# Disaster Risk Reduction

Methods, Approaches and Practices

**Series editor**
Rajib Shaw, Kyoto University, Japan

## About the Series

### Scope of the Series

Disaster risk reduction is a process, which leads to the safety of community and nations. After the 2005 World Conference on Disaster Reduction, held in Kobe, Japan, the Hyogo Framework for Action [HFA] was adopted as a framework of risk reduction. The academic research and higher education in disaster risk reduction has made/is making a gradual shift from pure basic research to applied, implementation-oriented research. More emphasis is given on the multi-stakeholder collaboration and multidisciplinary research. Emerging university networks in Asia, Europe, Africa and the Americas have urged for the process-oriented research in disaster risk reduction field. Keeping this in mind, this new series will promote the outputs of action research on disaster risk reduction, which will be useful for a wider range of stakeholders including academicians, professionals, practitioners and students and researchers in the related field. The series will focus on some of the emerging needs in the risk reduction field, starting from climate change adaptation, urban ecosystem, coastal risk reduction, education for sustainable development, community-based practices, risk communication, human security, etc. Through academic review, this series will encourage young researchers and practitioners to analyse field practices and link it to theory and policies with logic, data and evidences. Thus, the series emphasizes evidence-based risk reduction methods, approaches and practices.



More information about this series at http://www.springer.com/series/11575

Juha I. Uitto • Rajib Shaw
Editors

# Sustainable Development and Disaster Risk Reduction

Springer

*Editors*
Juha I. Uitto
Independent Evaluation Office
Global Environment Facility
Washington, DC, USA

Rajib Shaw
Graduate School of Global Environmental Studies
Kyoto University
Sakyo-ku, Kyoto, Japan

ISSN 2196-4106 ISSN 2196-4114 (electronic)
Disaster Risk Reduction
ISBN 978-4-431-55077-8 ISBN 978-4-431-55078-5 (eBook)
DOI 10.1007/978-4-431-55078-5

Library of Congress Control Number: 2015948258

Springer Tokyo Heidelberg New York Dordrecht London


Printed on acid-free paper

Springer Japan KK is part of Springer Science+Business Media (www.springer.com)

# Preface

This book's focus is on sustainable development and disaster risk reduction, two sides of the same coin. The editors firmly believe that effectively reducing the damage caused by natural disasters will be necessary for the achievement of sustainable development. Similarly, if our development does not reflect the sustainability principles, making a meaningful dent in disaster risk will be an uphill struggle. Reducing disaster risk, like moving towards sustainable development, not only depends on structural and engineering solutions but has important social, economic, political and cultural dimensions. The state of the environment underpins both endeavours. Global climate change is another factor playing into the equation, adding considerable risk and uncertainty. With the changing climatic conditions, disaster risk reduction and adaptation to climate change are natural priorities for sustainable development.

Yet disciplinary boundaries and silos are still hampering effective action and coordination between the two closely linked issues. Simply put, the professional and academic communities dealing with sustainable development, environment and disasters are still separate. They all address the same and related issues but often use different terminology and language. Key concepts that both must be concerned with include uncertainty, risk management, vulnerability, resilience and adaptation.

The year 2015 is a landmark in the coming together of the international community to forge global agreements aimed at addressing both our long-term future on the planet and disaster risk management. In March 2015, global leaders from governments, international organizations, civil society and the private sector came together for 5 days in the Japanese city of Sendai for the United Nations World Conference on Disaster Risk Reduction. It was symbolic to organize the gathering in Sendai, a city that suffered greatly from the Great East Japan Earthquake and Tsunami that struck the island nation on 11 March 2011 and the ensuing nuclear disaster. Sendai, like all of Japan, has shown great resilience in bouncing back from the disaster, largely because its development path has been rather sustainable. In contrast, major earthquakes which severely hit the mountainous country of Nepal in April 2015 have brought forward many sustainable development challenges that need to be addressed while the country is going through the recovery process. In September 2015, the world's nations are expected to approve the Sustainable Development Goals, a successor to the Millennium Development Goals. Finally, in November 2015, France will host the 21st Conference of the Parties of the United Nations Framework Conference on Climate Change and the 11th Session of the Meeting of the Parties of the 1997 Kyoto Protocol. The expectations for the Paris meetings are very high. Their aim is for the first time to reach a universal, legally binding agreement to combat climate change and to transition towards resilient low-carbon societies and economies.

In this context, we hope that this book makes a timely contribution to the debate and helps to integrate sustainable development and disaster risk management. The book brings together leading scholars and practitioners from around the world writing about their work and research. Contained here are 18 chapters covering a wide range of topics and sectors, ranging from climate change risk to water, food production and health. Many of the chapters outline conceptual issues in the context of practical case studies, primarily from Asia. There are also chapters that document lessons related to ecosystem-based disaster risk reduction, community-based approaches to sustainable development and disaster risk reduction, education and capacity building, integrating indigenous knowledge and how to evaluate sustainable development and disaster management efforts. The book emanated from discussions the editors had at the Graduate School of Global Environmental Studies at Kyoto University. Several of the authors are or have at one point been affiliated with the School, but the circle of authors is much wider.

The book has been written for researchers, students, practitioners and policymakers with an interest in sustainable development and disaster risk reduction. It is our sincere hope that they will find it stimulating and useful to their work.

Washington, DC, USA Juha I. Uitto
Kyoto, Japan Rajib Shaw

# Contents

# Contributors

**Md. Anwarul Abedin** Department of Soil Science, Bangladesh Agriculture University, Mymensingh, Bangladesh

**Rajarshi DasGupta** International Environment and Disaster Management Research Group, Graduate School of Global Environmental Studies, Kyoto University, Kyoto, Japan

**Rafaela Jane P. Delfino** The Oscar M. Lopez Center for Climate Change Adaptation and Disaster Risk Management, Foundation, Inc., Pasig City, Philippines

**Bharat Dhody** International Initiative for Impact Evaluation (3ie), New Delhi, India

**Glenn Fernandez** Environmental Education Laboratory, Graduate School of Global Environmental Studies, Kyoto University, Kyoto, Japan

Asian Disaster Preparedness Center, Bangkok, Thailand

**Richard Friend** Institute for Social and Environmental Transition (ISET), Bangkok, Thailand

**Adrienne I. Greve** Research Center for Disaster Reduction Systems, Kyoto University, Kyoto, Japan

Department of City & Regional Planning, California Polytechnic State University, San Luis Obispo, CA, USA

**Umma Habiba** Department of Agriculture Extension, Ministry of Agriculture, Dhaka, Bangladesh

**Kirsten Halsnæs** Management Engineering, Technical University of Denmark, Roskilde, Denmark

**Justin Henceroth** Institute for Social and Environmental Transition (ISET), Boulder, CO, USA

**Per S. Kaspersen** Management Engineering, Technical University of Denmark, Roskilde, Denmark

**Thi Kinh Kieu** Environmental Education Laboratory, Graduate School of Global Environmental Studies, Kyoto University, Kyoto, Japan

**Shimpei Kudo** Inter-Graduate School Unit for Sustainable Development and Survivable Societies, Kyoto University, Kyoto, Japan

**Kenneth MacClune** Institute for Social and Environmental Transition (ISET), Boulder, CO, USA

**M. Usman Mirza** Leadership for Environment and Development (LEAD) Pakistan, Islamabad, Pakistan

**Jesusa Grace J. Molina** Development Studies, School of Social Sciences, Faculty of Arts, University of Auckland, Auckland, New Zealand

**Daanish Mustafa** Department of Geography, King's College, London, UK

**Andreas Neef** Development Studies, School of Social Sciences, Faculty of Arts, University of Auckland, Auckland, New Zealand

**Erika Onagi** Department of International Studies, Graduate School of Frontier Sciences, The University of Tokyo, Kashiwa-shi, Chiba, Japan

**Juan M. Pulhin** Department of Social Forestry and Forest Governance, College of Forestry and Natural Resources, University of the Philippines Los Baños, College, Laguna, Philippines

**Jyotsna Puri** International Initiative for Impact Evaluation (3ie), New Delhi, India

**Rajib Shaw** Graduate School of Global Environmental Studies, Kyoto University, Sakyo-ku, Kyoto, Japan

**Nitin Srivastava** Graduate School of Global Environmental Studies, Kyoto University, Kyoto, Japan

**Akhilesh Surjan** Faculty of Engineering, Health, Science and the Environment, Charles Darwin University, Darwin, Australia

**Maricel A. Tapia** Department of Social Forestry and Forest Governance, College of Forestry and Natural Resources, University of the Philippines Los Baños, College, Laguna, Philippines

**Phong Tran** Institute for Social and Environmental Transition (ISET), Hanoi, Vietnam

**Sara Trærup** UNEP (United Nations Environment Programme), DTU (Technical University of Denmark) Partnership, Copenhagen, Denmark

**Juha I. Uitto** Independent Evaluation Office, Global Environment Facility, Washington, DC, USA

**Noralene Uy** Independent Researcher, San Mateo, Rizal, Philippines

**Minako Jen Yoshikawa** Inter-Graduate School Unit for Sustainable Development and Survivable Societies, Kyoto University, Kyoto, Japan

# About the Editors

**Juha Uitto** is the Director of the Independent Evaluation Office of the Global Environment Facility (GEF). His career has combined positions in international organizations and academia, including the United Nations Development Programme (UNDP), United Nations University (UNU), International Fund for Agricultural Development (IFAD) and the Nordic Africa Institute. He is Visiting Professor at Rutgers – the State University of New Jersey – and has held short-term visiting positions at the Graduate School of Global Environmental Studies, Kyoto University, in 2005 and 2013. His research interests pertain to assessing and measuring the effects and impact of environmental interventions, and he has written extensively on topics related to environment-poverty linkages, natural resources management, environmental hazards and evaluation. A geographer by training, he holds a PhD from the University of Lund and MSc from Helsinki University.

**Rajib Shaw** is a Professor in the Graduate School of Global Environmental Studies of Kyoto University, Japan. He worked closely with the local communities, NGOs, governments and international organizations, including United Nations, especially in the Asian countries. He is currently the Chair of the United Nations Asia Regional Task Force for Urban Risk Reduction and the President of Asian University Network of Environment and Disaster Management (AUEDM). His research interests are community-based disaster risk management, climate change adaptation, urban risk management and disaster and environmental education. He has published several books in the field of disaster and environmental management. He is also the Chief Editor of Asian Journal of Environment and Disaster Management.

# Chapter 1
# Sustainable Development and Disaster Risk Reduction: Introduction

**Juha I. Uitto and Rajib Shaw**

**Abstract** Sustainable development and disaster risk reduction are closely linked on many levels and the relationship cuts both ways. Disasters add often devastating costs to societies and communities in terms of financial losses, destroyed infrastructure and loss of life. They can set development back for years. Environmental destruction and lack of sustainable development exacerbate disaster risk and impact. Climate change is adding to the risk and uncertainty. Despite the obvious linkages, the sustainable development, climate change and disaster risk communities each approach the common problematique from different angles. Even in intergovernmental negotiations, the processes tend to be separate and on parallel tracks. This book is an attempt to address sustainable development and disaster risk reduction from an integrated perspective. The 18 chapters highlight issues from many angles and sectors covering them from theoretical and practical perspectives. A number of case studies, primarily from Asia, are highlighted.

**Keywords** Sustainable development • Disaster risk reduction • Climate change

## 1.1 Introduction

Sustainable development and disaster risk reduction are intrinsically intertwined. A single major natural disaster – be it an earthquake, storm, tsunami or a landslide – can undo progress made and set back development by years. At the same time, environmental factors and, more broadly, development that is not sustainable, contribute to the increase in disasters. Societies build infrastructure in places exposed to the forces of nature and that does not stand up to hazards. On every continent, people crowd onto coasts where cities grow uncontrollably and often without

J.I. Uitto (✉)
Independent Evaluation Office, Global Environment Facility, Washington, DC, USA
e-mail: juitto@thegef.org

R. Shaw
Graduate School of Global Environmental Studies, Kyoto University, Sakyo-ku, Kyoto, Japan
e-mail: shaw.rajib.5u@kyoto-u.ac.jp

 
J.I. Uitto, R. Shaw (eds.), *Sustainable Development and Disaster Risk Reduction*, Disaster Risk Reduction, DOI 10.1007/978-4-431-55078-5_1

planning. According to the United Nations, some 44 % of the world's people now live within 150 km from the coast. Individuals make decisions regarding settling into hazardous locations out of necessity or out of choice. In many rapidly growing cities, especially in the developing world, the places available for the poorest migrants from the countryside that still allow them access to employment and other urban opportunities are often in marginal and hazardous locations (Hewitt 1997). These may be on slopes that have been deforested by the migrants to make space for informal settlements, but at the same time making them unstable, exposed to storms and susceptible to landslides. On the other hand, many well-to-do citizens crave for a piece of waterfront property and face risks from coastal erosion and storms. Similarly, as we have seen in California too frequently in recent years, large homes are often built into forested hills that are susceptible to devastating fires (Simon 2014).

In some cases natural and technological hazards combine to compound the dangers to communities and the society at large. This was dramatically brought home by the Fukushima nuclear meltdown following the massive tsunami caused by the Great East Japan Earthquake on March 11, 2011 (Pritchard 2012; Shaw and Takeuchi 2012). On a more localized scale, such events are more frequent and often mostly affect the poor people whose dwellings in cities are often located near hazardous industrial facilities.

Human actions that degrade the environment are culpable for worsening the risks from natural hazards and increasing exposure leading to disasters. A case in point is the widespread removal of mangroves on the coasts of Southeast Asia to make way for aquaculture. It was clearly demonstrated in the case of the Indian Ocean tsunami on December 26, 2006, that the power of the tidal wave was more destructive in areas where the protective mangroves had been removed (Wun'Gaeo 2009). The mangroves also play other important ecosystem functions that get disrupted once the vegetation is removed. They act as spawning grounds for fish and other aquatic creatures, and they filter pollution from land-based sources that then runs directly into coastal waters when the mangroves are no longer in place.

Climate change is where human influence on the global environment is the most dramatic. While the exact mechanisms and the extent to which climate change is affecting weather patterns globally are still not known (Bouwer 2011; Pielke 2014), there is ample evidence to suggest that there is a correlation between climate change and the increased frequency and severity of extreme weather events, including storms, droughts, heatwaves and cold events. On regional and local scales these effects are even harder to predict, but their impact on local communities and economies is potentially devastating.

Apart from weather anomalies, climate change will result in rising sea levels that pose a severe threat to coastal settlements and infrastructure everywhere, from the richest cities in the world like New York, Miami and Tokyo, to vulnerable poor communities in Bangladesh and West Africa. Small islands are at the forefront of bearing the brunt of climate change induced sea level rise. Many of them, especially in the Pacific and Indian Oceans, consisting of little less than the coastal zone, are at risk of being entirely swallowed by the sea. This is particularly egregious, given that the small islands have been amongst those least contributing to climate change, and are now paying the price of globally unsustainable development (Pelling and Uitto 2001).

It is thus evident that sustainable development and disaster risk reduction must go hand in hand. Losses from disasters have been constantly on the rise and now place a huge burden on the global community in terms of lives lost, property damaged and opportunities missed. The resources consumed by natural disasters are directly away from development. The increases in losses, of course, cannot be attributed primarily to climate change and other environmental factors. Most of the increase is due to the growth of exposure because of, e.g., the great coastward movement of populations and the development of infrastructure in hazardous locations. When the losses are normalised taking into account trends in economic exposure, the upward trend in losses is obscured (Neumayer and Barthel 2011). It is also due to economic and political forces that render groups of people and communities powerless and thus vulnerable to natural hazards. Such outcomes often have deep historical roots within societies and in the international system (Hilhorst and Bankoff 2004).

Despite these obvious interlinkages between sustainable development and disasters, they still are treated separately in most of practice and literature. The professional and scientific communities are different, they speak a different language using different jargon about the same phenomena. In the international arena, there have been two parallel processes that seldom if ever meet. Even in areas where the overlap is most striking, communication and collaboration are almost entirely lacking: disaster risk managers do not talk to those dealing with climate change adaptation, and vice-versa.

This book arose from the recognition of the above fact and the need to bridge the gaps. The authors in the volume include primarily researchers who have their roots in one or the other community but who recognise the need for integrated approaches. We hope that the book will nudge the debate a notch further so that we can move towards a more sustainable development path. Sustainable development will require reducing vulnerability of societies and communities to natural hazards, enhancing adaptation to the impacts of the changing climate, and strengthening the resilience of societies, communities and individuals everywhere.

## 1.2 International Efforts: Parallel Processes

Sustainable development as a concept was conceived by the World Commission on Environment and Development – the so called Brundtland Commission named after its chair – that defined it as (WCED 1987): "Sustainable development is the kind of development that meets the needs of the present without compromising the ability of future generations to meet their own needs."

From the outset, sustainable development was intended to bring together the social, economic and political dimensions. The 1992 global Human Development

Report produced under the auspices of UNDP elaborated on the concept of sustainable human development (HDR 1992: 2):

> Global poverty is one of the greatest threats to the sustainability of the physical environment and to the sustainability of human life. Most of the poor live in the most ecologically vulnerable areas – 80 % of the poor in Latin America, 60 % in Asia and 50 % in Africa. They overuse their marginal lands for fuel wood and for subsistence and cash-crop production, further endangering their physical environment, their health and the lives of their children. In developing countries, it is not the quality of life that is at risk – it is life itself.

The 1992 United Nations Conference on Environment and Development[1] held in Rio de Janeiro, Brazil – the Earth Summit – sought to operationalize sustainable development on a global scale. Its main outcomes were the Rio Declaration and the Agenda 21, a massive blueprint for actions that would be needed in virtually all segments of society to move towards sustainable development. Other concrete outcomes from the Earth Summit included the global Convention on Biodiversity (CBD) and the UN Framework Convention on Climate Change (UNFCCC). The political deal was that the rich countries in the North would foot the bill to help the countries in the global South to switch their economic development patterns onto a sustainable path that would protect globally significant biodiversity and help the developing countries to avoid greenhouse gas emissions as they industrialise and develop their economies.

The Global Environment Facility[2] (GEF) was established as virtually the only source of funding for the entire sustainable development agenda and as financial mechanism for CBD and UNFCCC. Since then, new public funding sources have emerged, notably the Climate Investment Funds managed by the World Bank and four regional development banks, and the new Green Climate Fund (GCF) that is still to start operations, but these still are obviously insufficient to address the issues of global sustainable development. The overall performance evaluation of the GEF noted that the global environmental trends continue to decline and that global public funding allocated to environmental issues (around US$10 billion annually of which about one-tenth is distributed through the GEF) is dwarfed by the global public subsidies (about US$1 trillion) that lead to overexploitation of natural resources and environmental degradation (GEF IEO 2014).

Disaster risk reduction is visibly absent from the Rio documents and although this situation has improved in subsequent processes, the links are mostly at the levels of principles rather than action. The Rio+20 conference held in Johannesburg, South Africa, in 2012 identified ecosystems, climate change and disaster risk reduction among crosscutting issues. However, the UN sponsored Millennium Development Goals (MDGs; 2000–2015) were silent on disaster risk.

Instead, there has been a parallel process focusing on disaster risk reduction. The United Nations General Assembly designated the 1990s as the International Decade for Natural Disaster Reduction (IDNDR) that led to the establishment of a permanent secretariat in the United Nations to promote disaster risk reduction worldwide.

---

[1] http://www.un.org/geninfo/bp/enviro.html

[2] http://www.thegef.org/gef/

The Hyogo Framework for Action, 2005–2015, was the outcome of the World Conference on Disaster Reduction held in Kobe, Hyogo, Japan, in 2005. It focused on building the resilience of nations and communities to disasters. It detailed the requirements for different sectors and actors, including governments, international agencies, disaster experts and others, to reduce disaster losses. It outlined five priorities for action: (1) Ensure that disaster reduction is a national and local priority with a strong institutional basis for implementation; (2) Identify, assess and monitor disaster risks and enhance early warning; (3) Use knowledge, innovation and education to build a culture of safety and resilience at all levels; (4) Reduce underlying risk factors; and (5) Strengthen disaster preparedness for effective response at all levels (UN 2005). The Hyogo Framework identifies environmental conditions and vulnerabilities as contributing to disaster risk. It further identifies activities related to environmental and natural resource management, including land-use planning and development, integrated flood management and management of fragile ecosystems as part of the package for disaster risk reduction. It also promotes the integration of risk reduction associated with existing climate variability and future climate change, and the identification of climate-related risks.

The year 2015 is seen as a watershed with numerous important events taking place. In March 2015, the disaster community met in the Japanese city of Sendai for the Third UN World Conference on Disaster Risk Reduction. The Sendai Declaration mentions neither the environment nor climate change, although climate change and variability feature rather prominently in the final document (UN 2015). Sustainable development comes in one of the key guiding principles of the disaster risk reduction as: "The development, strengthening and implementation of relevant policies, plans, practices and mechanisms need to aim at coherence, as appropriate, across sustainable development and growth, food security, health and safety, climate change and variability, environmental management and disaster risk reduction agendas. Disaster risk reduction is essential to achieve sustainable development". In the conference, the secretary general of the UN Ban Ki-moon expressed that "an ambitious outcome at the WCDRR will put the world on a path to a new sustainable development agenda in 2015, together with the forthcoming Sustainable Development Goals (SDGs) and a meaningful climate change agreement". The Sendai Declaration also calls for collaboration across global and regional mechanisms and institutions relevant to disaster risk reduction, including those for climate change, biodiversity, sustainable development, poverty eradication, environment and others.

The UN is leading the development of a post-2015 agenda and the new Sustainable Development Goals to replace the MDGs. At the time of this writing, there are 17 SDGs that will be presented to the UN General Assembly in September 2015 for adoption. These have been produced through a lengthy and inclusive negotiation process. Four of the proposed goals make the connection to disaster risk.[3] Goal 1, End poverty in all its forms everywhere, refers to the need to build resilience of the poor and those in vulnerable situations, and reduce their exposure and vulnerability

[3] https://sustainabledevelopment.un.org/sdgsproposal

to climate-related extreme events and other economic, social and environmental shocks and disasters. Goal 2, End hunger, achieve food security and improved nutrition, and promote sustainable agriculture, refers to resilient agricultural practices that increase productivity and production, that help maintain ecosystems, that strengthen capacity for adaptation to climate change, extreme weather, drought, flooding and other disasters. The most extensive references to disaster risk pertain to Goal 11, Make cities and human settlements inclusive, safe, resilient and sustainable, which calls for significantly reducing the number of deaths and the number of affected people and economic losses caused by disasters, including water-related disasters, with the focus on protecting the poor and people in vulnerable situations. It further calls for cities to adapt to climate change and build resilience to disasters. Finally, Goal 13, Take urgent action to combat climate change and its impacts, urges strengthening resilience and adaptive capacity to climate related hazards and natural disasters in all countries.

In November 2015, the negotiators will gather in Paris for a landmark meeting on climate change. The UN Climate Change Conference COP21 is intended to produce a universal climate agreement to replace the Kyoto Protocol adopted in 1997.

All of these international processes are critical to the future of sustainable development and disaster risk reduction on a global scale. Despite the increasing references to links between the spheres, the practical challenge of producing a coherent and integrated framework that takes into account sustainable development, disaster risk reduction and environmental issues remains. Partly this is due to the compartmentalisation of the issues within the international negotiations and national governments, which poses limits to not only the political agreements, but also their implementation.

## 1.3 Importance of Implementation, Monitoring and Evaluation

The international processes and agreements, while important, obviously are not sufficient by themselves. It is essential to move from political proclamations and strategies to implementing actions on the ground. Given the daunting challenges, no one actor or sector will be adequately powerful and resourced to address the issues of sustainable development and disaster risk reduction. It is important to mobilise public action and funding, but the financing available through mechanisms such as the GEF and GCF can only be catalytic. It is far from sufficient for directly solving the problems. It is important that environmental and disaster risk reduction concerns be mainstreamed into all development endeavours.

It is also imperative that the private sector participates actively. The decisions made by companies are much more powerful drivers than the direct funding from public sources can ever be. The good news is that companies are already recognising the burden that natural disasters place on their operations and profits. They also see climate change as a direct risk. Companies are already starting to hire a new cadre of professionals to help them to manage climate risk and to adapt their operations to the changing climate (Westervelt 2015). When motivated, the private sector can move much faster than the political process.

Cities are also important actors and many have taken decisive action to reduce their vulnerabilities and enhance their resilience in the face of climate change. For New York, the 2012 Superstorm Sandy was a rude wake-up call that led to concrete actions and changes in policies in the coastal mega-city.

We will also need to know whether we are achieving results and whether we are doing so in a way that is cost-effective. Monitoring and evaluation are important tools for this purpose and will be needed at multiple levels.

It is important to monitor implementation of the agreed policies and strategic frameworks, as well as the individual programmes and projects so that the various stakeholders ranging from the funders and tax payers to the people on the ground that are intended to benefit from these actions can be assured that implementation is on track; or if it isn't, that corrective action can be taken. In developing the SDGs, the international community has placed significant emphasis on results frameworks, monitoring and indicators.

Given that we call for integration of sustainable development and disaster risk reduction goals, there is a need for developing new metrics that capture these dimensions. For one, climate change has increased the importance of risk and uncertainty that need to be also built into the monitoring and evaluation frameworks (Picciotto 2007). It is not possible to continue business as usual and assume continued linear changes in complex systems that are characterised by uncertainty, discontinuities and unknown tipping points.

Although monitoring is essential for the international community to know that strategies, programmes and projects are proceeding on target, it is not enough. Monitoring should be a routine management task and indicators can only measure change in a limited number of areas. They cannot explain why things change, what are the causal mechanisms and conditions where interventions are effective. For this, evaluation is required.

Evaluation involves an objective and rigorous analytical process using different types of data and methodologies to enhance our understanding of the causal mechanisms and underlying factors of why an intervention works or does not work, under what circumstances, and for whom. Unlike monitoring, evaluation does not take the intervention as a given, but will question whether the strategy or approach chosen was the correct one or should be abandoned. Evaluation should assess the relevance of the intervention, not only based on whether it fits into an agreed national and/or international framework, but also whether it is making a difference on the ground. In this sense, relevance approaches impact (Van den Berg 2011).

There are many approaches and ways of evaluating the performance and impact of a policy, strategy, programme or project. Impact evaluations often utilize econometric tools and experimental and quasi-experimental methods, such as randomised controlled trials. Such methods can be useful in certain contexts, but have their limitations (see Puri and Dhody in Chap. 15 of this volume). Other methods can be used rigorously to enhance our understanding of what works and how to improve performance. Most often theory-based approaches utilising multiple methods are the most feasible way of triangulating evidence from different sources and arriving at useful results. To the extent possible, the intended beneficiaries should be involved in setting the evaluation questions and participating in the evaluation. After all, the goal of the interventions is to benefit the people on the ground and there is a strong element of downward accountability to them.

When designing the evaluation framework and evaluation questions, it is important to focus on the ultimate goals of what the policy, strategy, programme or project is trying to achieve. There may be multiple goals and when multiple organizations are involved, it is not always clear that the goals are fully aligned (Uitto 2014). It is important to be fully cognizant of these tensions and also to look out for unintended consequences that virtually every intervention has. As we are moving into new territory of integrating disaster risk reduction and sustainable development, rigorous evaluation can be truly helpful in ensuring that stakeholders benefit from the intervention and helping proponents learn and improve performance.

## 1.4 Structure of This Book

The book contains 17 chapters apart from this introductory one. They explore the linkages between sustainable development and disaster risk reduction from a variety of angles, addressing theoretical and conceptual issues as well as practical lessons from the ground. They also cover a variety of sectors.

In the next chapter, Adrienne Greve provides a summary of the state of the art on the ways in which climate change affects disaster type, location, frequency and severity. She uses this background to explore how disaster management procedures must adjust to accommodate progressive climate change, and discusses the characteristics of effective climate adaptive disaster management procedures and strategies.

In Chap. 3, Akhilesh Surjan, Shimpei Kudo and Juha Uitto deconstruct risk and vulnerability to natural disasters, recognising that they are not evenly distributed. Risk varies geographically dependent on geographical location, but vulnerability is dependent on social, economic and political factors. In many ways, poor people are more vulnerable to hazards, often living in exposed areas and substandard housing, having inadequate means to prepare for and recover from shocks brought about by natural disasters, including slow-onset disasters. The chapter also explores psychological dimensions and trauma caused by natural disasters that can lead to long-standing damage.

Urban areas are particularly vulnerable to the impacts of climate change, but there is no agreed method for assessing urban vulnerability. Tran Phong and Nguyen Huy argue, in Chap. 4, that to overcome this constraint and to make the concept of vulnerability operational, it is useful to use a resilience approach that allows for consideration of complex systems and their interactions in cities. The chapter reviews key challenges to disaster risk management in the context of rapid urbanisation and impacts of climate change in light of experience from practice. It presents a new vulnerability assessment approach based on a climate resilience framework. It concludes with key findings, experiences and lessons learned from the application of urban climate vulnerability assessment in Hue City, Vietnam.

Climate change is most often presented as a long-term phenomenon and little attention is given to assessments of how extreme weather events already today cause serious losses, particularly in the least developed countries that are vulnerable due to low incomes, weak infrastructure and institutions, and low capacity for coping with climate change. Despite data limitations and uncertainties, Kirsten Halsnæs, Per Kaspersen and Sara Trærup develop a methodological framework for damage cost assessment that reflects key assumptions regarding the specific vulnerabilities in a developing country context. Presenting the framework in Chap. 5, they apply it to the assessment of the consequences of severe storms in Cambodia based on 18 years of statistical records of events.

Natural disasters and climate change impacts are a leading cause of hunger, affecting all dimensions of food security, including access to food, availability and stability of supplies, and nutrition. Food security, climate change adaptation and disaster risk are the focus of Chap. 6 by Umma Habiba, Md. Anwarul Abedin and Rajib Shaw. Most food insecure people live in areas prone to natural hazards and they are the least able to cope with shocks. Poor households are often trapped in a downward spiral of food insecurity and poverty. The chapter focuses on the governance of food systems in order to understand their vulnerability to environmental change and to identify solutions.

The extent to which sustainable development benefits a community is closely tied to its level of health, argue Minako Jen Yoshikawa and Akhilesh Surjan in Chap. 7. Health is a product of economic, social, political and environmental factors, as well as of health services. Sustained improvements in health must been seen as an integral part of sustainable development. Health in turn contributes to economic, social and environmental development through multiple pathways. There is thus a virtuous cycle between improved health and sustainable development.

Ecosystems, climate change and disaster risk reduction were among the cross-cutting issues highlighted in Rio+20. With this in mind, Noralene Uy Rafalea, Jane Delfino and Rajib Shaw discuss the important role of ecosystem-based disaster risk reduction (Eco-DRR) in sustaining ecosystems and building disaster-resilient communities. Chapter 8 describes ecosystem management practices that link ecosystem protection and disaster risk reduction. The authors further analyse trends in Eco-DRR elucidating the challenges in advancing its use and linking it to policy.

In Chap. 9, M. Usman Mirza and his co-author Daanish Mustafa review water research literature through the tri-focal lens of access, equity and hazards. Building resilience and adaptation capacity to guard against water-related hazards must be an integral part of water resource planning. With the increasing awareness of water-related hazards, their impacts and associated risks, it is no longer possible to forecast based on averages.

Hydro-meteorological disasters, such as flooding, storm surges and wet mass movement account for nearly 90 % of total catastrophic events in the world. At the same time, more than 60 % of economic damage due to natural disasters has been in coastal areas. In Chap. 10, Rajarshi DasGupta and Rajib Shaw highlight key linkages between sustainable development and disaster risk reduction in coastal areas, addressing the emerging challenges in risk sensitive coastal zone management in the face of climate change and coastward migration, land development, urbanization and loss of ecosystem services.

Integrated approach to water resources management under conditions of climate change is the focus of Chap. 11 by Erika Onagi. Her research focuses on the Murray-Darling Basin in Australia and the basin plan negotiated under a federal political system. She raises several questions in order to provide lessons from the case study and to suggest implications to other situations of transboundary river management.

Using the Philippines, one of the most climate vulnerable countries, as a case, Juan Pulhin explores how the current and future potential impacts of climate change threaten the contribution of agriculture and water sectors to the country's economic development. In Chap. 12, issues and challenges facing the sectors are also analysed and potential solutions explored to reduce the adverse impacts of climate change with the aim of helping achieve the country's quest of sustainable development.

In Chap. 13, Rajib Shaw explores sustainability elements of community-based approaches in development projects and links them to the risk reduction paradigm. His argument from a historical point of view is that communities have been active in development activities even before states were formed. However, after state formation and governmental control of most of development activities, community-based approaches must remain central.

Chapter 14 by Glenn Fernandez and Rajib Shaw reviews the achievements of the UN Decade for Sustainable Development in the area of Disaster Risk Reduction Education (DRRE). Specific examples of successful education, training and capacity building initiatives in formal and informal DRRE are presented. The remaining challenges of utilising DRRE as a tool to build a culture of disaster resilience are discussed to explore how DRRE can be further enhanced.

In Chap. 15, Jyotsna Puri and Bharat Dhody lay out a paradigm for evaluating adaptation in forestry projects, policies and programmes. The authors examine how experimental and quasi-experimental methods can be used to understand the effectiveness of adaptation projects in the forestry sector. There have been few studies that have used robust attribution methods to assess the impacts of programmes on how well forests are adapting and are sustainable. Impact evaluation methodologies can make big contributions to the field, but there are also many limitations in traditional methods that can limit the understanding of impacts in multi-intervention and multi-sectoral contexts. The authors present possibilities in methodology and data that represent an important way forward.

Jesusa Grace Molina and Andreas Neef in Chap. 16 make a case for integration of indigenous knowledge into disaster risk reduction and management policies for sustainable development. Due to a combination of physical, socio-economic and political factors, the Agta, an indigenous group in Casiguran, Philippines, are highly susceptible to the threat of natural hazards. Despite their vulnerabilities, they possess valuable knowledge generated through practical and long-standing experiences, culture and local resources, which should be brought to bear on the decision-making and planning, and policy formulation processes of the local government. The authors recommend mechanisms for ensuring the Agta's inclusion.

Chapter 17 by Thi Kinh Kieu, Glenn Fernandez and Rajib Shaw traces the history, development and purpose of the Sustainability Literacy Test (SLT) promoted by several universities worldwide to ensure that they are producing sustainability literate graduates. A comparison between SLT and several similar tests is made to offer insights regarding lessons learned from experiences and to provide suggestions for improving SLT. The chapter also presents initial student feedback on SLT and their recommendations to enhance its usefulness.

To conclude the book with Chap. 18, Nitin Srivastava, Glenn Fernandez, Rajarshi DasGupta, Akhilesh Surjan and Rajib Shaw explore the inclusion of disaster risk reduction and resilience in the post-2015 SDGs. The chapter also focuses on the role various stakeholders can play in disaster risk reduction through social inclusion. Yet, investment for disaster risk reduction, enhancement of disaster knowledge and access to information, and a conducive international environment still pose challenges in the post-2015 era.

**Acknowledgements** The idea for the book was conceived at the Graduate School of Global Environmental Studies at Kyoto University, Japan, where Juha Uitto was visiting professor in 2013 hosted by the International Environment and Disaster Management laboratory headed by Rajib Shaw. Juha Uitto is grateful to the Kyoto University GSS program [Sustainability and Survivability Studies program] for supporting his visit.

## References

Bouwer LM (2011) Have disaster losses increased due to anthropogenic climate change? Bull Am Meteorol Soc 92(1):39–46

GEF IEO (2014) Final report: at the crossroads for higher impact. Fifth overall performance study of the GEF. Global Environment Facility Independent Evaluation Office, Washington, DC

HDR (1992) Human development report 1992. United Nations Development Programme. Oxford University Press, New York

Hewitt K (1997) Regions of risk: a geographical introduction to disasters. Longman, Harlow

Hilhorst D, Bankoff G (2004) Introduction: mapping vulnerability. In: Bankoff G, Frerks G, Hilhorst D (eds) Mapping vulnerability: disasters, development and people. Earthscan, London, pp 1–9

Neumayer E, Barthel F (2011) Normalizing economic loss from natural disasters: a global analysis. Glob Environ Chang 21:13–24

Pelling M, Uitto JI (2001) Small island developing states: natural disaster vulnerability and global change. Environ Hazards 3(2001):49–62

Picciotto R (2007) The new environment for development evaluation. Am J Eval 28(4):509–521

Pielke R Jr (2014) The rightful place of science: disasters and climate change. Arizona State University: Consortium for Science, Policy and Outcomes, Tempe, Washington, DC

Pritchard SB (2012) Japan forum – an envirotechnical disaster: nature, technology, and politics at Fukushima. Environ Hist. First published online February 23, 2013 doi:10.1093/envhis/ems021

Shaw R, Takeuchi Y (2012) East Japan earthquake and tsunami: evacuation, communication, education, volunteerism. Research Publishing, Singapore
Simon GL (2014) Vulnerability-in-production: a spatial history of nature, affluence, and fire in Oakland, California. Ann Ass Am Geogr 104(6):1199–1221
Uitto JI (2014) Evaluating environment and development: lessons from international cooperation. Evaluation 20(1):44–57
UN (2005) Hyogo framework for action 2005–2015: building the resilience of nations and communities to disasters. Extract from the final report of the World Conference on Disaster Reduction (A/CONF.206/6). International Strategy for Disaster Reduction. United Nations, Geneva
UN (2015) Sendai framework for disaster risk reduction 2015–2030. A/CONF.224/CRP.1. United Nations, Sendai
Van den Berg RD (2011) Evaluation in the context of global public goods. Evaluation 17(4):405–415
WCED (1987) Our common future. World Commission on Environment and Development. Oxford University Press, Oxford
Westervelt A 2015 How climate adaptation experts help companies to prepare for disaster. Ensia. http://ensia.com/features/how-climate-adaptation-experts-help-companies-prepare-for-disaster/
Wun'Gaeo S (2009) Environment as an element of human security in southeast Asia: case study of the Thai Tsunami. In: Brauch H-G, Oswald Spring U, Grin J, Mesjasz C, Kameri-Mbote P, Behera NC, Chourou B, Krummenacher H (eds) Facing global environmental change: environmental, human, energy, food, health and water security concepts. Springer, Berlin, pp 1131–1142

# Chapter 2
# Sustainable Development, Climate Change Adaptation and Disaster Management

**Adrienne I. Greve**

**Abstract** Climate change necessitates changes to the methods and procedures of disaster management. A triggering event results in disaster due to the severity of impacts and sequential, secondary consequences. These consequences result from factors such as the biophysical, social, political, economic, and structural context of the community experiencing the event. Climate change adds an additional progressive factor that increases the potential severity and frequency of triggering events and the vulnerability of communities experiencing the impact. Projecting future climate change impacts is imperfect, with a high degree of uncertainty. This uncertainty combined with the self-organizing, emergent properties that characterize urban systems, demands evaluation of the question, what qualifies as recovery? The manner in which policy is formulated must be adjusted to accommodate the dynamic contexts that may be subject to disaster triggering events. To address this need, disaster management must make adaptive capacity one of its primary goals. This is achieved through a process that combines risk assessment, comprehensive vulnerability assessment, iterative policy development and implementation, and ongoing public engagement. One of the critical aspects to long-term disaster recovery in the context of climate change is consideration of both local and regional contexts. Future adaptive capacity is closely tied with regional sustainability. This connection to sustainability is due to the fact that progressive climate change may further stress systems already being pushed by human development such as water and air pollution, habitat loss, and other degraded ecosystem services. Disaster events can result in the breaking of already stressed system connections. This confluence of events must be considered as part of disaster management. Further, ongoing monitoring of these systems and clear acknowledgement of their role in the ongoing functioning and vitality of a city is critical. Without continuous monitoring and iterative update of management procedures and local policy, communities will not be adequately prepared for an uncertain future.

**Keywords** Climate change adaptation • Disaster management • Hazards mitigation • Resilience • Climate action planning

---

A.I. Greve (✉)
Research Center for Disaster Reduction Systems, Kyoto University, Kyoto, Japan

Department of City & Regional Planning, California Polytechnic State University, San Luis Obispo, CA, USA
e-mail: agreve@calpoly.edu



J.I. Uitto, R. Shaw (eds.), *Sustainable Development and Disaster Risk Reduction*, Disaster Risk Reduction, DOI 10.1007/978-4-431-55078-5_2

## 2.1 Introduction

The aim is simple: to leave a more resilient world to future generations.
- UN Secretary-General Ban Ki-moon (UN News Centre 2014, 2)

The International Panel on Climate Change (IPCC) defines risk as the intersection of hazards, vulnerability, and exposure (2012, 2014). These three broad categories, while fairly simple to understand as concepts, are much more difficult to determine specifically enough to serve as the basis for the formation of public policy intended to guide future development, protect human health and safety, and assure a healthy, sustainable biophysical context. This chapter focuses on the integration of climate change into disaster management. This integration is a critical component of comprehensive climate change adaptation policy. While the this chapter is concentrates on a fairly narrow policy aim, these goals fit within and must be informed by the much broader goal of sustainable development.

Disaster management refers to efforts to reduce long-term risk posed by disaster events to life and property. While many different terms are used for disaster management, such as hazard mitigation or risk reduction, the activities that these terms describe are generally the same. They can be broken into three broad, closely-related phases: prevention and preparedness, emergency response, and recovery and rehabilitation (Director General for Disaster Management, Government of Japan 2011; FEMA 2013). Incorporation of climate change into disaster management occurs primarily in the prevention and preparedness phase, though it should also figure prominently in the recovery and rehabilitation process.

Prevention and preparedness is a phase occurring well before a disaster event. Traditionally, this stage has focused on strengthening built structures (buildings, infrastructure, and protective elements such as flood walls), adjusting land use to move development away from disaster-prone areas, and maintaining or restoring protective natural features such as forests, wetlands, or other natural elements that dampen disaster impacts (Godschalk et al. 1999). Planning and implementing these actions requires estimates of the recurrence and subsequent consequences of disaster events, this is termed a risk assessment (FEMA 2013; IPCC 2014). Climate change alters the manner in which these estimate are made, the certainty of the estimates, and how recovery is defined. Addressing these challenges, particularly those focused on long-term recovery, requires broader consideration of how choices regarding future development relate to biophysical context and regional ecological viability.

Alesch et al. (2009) and Alesch and Siembieda (2012) explain that a disaster occurs due to a primary, triggering event that causes a series of cascading consequences that unfold over time. These secondary consequences can cause discontinuities in critical urban system elements. The cascade effect of disaster impacts continues through time and is bound up with regional relationships that also may be broken or interrupted. These regional relationships may result in consequences being experienced well beyond the spatial extent of the directly impacted areas.

What climate change adds to disaster management are additional dynamic components. The severity and frequency of some disaster triggering events may be affected. In addition, progressive climate change may slowly affect the viability of certain parts of an urban system and its regional relationships that will influence local capacity to cope with and recover from disaster. The reduced predictability of future events forces an examination of the tools cities have available to address the threats posed by climate change and disaster events. When viewing the larger biophysical and regional context of a city, the shifts resulting from climate change must be assessed, as well as the changes resulting from human development. Climate change, resulting from greenhouse gas emissions, is itself a consequence of the human development. In addition, climate change has the potential to exacerbate some of the other negative consequences of human development that have long been the focus of sustainability such as habitat loss, intensive agricultural cultivation and grazing, and air and water pollution.

### *2.1.1 Disaster Management and Dynamic Systems*

Disaster management has traditionally used historic recurrence intervals to estimate the likelihood of future events (e.g. FEMA 2001). This method is based on the assumption that the likelihood of disaster events is largely static. The consequences may evolve through time due to shifts in factors such as development patterns, infrastructure, and cultural norms, but the anticipated frequency and severity of natural disaster events has been presumed constant in the past. Climate change results in history being an inaccurate predictor of future disaster recurrence and severity. Accommodating this change requires a new approach to disaster management and entails more than simply altering the statistical methods used to estimate recurrence intervals. Two factors that require consideration are as follows:

1 *Uncertainty*. Climate change science has a high level of uncertainty that increases at smaller regional and local spatial scales. In addition, the estimates of future climate impacts are reliant on current and future greenhouse gas (GHG) emissions, which are also changing.
2 *Gradual climate change*. Climate change influences both episodic disaster events such as flooding and storms, but also results in slow changes such as the shifting of seasonal temperature and precipitation patterns. These slow shifts interact with disaster events and have the potential to alter both local vulnerability and regional consequences.

Researchers often conceptualize cities as open systems comprised of biophysical and human elements. Increasingly such systems are viewed as self-organizing and dynamic, exhibiting emergent behavior (e.g. Folke 2006; Manyena 2006; Djalante et al. 2011; Alesch and Siembieda 2012). While this theoretical conceptualization is appealing when considering urban spaces in the abstract, it can be difficult to use as a basis for the formulation of policy. The interaction between social, economic, ecological, and political contexts cannot be precisely predicted, particularly in the context of progressive climate change. If the future cannot be reliably projected, how can policy be devised to assure local capacity to recover and adapt? Climate change may alter the character of disasters, but also has the potential to alter the context in which the disasters are occurring. As a result, disaster management measures that seek to bolster adaptive capacity must be iterative, allowing room to address system unpredictability and dynamic change.

## 2.2 Climate Change and Projected Impacts

There is a fairly small number of what are often termed *direct* climate impacts. These include altered temperature and precipitation patterns, sea level rise, ocean acidification, and increased extreme events (IPCC 2013). Similar to the disasters discussed above, these direct impacts result in a much larger set of secondary outcomes. The summary of climate impacts presented below is broken into episodic events (e.g. disasters) and changes likely to occur more slowly (often termed progressive climate change).

Ongoing temperature, precipitation, sea level, and other weather-related measurements reveal that the projected changes to the climate are already occurring (IPCC 2013). In the Northern Hemisphere, the last 30-years were one of the warmest periods in the last 1,400 (IPCC 2013, p. SPM-5). This warming, driven by anthropogenic greenhouse gas (GHG) emissions, is unlikely to slow in the near future despite increasing global focus on emissions reduction. Current climate observations and GHG emissions are used to calibrate and refine the models used to project future climate change. This results in a series of scenarios reflecting possible futures from best case to worst case. Planners, managers, or any others working towards a sustainable, resilient future must use such scenarios as a starting point for making decisions.

The projected impacts of climate change have varying levels of statistical confidence (Table 2.1). An important characteristic to note about the estimates of certainty is that probability is higher for broader trends, such as average daily temperature, as opposed to phenomena that are discrete events that occur over smaller spatial scales (e.g. drought, cyclone). Even those impacts that have high levels of certainty for occurring, precise estimates of the impact in a particular location are difficult and include a high level of uncertainty. Methods for handling uncertainty in a policy context are summarized in subsequent sections. Uncertainty

**Table 2.1** Climate events and the likelihood of future change

| Climate event | Likelihood of future changes | | Regions with highest confidence level[a] |
|---|---|---|---|
| | Early twenty-first century | Late twenty-first century | |
| Warmer and/or fewer cold days and nights | >90 % probability | >99 % probability | N. America, C. America, Europe and Mediterranean, Southern Africa, Asia, SE Asia, Oceania |
| Increased frequency and/or duration of heat waves | *Not assessed* | >90 % probability | Europe and Mediterranean, Asia, Australia |
| Increase in the frequency, intensity, and/or amount of heavy precipitation | >66 % probability | >90 % probability | North and Central America, Europe and Mediterranean |
| Increases in intensity and/or duration of drought | *Low confidence* | >66 % probability | North America, Europe and Mediterranean, West Africa, Australia |
| Increases in intense tropical cyclone activity | *Low confidence* | 33–66 % probability | *Regional confidence levels unavailable* |
| Increased incidence and/or magnitude of extreme high sea level | >66 % probability | >90 % probability | *Regional confidence levels unavailable* |

Source: IPCC (2013, 2014)
[a]Listed regions were included if assessments were labeled "high confidence" (IPCC 2013); within any one region there is spatial variation in the listed climate event

should not be viewed as a means for justifying inaction on lower probability events, but it is a critical factor when evaluating and prioritizing adaptation strategies.

### *2.2.1 Climate Change and Episodic Disaster Events*

Climate change has the potential to alter the location, spatial extent, severity, and frequency of disaster events. The disaster triggering events most likely to be affected by climate change are briefly described below.

#### 2.2.1.1 Flooding

Increased severity and/or frequency of flooding can be caused by a variety of climate change impacts including sea level rise, intense rain, and rapid snowmelt. This is projected to pose particular challenges in Europe, parts of Asia, Australia, and small island nations (IPCC 2014). In coastal areas, sea level rise (SLR) is projected to result in increased coastal flooding and inundation. Coastal communities must also consider what increases in mean sea level will mean for extreme high tide events, coastal storms, and coastal erosion.

Three-quarters of large cities are located on the coast, with half of the world's population living within 100 km of the shore (UNEP and UN-HABITAT 2005). In addition to threats to human life, the proximity of population centers to the coast poses a particular threat to coastal development and infrastructure. Many industrial, port, and water reclamation facilities are located in low-lying, coastal areas. These facilities are often vulnerable to rising seas and could result in long-term disruption of critical community functions. In addition, sea level rise may also result in loss of tourist and recreational resources, which may have far-reaching economic consequences.

Intense rainfall and/or rapid snowmelt have the potential to result in increased loss of life, structures, and aquatic ecosystem resources above already identified disaster risks. Flood risk associated with climate change is progressive. Meaning, current strategies must to be assessed to determine if they will continue to be adequate, require strengthening, or necessitate a new approach entirely.

#### 2.2.1.2 Storm Events

Climate change may alter the severity and/or frequency of large storm events such as cyclones, hurricane, typhoon, and tornado. However, these projections are considered *low confidence* (Knutson et al. 2010; IPCC 2012, 2014). This is partly due to the complexity of climate events, which makes them difficult to model, and the relatively small scale of individual weather events. Nonetheless, in areas already victimized (however infrequently) by these events, consideration of current and future risk must be conducted.

#### 2.2.1.3 Fire

Climate change is projected to alter wildfire timing (seasonality and frequency), spatial distribution (fire size and complexity), and magnitude (intensity, severity, and type). This is projected as a particularly important consideration for North America (IPCC 2014). Climate change has the potential to alter wildfire occurrence due alteration of species mix, moisture and fuel load, wind patterns, and number of ignitions (Price and Rind 1994; Miller and Schlegel 2006; IPCC 2014).

#### 2.2.1.4 Landslide

Climate impacts cannot be considered in isolation. Climate change influenced landslides are most likely due to a combination of wildfire and intense rainfall events (Cal EMA and CNRA 2012). Further compounding this threat are the changes to species mix described below. A shift in vegetative species, due to seasonal temperature and precipitation impacts, can alter slope stability. These are critical considerations, particularly in communities already prone to periodic landslides.

#### 2.2.1.5 Drought and Extreme Heat

Drought results from prolonged periods of reduced rainfall. Extreme heat events, or heat wave, pose threats to public health and safety and can exacerbate the consequences of drought. Drought and/or heat wave are not always considered an episodic disaster event, but given the potential consequences to human health and safety, are increasingly being considered in a disaster management context. Drought also can affect the provision of basic needs such as food and water. Heat threats to human health may be increased for particular populations disproportionately sensitive due to intrinsic factors such as age, gender, ethnicity, and health, or extrinsic factors such as occupation and financial status (Parry et al. 2007; Portier et al. 2010; Cal EMA and CNRA 2012).

### *2.2.2 Non-disaster Climate Change Impacts*

The risk associated with climate change exacerbated disasters is a product of both episodic weather events, as well as slow shifts that may occur due to the progressive nature of climate change. This can include shifts in vegetative cover, altered disease vectors, reduced water supply, and changes to agricultural productivity. The progressive consequences of climate change have been described as a slow-onset disaster (e.g. IRIN 2012).

The consequences of a disaster event is related both the characteristics of the particular event (severity, duration, etc.) and the characteristics of the location being impacted. These characteristics can include the development pattern, local culture, economic base, as well as conditions of the biophysical context. All communities rely, to some degree, on the ecosystem services provided by the biophysical setting. This includes many factors that have been the focus of sustainability advocates for decades such as water, air, food, pollination, soils, nutrient cycling, and more. Progressive climate change may alter the provision of these services, which may affect a community's resilience in the face of disaster. Two of the changes most likely to affect disaster management are described below.

#### 2.2.2.1 Public Health and Social Equity

In addition to public health and safety threats of episodic events, climate change can result in long-term impacts such as food-, water-, and vector-borne diseases, food insecurity, water availability, water quality, and cardiorespiratory morbidity and mortality (Maibach et al. 2011; Cal EMA and CNRA 2012). Many of these impacts are produced due to changes in seasonal temperature and precipitation patterns. The risks to public health are directly related to intrinsic and extrinsic factors. Intrinsic factors such as age, pre-existing health status, and gender can influence susceptibility to health impacts. For example, children under the age of 5 and elderly over the age of 65 are more likely to be adversely affected by extreme heat events. Additionally, prolonged periods of elevated heat can affect air quality, which can further respiratory ailments. The slow decline in public health for particular populations can result in those populations being more heavily impacted or less able to respond in a disaster situation.

Extrinsic factors such as knowledge, occupation, geographic location, and financial status also affect individual risk. Climate change does not affect all populations equally. The extrinsic factors result in some populations being disproportionately vulnerable (Morello-Frosch et al. 2010). This is particularly true for minorities, the poor, and other socially marginalized populations (Cutter et al. 2009; Shonkoff et al. 2011; Cal EMA and CNRA 2012). Factors to consider include physical location, structural quality of residence, access to basic lifelines (e.g. communication, power, water), pre-existing knowledge, and occupation. Each of these can increase individual risk to slowly progressing climate change, which can, in turn, increase risk with respect to disaster events.

#### 2.2.2.2 Terrestrial and Aquatic Ecosystems

Seasonal temperature and precipitation pattern changes have the potential to alter the distribution of native habitat and species, which further exacerbates the impacts of urban development and environmental pollution. In addition, invasive species and pest ranges are projected to both expand and move (Snover et al. 2007; CNRA 2009). These changes in vegetative cover and associated species directly impact hazard events such as fire or landslide. The same climate shifts that stress native flora and fauna, also pose a threat to cultivated agriculture and livestock. These changes are particularly important in areas where crops are already near the upper end of their preferred temperature range or the lower end of the moisture range. In a disaster context, such impacts can limit local provision of food in recovery situations.

Climate change will also affect aquatic ecosystems. These ecosystems not only support fisheries, but also serve to provide water supply and hydroelectric power. Reduced precipitation and alteration in the seasonal distribution of rain will alter the flow level in rivers and streams. Water temperature may also increase, which affects water quality and the health of aquatic species. The reduced water levels during certain seasons and altered water quality will affect communities relying on these systems for water supply, food, and power. Again, food, water, electricity, and associated employment are all factors that must be considered when evaluating local capacity to withstand disaster events.

Ecosystems such as wetlands, forests, and estuaries provide natural disaster protection, particularly from coastal storms and flood events. The changes described above have the potential to affect the health of these protective ecosystems. Communities that rely on the protective ecosystem services provided by these habitats should include potential climate change impacts to their health as part of disaster management.

## 2.3 Addressing Climate Impacts on Disaster Events

> Heat stress, extreme precipitation, inland and coastal flooding, landslides, air pollution, drought, and water scarcity pose risks in urban areas for people, assets, economies, and ecosystems (very high confidence). Risks are amplified for those lacking essential infrastructure and services or living in poor-quality housing and exposed areas. Reducing basic service deficits, improving housing, and building resilient infrastructure systems could significantly reduce vulnerability and exposure in urban areas.
>
> - (IPCC Working Group II Summary for Policymakers 2014, p. 18)

Climate change is often perceived as occurring in the future. This can result in political reluctance to allocate resources that could be used to meet current needs. One of the most important things for urban planners, decision makers, and other community stakeholders to realize is that climate change, disaster management, and sustainable development are not distinct spheres of policy development; nor are they in conflict with short-term community needs. All areas of urban policy development are focused on sustaining long-term vibrancy and resilience. Each focus area within planning and policy offers considerations that must be included in any policy development process, regardless of which happens to be the focus at a particular time. In this discussion, it begins with disaster management, but effective disaster management integrates the needs and considerations required in the other planning and policy areas.

Disaster events are spatially specific. The communities in the location of the disaster most directly experience the impacts. While action to address climate change exacerbated disasters must be taken at all levels of government, the local level is a critical scale for policy development (IPCC 2012). There are three phases to developing strategies to address climate change impacts on disaster events: preliminary activities, risk assessment, and strategy development (Fig. 2.1). There is one action that must take place throughout all three phases, public outreach and education. The outreach and education lays the foundation for ongoing collaboration. This process should include residents, the various departments of the jurisdiction taking action, community groups and other non-governmental organizations, and science entities.

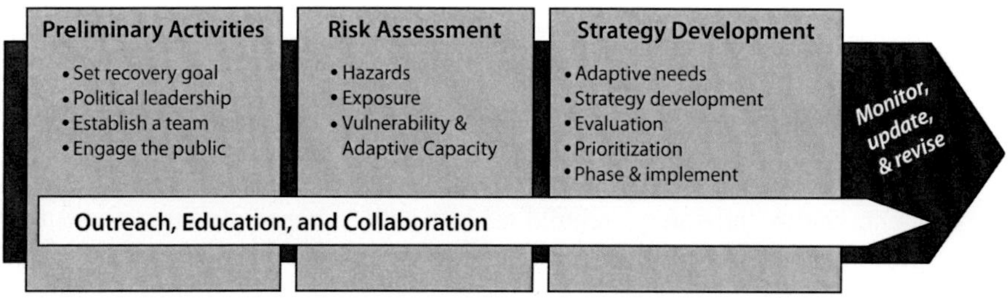


**Fig. 2.1** Steps in climate change adaptation policy development

Local government and other organizations at this scale are most aware of the specific situation in a given location from social setting, economic conditions, community networks, and cultural context. These factors are all critical to assessing vulnerability and identifying strategies most likely to be effective (Boswell et al. 2012). For this reason, the policy development steps presented in this section will refer primarily to local action. Many of the steps are appropriate for larger scales, but specific considerations may require adjustment.

## 2.3.1 *Preliminary Activities*

Disaster management and climate change adaptation overlap. This section details the preliminary steps to taking action that are equally applicable in both fields. Prior to conducting a risk assessment, decision makers and planners should complete a series of steps to assure that the risk assessment is comprehensive, that the strategies devised to address the identified risks are likely to be effective, and that implementation will be ongoing and iterative (Fig. 2.1). Binder et al. (2010) reviewed characteristics of communities with demonstrated success in developing climate change adaptation strategies and found that these communities shared several common characteristics including committed political leadership, establishment of a team of staff and local stakeholders, and community engagement throughout the process from initiation to implementation.

A critical goal for these preliminary steps is development of a feasible process. Too often, a lack of resources or a perceived need for precision is cited as a barrier to climate change adaptation policy development. It is critical communities view adaptation as both important and feasible. In particular, it is the most resource poor communities that are often disproportionately vulnerable to climate change exacerbated disasters. The steps described below can be pursued as part of a long-term, cost-intensive, and often quantitative process. Conversely, the same steps can be completed much more quickly with a small budget, where analysis is primarily qualitative. Communities can fall anywhere between these two extremes based on their defined community goals, resources, and needs. Regardless of approach, a prioritized set of community needs, a suite of strategies to address them, and a monitoring and implementation process can be developed.

### 2.3.1.1 Set Recovery Goals

In disaster situations, the term recovery often refers to the restoration of basic services; public health and safety; rebuilding lost and damaged structures; and the return of economic stability. In the past, recovery could be most easily defined as *back to the way things were before*. Climate change increases the chance that disasters, episodic or slow-onset, will alter the character of the affected community and associated biophysical system. What if going back is impossible? This can be due to land being no longer suitable for habitation due to flooding, subsidence, or other impacts. It can also be due to changes that limit the viability of certain economic enterprises such as ocean acidification or seasonal shifts that impact the productivity of fisheries or agriculture. In addition, factors such the availability or cost of water for residential or commercial use may limit the long-term viability of past land use patterns, building types, or businesses.

If the disaster itself, the resulting impacts, and the context for recovery cannot be precisely projected, how can communities prepare effectively for disaster? The answer to this question must start with a definition of recovery that does not use the past as its sole point of reference. Communities facing disasters want to assure that community members have access to food, water, and shelter. In addition, public health and safety, community function, and economic viability must be assured. These goals can be pursued without returning to the way in which they were met prior to the disaster event. Instead, a series of clearly defined desired future conditions must be developed. These goals can then be used as a point of evaluation for strategies to address projected impacts.

For example, one such goal could be community well-being. Well-being is a comprehensive concept that can be applied to both an individual and a community. What makes it useful when considering post-disaster recovery in the context of climate change is that well-being can be recovered, but the factors that contribute to the sense of well being do not have to be the same. This frees communities to focus on a preparing for disasters without being beholden to maintaining prior conditions. If the past is considered a starting point, rather than a desired endpoint, it begs the question of how do you plan for the future? The planning for the future begins with collaborative definition of the most important components of overarching goals such as well-being. These locally defined priorities add specificity to the goals.

Defining a set of specific community characteristics that should be maintained post-disaster is a critical step prior to engaging in risk assessment or strategy development. This definition of community characteristics that should be maintained or recovered can become the focal point when evaluating the potential level of community disruption posed by individual disaster events, as well as for prioritizing both adaptive needs and potential strategies. There is no standard rule or accepted set of community characteristic above that of health, shelter, and access to basic services. Communities need an economic and employment base, as well as other community resources. The desired nature of how these needs are met must be defined locally.

#### 2.3.1.2 Political Leadership

Regardless of scale, climate change can be a politically difficult issue to address due partially to temporal distance. Disaster management traditionally addresses hazards that are just as likely tomorrow as 10 years in the future. Though there is ample evidence that climate change is occurring now, the most damaging potential impacts to a community are often several decades into the future. This distance can lead decision makers to be hesitant to take action because it requires an allocation of funds that is perceived as not yielding short-term benefit (Binder et al. 2010). Education and outreach must begin with decision makers; the temporal distance should be viewed as an opportunity. The longer the timeframe, the greater the number of viable options for addressing climate change exacerbated disasters. Taking action in the short term allows strategies that take long periods of time, such as land use change or alteration of linear infrastructure, to be pursued. It also makes fiscal sense, as the relative cost of taking action in the short term will be considerably less than reactively addressing impacts after they occur.

Once support from decision makers has been obtained, a formalized commitment to address climate change and its influence on disasters is critical to long-term implementation. The commitment can be in the form of a resolution, added language to governing documents such as a comprehensive plan, and many other options depending on the entity pursuing these policies. A formal commitment clearly communicates to departments, residents, and other stakeholders that climate change and disaster management is a priority (Smit et al. 2000, 2009). Dedication of funds and/or staff to the effort also is more likely with a formal commitment.

#### 2.3.1.3 Establish a Team

Climate change has the potential to affect nearly every aspect of city function. A comprehensive risk assessment, even if the focus is solely on disaster events, will require an assessment that includes all structures, functions, and populations of a community. Accounting for this diverse set of information requires the assembly of a team (Smit and Wandel 2006; Boswell et al. 2012). Prioritization of community needs and development of effective policy also relies on specific information about potential impacts to a community. This information will be most accurate if developed by those most familiar with the structures, functions, and populations that may be impacted. In addition, this information is also critical to identifying current capacity to adapt and strategies most likely to be locally effective. A team comprised of individuals from the community and government can also serve to facilitate integration across departmental boundaries and strengthen community connection.

#### 2.3.1.4 Engage the Public

Effective, ongoing policy development and implementation requires support and input from the affected populations. It can be difficult for decision makers to take action without support and acceptance of the both the potential impacts and proposed measures to address them. This support relies on development of an informed and involved community. An engaged community also is critical to maintaining momentum over time when faced with budget cuts and/or staff turnover. An involved public helps ensure that policy measures address community needs and can be efficiently implemented (Boswell et al. 2012; Cal EMA and CNRA 2012).

## *2.3.2 Risk Assessment*

The intent of a risk assessment is to determine the likelihood of a disaster occurring, the expected severity, and potential consequences. This assessment serves as the basis for the development of strategies to address the identified risks (Fig. 2.1). The IPCC (2012) defines risk as being produced by a combination of hazards, exposure, and vulnerability. These elements can be determined in varying levels of detail depending on the resources (staff and funding) available to dedicate to the task and the timeframe for completing the assessment.

Risk assessment is presented in a series of steps, but should be viewed as an iterative process. Each subsequent step may yield a need for additional information from prior steps. Completion of each step relies on the team described above.

### 2.3.2.1 Hazards

Climate change exacerbated natural hazards are produced through two interacting processes: natural weather patterns and anthropogenic climate change (IPCC 2014). This goal of this step is to identify which hazards may impact a particular location or jurisdiction and how climate change is projected to alter the frequency, severity, location, duration, or spatial extent.

Determining this information requires climate science. The availability of high-resolution climate projections for any particular location varies. The IPCC provides a global and regional perspective, which in the absence of any other information can still be used as a basis for risk assessment and policy development (2013, 2014). Many national and regional governments, as well as academic institutions and non-governmental organizations, have developed climate projections, often with higher spatial resolution. However, even these downscaled projections have limited resolution, between 5 and 12 km resolution (e.g. Solomon et al. 2007; Cayan et al. 2011). Communities should seek out the best data available to serve as a basis for risk assessment.

Most scientific projections of climate change will include temperature, precipitation, and sea level rise. In some cases, additional changes also will be included such as wildfire, flooding, tropical storms, drought, and heat wave. There is a suite of secondary consequences that can occur due to the interaction of direct changes to the climate (Table 2.2). While specific projections may not be available for these events, the climate changes that contribute will be and should be assessed with them in mind.

**Table 2.2** Climate change impacts and selected secondary impacts

| Climate change impact | Associated secondary impact |
|---|---|
| Sea level rise | Inundation or long-term waterline change |
| | Extreme high tide |
| | Coastal erosion |
| | Saltwater intrusion |
| Changed temperature and/or precipitation patterns | Changed seasonal patterns |
| Increased temperature | Heat wave |
| Increased temperature and/or reduced precipitation | Drought |
| | Wildfire |
| | Reduced snowpack |
| Increased temperature and/or changed precipitation | Intense rainstorms |
| | Flooding |
| Wildfire and/or increased precipitation | Landslide |

Source: IPCC (2013) and Cal EMA and CNRA (2012)

The informational goals for defining the climate change influence on hazard events focus on the extent and speed of change expected as compared to current conditions. In addition, a similar evaluation of non-hazard climate changes should also be conducted because these changes will influence local vulnerability to hazard events and the capacity for recovery. The key questions to ask for all potential climate changes are listed below. A jurisdiction can make these evaluations qualitatively by simply ranking responses on a high, medium, and low scale, or spend considerably more time and money for increased precision and quantitative evaluation.

1. *How different are conditions projected to be from current conditions*? This evaluation should match the time horizon for planning (e.g. 2030). Communities can also choose to use 2050 and 2100 as evaluative benchmarks due to the frequency of use in scientific projections.
2. *How quickly are the changes projected to occur*? This does not have to be quantitatively specific. Simply classifying changes in categories such as near term (20 years or less), mid-term (20–50 years), or long-term (greater then 50 year) can be enough.
3. *Over how large an area are these changes projected to occur*? This question is about determining how much a community may be impacted. This too can be qualitative (high, medium, or low) or quantitative with specific estimates such as potentially inundated land areas.
4. *How certain are the projections*? As described above and listed in Table 2.1, climate projections have varying levels of statistical certainty associated with them. This is a measure of how likely the projected impacts are to occur based on best available science. Climate projections are updated and refined regularly. This measure, along with the others, can and should be revised as additional data come available.

#### 2.3.2.2 Exposure and Vulnerability

This step asks three key questions: (1) What aspects of the community will be exposed to the identified disaster triggering events?; (2) Who or what will be affected by this exposure?; and (3) How prepared is the community for these points of exposure? Answering these questions in a systematic manner requires involvement from the team described above. Detailed, local knowledge of the land development pattern, local economy, social interaction, political context, regional ecological health, and other community functions is required to comprehensively and accurately complete this step.

*Exposure*, also referred to as sensitivity, simply evaluates which aspects of a community may be affected by the hazards identified above. It can be a simple checklist that identifies whether or not a particular change is likely to be experienced by certain aspects of a community. This is a standard part of disaster management where structures are evaluated for vulnerability to projected hazard events. This process should also extend to populations, community functions, and ecosystems, particularly those providing ecosystem services such as wetlands or floodplains. Populations can be specific groups that may be disproportionately vulnerable to projected hazards due to factors such as health, age, employment, residence location, structural quality of residence, language spoken, and transportation access. Community function can be factors such as economic continuity or community cohesion. Disasters have the potential to disrupt transportation infrastructure, which may disrupt supply chains for local businesses, or communication networks, which has the potential to limit local communication. Ecosystems provide several services to communities that can be disrupted by disaster events such as local capacity to sustain community function and the ability to withstand future events. These impacts must all be identified in this step, though the extent of the potential risk is not determined here.

Particular attention should be paid to critical nodes or particularly vulnerable populations. This can include locations that hold hazardous materials, where damage could endanger local populations, or critical infrastructure nodes that could disrupt the provision of basic services such as water or power. Also in need of consideration are locations of particular cultural value such religious structures or community services such as schools and hospitals. A team comprised of multiple jurisdictional departments and community stakeholders must define the most important elements in a given community.

*Vulnerability* assessment includes evaluation of potential impacts and existing adaptive capacity. This evaluation answers two questions: (1) What and who will be affected for each point of exposure; and (2) How prepared is the community? In a disaster management process, this step is often data intensive and detailed. As compared to historic disaster recurrence estimates, climate change projections have a relatively high level of uncertainty that escalates at smaller spatial scales and grows yet again when identifying consequences that may be several links down a causal chain. Due to this uncertainty, vulnerability assessment is often qualitative. Similar to the other steps, this evaluation relies heavily on the team of staff and community members to assess risk and devise strategies.

It is important that this analysis be comprehensive for each community resource affected. For each point of exposure the following should be determined (Cal EMA and CNRA 2012). Based on the answer to these questions a rating of high, medium, or low can be determined. This rating must reflect local context and community needs.

(a) The temporal extent of the impact
(b) The spatial extent of the impact
(c) The permanence of the impact
(d) The level of disruption to normal community function

The evaluation of each point of exposure should be as comprehensive as possible. For example, if a hospital or water treatment facility is projected to be impacted by climate change exacerbated hazards, all residents reliant on these facilities must also be identified as potentially vulnerable. The analysis then must assess how long the facility, and those reliant on it, would be impacted (e.g. How long would floodwater inundation persist?), the spatial extent (e.g. Is it one portion of the facility or the whole site?), how permanent (e.g. Is repair possible and/or advisable?), and how much the disruption of the facility would disrupt other community processes (e.g. Are there other community functions that would be disrupted such as industrial uses reliant on the water? or Are there threats such as health consequences to those reliant on the facility if the disruption persists?).

It is also critical that the potential interaction between impacts be evaluated. For example, recovery from a flood disaster may be hampered by limited access to food and water due to progressive climate change that has affected water supply and agricultural productivity. Similarly, health threats from hazards may be amplified by non-disaster climate impacts such as altered disease vectors. These potential points of interaction must be included the assessment of potential impacts and the existing capacity to adapt, the second component of vulnerability assessment.

The other part of an assessment of vulnerability is the capacity to meet the challenges posed by the identified impacts. For each potential point of exposure that has been evaluated for who and what it will be affected, the local capacity to address the threat must be assessed. This final step helps determine the areas most in need of adaptation strategy development. Existing local plans and policies should be evaluated. In addition, proposed or not yet implemented measures should also be included. In a community that already has robust measures in place to address a particular hazard, there may be little change necessary. Conversely, a community may find that it is currently underprepared for some projected impacts. This step seeks to identify the extent to which a community is ready.

For each potential impact evaluated, the following questions should be addressed (Cal EMA and CNRA 2012).

- To what extent does existing policy addresses the potential impacts?
- Are there actions in progress, planned, or readily implemented to address projected impact?

- If the policy or program is not yet implemented, how long will it take and what are the resources necessary for implementation?
- Can existing policy or programs be strengthened to address the projected impacts?

The identification of existing policy that have demonstrated local effectiveness also lays a foundation for future strategy development. The level or readiness can also be categorized simply in into low, medium, and high. However, this process requires that a local team of stakeholders and governmental staff has specifically defined how low, medium, and high were determined.

### 2.3.3 *Strategy Development*

Strategy development and implementation should receive immediate action, partly due to the lag in time before the most damaging impacts are projected to occur. While this statement may seem contradictory, it is critical that decision makers understand the potential benefits of pre-emptive action. The earlier an action is taken to address potential climate change impacts, the greater the number of viable options to address the risk and the lower the cost. Some strategies such as changes to land use pattern and/or diversification of economic base can take a long period of time to implement. Similarly, for strategies that utilize new approaches or rely on technology still under development, it is critical to have time to implement, evaluate, and adjust for long-term effectiveness.

One of the reasons decision makers can be resistant to taking action on climate impacts is that they will prioritize more immediate needs over those in the future. One remedy to this apparent conflict is policy integration. Effective adaptation policy is integrated into existing operations and governing plans. Public safety, environmental health, GHG emissions reduction, and economic development are all potential short-term co-benefits of adaptation and disaster management. Measures that address potential climate change impacts as part of standard planning practices have been identified as a critical characteristic of successful strategy implementation (Urwin and Jordan 2008; Binder et al. 2010).

The risk assessment yields a set of community needs and an assessment of the current level of preparedness. The first step in developing policy is prioritizing these needs. Often this assessment is qualitative and can be difficult to use by decision makers to justify the allocation of resources. To address this difficulty many organizations, jurisdictions, and guidance resources advocate the use of decision support tools. One such tool, decision matrices, provide a defensible way to balance the qualitative evaluation of risk and the subsequent evaluation of strategies developed to address the risk (e.g. DEH 2006; Snover et al. 2007; NRC 2010; Cal EMA and CNRA 2012; FEMA 2013).

The first step in developing strategies is prioritization of the community needs identified in the risk assessment. This prioritization should balance the certainty of

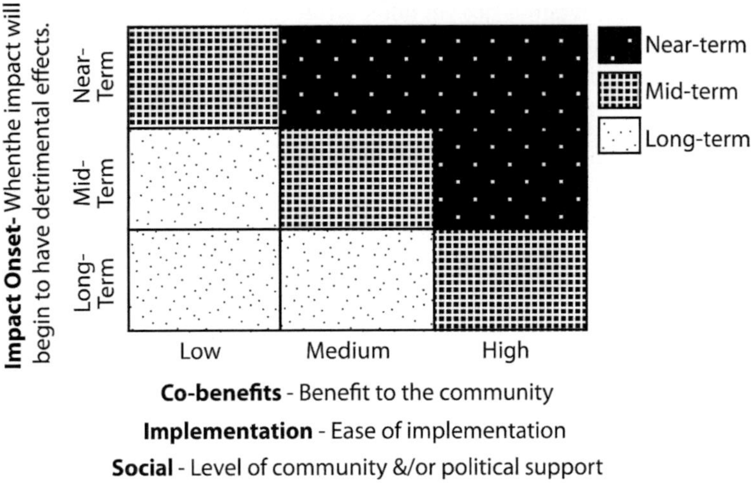


**Fig. 2.2** Example decision matrix (Source: Adapted from Cal EMA and CNRA 2012)

the projected impact, the level of vulnerability, speed of onset, adaptive capacity and other characteristics determined as part of the risk assessment. The team that conducted the risk assessment should be used to determine how the categories in a decision matrix are defined. The matrices, or other decision support tools, not only provide transparency for the process, but also can be easily communicated to stakeholders.

The matrices can be used to assist the local policy development team with the following decisions (Fig. 2.2 provides an example):

- Identifying impacts that deserve immediate action and which warrant further monitoring and investigation based on factors such as the following:
  - Speed of onset
  - Certainty
  - Vulnerability measures
  - Adaptive capacity
- Determining which strategies should be identified for implementation based on factors such as the following:
  - Cost
  - Co-benefits
  - Political support
  - Period of time needed for implementation
- Phasing of strategies (near-, mid-, and long-term) based on factors such as the following:
  - Speed of impact onset
  - Associated vulnerability

- Period of time to begin implementation
- Envisioned duration of implementation

#### 2.3.3.1 Adaptation Strategies Characteristics

While the decision matrices provide a tool for analyzing qualitative categorizations, it does not help formulate specific policy recommendations. From the risk assessment, community needs should be prioritized. Those deemed worthy of attention should have strategies developed. The following list has characteristics of effective adaptation measures (Boswell et al. 2012; Cal EMA and CNRA 2012).

- *Flexible*: Climate change means devising strategies in an uncertain context. As a result, effective measures are applicable in a wide range of conditions (Smith 1997). de Loe et al. (2001) takes the definition of flexibility even further, advocating for reversibility if a strategy's outcome does not turn out as anticipated.
- *Cost Effective*: Balancing immediate community needs and long-term adaptation needs is often a challenge for decision makers, particularly in economically trying times. Co-benefits is a term often used to describe the short-term community benefits of long-term adaptation policies. Such dual-purpose policies can be viewed as economically efficient and politically advantageous.
- *Targets Irreversible Impacts*: The oft repeated phrase of 'no regrets' adaptation planning refers to taking action when potentially catastrophic consequences are projected even if it is a low probability event. Smith (1997) provides a bit more detail to what could be considered 'no regrets' situations: (1) irreversible impacts such as extinction, loss of an ecosystem, or extreme weather (i.e. hurricanes); (2) unfavorable trends where enacting adaptation measures now is more feasible now than it is likely to be in the future (i.e. limiting population density in coastal areas); and (3) decisions, such as those regarding infrastructure, that have long life spans.
- *Specific*: One component to developing policy in an uncertain context is specifically target an impact in a particular context (Smit and Wandel 2006). Through the risk assessment process, impacts will have an identified speed of onset, rate of change, and anticipated scale (Smit et al. 2000). This information is critical to tailoring policy to address particular anticipated impacts.
- *Feedback loops and adaptive management*: In an uncertain context, strategy effectiveness is not assured. Long-term implementation requires ongoing monitoring to evaluate the success of adaptation measures. In addition, climate science is regularly updated and new technology can be developed. These factors should also inform revision of adaptation strategies.

#### 2.3.3.2 Implementation

The final stage in policy development is assuring that strategies are effective. Climate change impacts to both episodic disasters and slow-onset changes occur over the course of many decades, meaning success or effectiveness is only observable in the long-term. As a result, implementation requires sustained effort. There are a few critical components that can bolster implementation effectiveness.

- *Responsible entity*: A department, agency, or staff person, should be identified for each strategy. This entity should be funded specifically for the task. To complement the funded staff person, department, or organization, a sustainable funding source for implementation should also be defined and allocated.
- *Monitoring*: The single most important contributor to long-term effectiveness is the identification of indicators for each strategy and the establishment of monitoring systems to track them. The collection of data to track implementation is critical to creating feedback loops that allow for adjustment and update. Climate adaptation is a fairly new area of policy development and climate science is updated regularly. All strategies intended to address climate impacts must be regularly evaluated to assure that they are working as intended and continue to be adequate measures for the projected impacts. If or when updated climate change data indicates impacts may be more severe than expected or that strategies are not as effective as envisioned, this data should be incorporated into the steps of the adaptation strategy development process to revisited and adjust the strategies necessary to assure long term resilience.

## 2.4 Conclusion

The ultimate goal should be for climate change, disaster management, and sustainability, as distinct areas of policy development, to disappear because they have become standard considerations in all areas of urban planning and management. Until that occurs, there are some overarching themes that should be kept in mind when considering preparation for climate change exacerbated disasters on a local, regional, or national scale. The first is that climate change, disaster management, and sustainable development are inseparable. For example, while urban development and economic growth of the past may have contributed to climate change, increased disaster risk, and lower regional sustainability; they are also critical components of long-term resiliency. The difference is the revision of the goals that guide policy development. These goals that guide future growth should keep the following in mind.

- *Diversity*. Resilient systems have diversity and redundancy. In a climate change and disaster context that means bolstering local resilience and reducing risk by assuring that that any external change, such as a disaster triggering event, will not undermine the viability of most community functions. For example, if a city's economic and employment base is predominantly dependent on a single industry or a single resource, disaster events affecting these resources will have far reaching consequences. Diversity in a local economy is one factor that strengthens capacity to recover. This is because, while some industries will be

dealing with severe impacts, others are less likely to be, which dampens the impact on the overall economy. Similar to economic and employment bases, diversity is also critical in other areas of disaster management, particularly in the context of climate change. The types of strategies developed to address climate change exacerbated disasters should also be diverse (e.g. mandate, incentive, structural, educational, etc.). This diversity reduces the likelihood of gaps occurring in the overall management plan and provides a good process by which to test and refine policy making. The final area in which diversity should be a focus is in the participants in the policy development and implementation process. Communities, cities, regions, and more are diverse. The more voices informing the process the more likely support for the strategies and collaborative implementation will occur. In situations where success is partially reliant on dynamic self-organization, developing a network of engaged, aware stakeholders can be a good predictor of adaptive capacity.
- *Short-term Benefits and Long-Term Goals.* Disaster management and climate change adaptation, in particular, require long time horizons. In many cases, the achievement of policy goals will be distant and require ongoing momentum to achieve full implementation. This momentum requires identification of short-term benefits and clear communication to decision makers and community members to raise awareness. Not all disaster management and climate adaptation strategies lend themselves equally well to the dual purpose, but those that do should be prioritized. Questions to consider when evaluating policy choices are the following: Does the strategy bolster long-term economic viability?; Will a strategy reduce GHG emissions?; Will local ecosystem health (and associated ecosystem services) be strengthened as a result?; and Does the strategy contribute to regional, national, and global sustainability efforts by not creating undue burden in other areas (externalities)?
- *Regional Connections.* Climate change is a global phenomena with spatially specific consequences. The challenge for cities is in managing their own risks in the context of regions also experiencing climate change impacts. For this reason, regional connections must be established or bolstered and should be considered in the formulation of policy. Local economies are reliant on resource supply chains, infrastructure networks, and more that have regional, national, or even global connections. This context must be included in vulnerability assessment as potential points of weakness and identified in policy development as focus areas for strengthening. This means careful examination of a local economic base and identifying those industries most likely to thrive in the future with most local, regional, and global benefits. Two other areas where regional collaboration can be particularly important is in the development of higher resolution climate science and in development of policy options for impacts that have large spatial extents. Development of high resolution climate change projections can be costly, collaboration between regional partners can be beneficial to all jurisdictions. Similarly, some climate change impacts, such as those related to sea level rise, require regionally complementary strategies.

Disasters are events that occur infrequently and can result in extensive disruption to all aspects of an urban system. Climate change has the potential to further the potential consequences by altering both the characteristics of the disaster-triggering event itself and by altering the capacity of impacted communities to recovery, adapt, and transform. These changes cannot be precisely projected, but the risks associated with climate change exacerbated disasters are high enough that measures must be developed to address them despite the associated uncertainty. Effective policy development in this context requires measures that are flexible and monitored for effectiveness. Too often monitoring is sacrificed for budgetary reasons. In a dynamic context, where new data is regularly available, monitoring to assure that strategies are performing as anticipated and continue to adequately address community disaster management needs is critical. More than other forms of policy that are updated every 5–10 years, climate change policy must be reviewed, revised, and updated on a more frequent basis.

## References

Alesch DJ, Siembieda W (2012) The role of the built environment in the recovery of cities and communities from extreme events. Int J Mass Emerg Dis 32(2):197–211

Alesch DJ, Arendt LA, Holly JN (2009) Managing for long-term community recovery in the aftermath of disaster. Public Entity Risk Institute, Fairfax

Australian Greenhouse Office of the Department of the Environment and Heritage (DEH) (2006) Climate change impacts & risk management – a guide for business and government. Retrieved from http://ccsl.iccip.net/risk-management.pdf

Binder LCW, Barcelos JK, Booth DB, Darzen M, Elsner MM, Fenske R, Graham TF et al (2010) Preparing for climate change in Washington State. Clim Change 102(1–2):351–376. doi:10.1007/s10584-010-9850-5

Boswell MR, Greve AI, Seale TL (2012) Local climate action planning. Island Press, Washington, DC, 284 p

California Natural Resources Agency (CNRA) (2009) 2009 California climate adaptation strategy. Author, Sacamento, 200 p

California Emergency Management Agency (Cal EMA) & California Natural Resources Agency (CNRA) (2012) California climate adaptation planning guide. Retrieved on December 13 2013 from http://resources.ca.gov/climate_adaptation/local_government/adaptation_policy_guide.html

Cayan D, Tyree M, Dettinger M, Hidalgo H, Das T, Maurer E, Bormirski P, Graham N, Flick R (2011) Climate change scenarios and sea level rise estimate for the California 2009 climate change scenarios assessment. California Climate Change Center, Sacramento. CEC-500–2009–014-F. Retrieved from http://www.energy.ca.gov/2009publications/CEC-500–2009–014/CEC-500–2009–014-F.PDF

Cutter S, Emrich C, Webb J, Morath D (2009) Social vulnerability to climate variability hazards: a review of the literature. Final Report to Oxfam America. Retrieved from http://adapt.oxfamamerica.org/resources/Literature_Review.pdf

de Loe R, Kreutzwiser R, Moraru L (2001) Adaptation option for the near term: climate change and the Canadian water sector. Glob Environ Chang 11(3):231–245

Djalante R, Holley C, Thomalla F (2011) Adaptive governance and managing resilience to natural hazards. Int J Disaster Risk Sci 2:1–14

Federal Emergency Management Agency (FEMA) (2001) Understanding your risks: identifying hazards and estimating losses. FEMA 386-2. Retrieved from http://www.fema.gov/library/viewRecord.do?id=1880

Federal Emergency Management Agency (FEMA) (2013) Local mitigation planning handbook. Author, Washington, DC, 162 p, http://goo.gl/xPqbEL
Folke C (2006) Resilience: the emergence of a perspective for social-ecological systems analyses. Global Environ Change 16:253–267
Godschalk DR, Beatley T, Berke P, Brower DJ, Kaiser EJ (1999) Natural disaster mitigation: recasting disaster policy and planning. Island Press, Washington, DC, 591 p
Government of Japan (2011) Disaster management in Japan. Retrieved from http://www.bousai.go.jp/1info/pdf/saigaipanf_e.pdf
Intergovernmental Panel on Climate Change (IPCC) (2012) In: Field CB, Barros V, Stocker TF, Qin D, Dokken DJ, Ebi KL, Mastrandrea MD, Mach KJ, Plattner G-K, Allen SK, Tignor M, Midgley PM (eds) Managing the risks of extreme events and disasters to advance climate change adaptation. A special report of Working Groups I and II of the IPCC. Cambridge University Press, Cambridge, 582 pp
Intergovernmental Panel on Climate Change (IPCC) (2013) In: Stocker TF, Qin D, Plattner G-K, Tignor M, Allen SK, Boschung J, Nauels A, Xia Y, Bex V, Midgley PM (eds) Climate change 2013: the physical science basis. Contribution of Working Group I to the fifth assessment report of the Intergovernmental Panel on Climate Change. Cambridge University Press, Cambridge, 1535 pp
Intergovernmental Panel on Climate Change (IPCC) (2014) In: Field CB, Barros VR, Dokken DJ, Mach KJ, Mastrandrea MD, Bilir TE, Chatterjee M, Ebi KL, Estrada YO, Genova RC, Girma B, Kissel ES, Levy AN, MacCracken S, Mastrandrea PR, White LL (eds) Climate change 2014: impacts, adaptation, and vulnerability. Part A: global and sectoral aspects. Contribution of Working Group II to the fifth assessment report of the Intergovernmental Panel on Climate Change. Cambridge University Press, Cambridge
IRIN (2012) Disasters: slow-onset disasters take toll. Retrieved from http://www.irinnews.org/report/96989/disasters-slow-onset-disasters-take-toll
Knutson TR, McBride JL, Chan J, Emanuel K, Holland G, Landsea C, Held I, Kossin JP, Srivastava AK, Sugi M (2010) Tropical cyclones and climate change. Nat Geosci 3:157–163
Maibach E, Nisbet M, Weathers M (2011) Conveying the human implications of climate change: a climate change communication primer for public health professionals. George Mason University Center for Climate Change Communication, Fairfax
Manyena SB (2006) The concept of resilience revisited. Disasters 30:434–450
Miller NL, Schlegal NJ (2006) Climate change–projected Santa Ana fire weather occurrence. Sacramento: California Climate Change Center-California Energy Commission. Publication # CEC-500-2005-204-SF
Morello-Frosch F, Pastor M, Sadd J, Shonkoff S (2010) The climate gap: inequalities in how climate change hurts americans & how to close the gap. University of Southern California, Program for Environmental and Regional Equity (PERE). Retrieved from http://college.usc.edu/pere/documents/ClimateGapReport_full_report_web.pdf
National Research Council (NRC) (2010) Adapting to the impacts of climate change. National Academies Press, Washington, DC, 293 p
Parry ML, Canziani O, Palutikof J, van der Linden P, Hanson C (2007) Climate change 2007: impacts, adaptation and vulnerability. Contribution of Working Group II to the fourth assessment report of the Intergovernmental Panel on Climate Change. Cambridge University Press, Cambridge/New York
Portier CJ, Thigpen Tart K, Carter SR, Dilworth CH, Grambsch AE, Gohlke J, Hess J, Howard SN, Luber G, Lutz JT, Maslak T, Prudent N, Radtke M, Rosenthal JP, Rowles T, Sandifer PA, Scheraga J, Schramm PJ, Strickman D, Trtanj JM, Whung P (2010) A human health perspective on climate change: a report outlining the research needs on the human health effects of climate change. Environmental Health Perspectives/National Institute of Environmental Health Sciences, Research Triangle Park. doi:10.1289/ehp.1002272. Available at www.niehs.nih.gov/climatereport
Price C, Rind D (1994) Possible implications of global climate change on global lightning distributions and frequencies. J Geophys Res 99(10):823–831

Shonkoff S, Morello-Frosch R, Pastor M, Sadd J (2011) The climate gap: environmental health and equity implications of climate change and mitigation policies in California – a review of the literature. Clim Change 109(Suppl 1):S485–S503. doi:10.1007/s10584-011-0310-7
Smit B, Wandel J (2006) Adaptation, adaptive capacity and vulnerability. Glob Environ Chang 16(3):282–292
Smit B, Burton I, Klein RJT, Wandel J (2000) An anatomy of adaptation to climate change and vulnerability. Clim Change 45:223–251
Smith JB (1997) Setting priorities for adapting to climate change. Glob Environ Chang 7(1):251–264
Smith JB, Vogel JM, Cromwell JE III (2009) An architecture for government action on adaptation to climate change: an editorial comment. Clim Change 95(1–2):53–61
Snover AK, Whitely Binder L, Lopez J, Willmott E, Kay J, Howell D, Simmonds J (2007) Preparing for climate change: a guidebook for local, regional, and state government. In association with and published by ICLEI – Local Governments for Sustainability, Oakland
Solomon S, Qin D, Manning M, Marquis M, Averyt K, Tignor MMB, Miller HL, Chen Z (2007) Climate change 2007: the physical science basis: contribution of Working Group I to the fourth assessment report of the Intergovernmental Panel on Climate Change. Summary for Policymakers. Cambridge University Press, Cambridge
UN News Centre (2014) Ahead of Sendai conference, Ban pushes for 2015 global agreement on disaster risk. Retrieved from http://goo.gl/7hjKWG
United Nations Environment Programme (UNEP), United Nations Human Settlements Programme (UN-HABITAT) (2005) Coastal area pollution – the role of cities. Retrieved from http://www.unep.org/urban_environment/PDFs/Coastal_Pollution_Role_of_Cities.pdf
Urwin K, Jordan A (2008) Does public policy support or undermine climate change adaptation? Exploring policy interplay across different scales of governance. Glob Environ Chang 18(1):180–191. doi:10.1016/j.gloenvcha.2007.08.002

# Chapter 3
# Risk and Vulnerability

**Akhilesh Surjan, Shimpei Kudo, and Juha I. Uitto**

> *Risk in the modern world is confronted and dealt with in three fundamental ways. Risk as feelings refers to our fast, instinctive, and intuitive reactions to danger. Risk as analysis brings logic, reason and scientific deliberation to bear on hazard management. When our ancient instincts and our modern scientific analyses clash, we become painfully aware of a third reality – risk as politics. (Paul Slovic in 'The Feeling of Risk', Page 21, 2010)*

**Abstract** The authors argue that sustainable development cannot be achieved without consideration of risk and vulnerability. Losses due to natural disasters, including those related to extreme climatic events, have been on the rise but risk and vulnerability are not equally distributed. Climate risk affecting both natural and human systems affects geographical regions to differing extents. Coastal areas where half of the world's population and many major urban areas are located will bear the brunt of storms and sea level rise. Similarly, risks of flooding, drought etc. are unevenly distributed. Vulnerability is dependent on social, economic and political factors. In many ways, poor people are more vulnerable to climatic hazards, often living in exposed areas and substandard housing, having inadequate means to prepare for a recover from shocks. They are also vulnerable to slow-onset disasters. Vulnerability has also a psychological dimension and trauma caused by natural disasters can lead to long-standing psychological damage and a changed perception of the external world.

**Keywords** Disaster risk • Environmental risk • Psychological vulnerability • Human vulnerability

---

A. Surjan (✉)
Faculty of Engineering, Health, Science and the Environment, Charles Darwin University, Darwin, Australia
e-mail: akhilesh.surjan@cdu.edu.au

S. Kudo
Inter-Graduate School Unit for Sustainable Development and Survivable Societies, Kyoto University, Kyoto, Japan

J.I. Uitto
Independent Evaluation Office, Global Environment Facility, Washington, DC, USA



J.I. Uitto, R. Shaw (eds.), *Sustainable Development and Disaster Risk Reduction*, Disaster Risk Reduction, DOI 10.1007/978-4-431-55078-5_3

## 3.1 Introduction

History of human evolution shows that risk and vulnerability have always been associated with humankind, although the origins and sources of risk may be changing over the period of time. The focus of this chapter is to present risk and vulnerability in the context of sustainability. Throughout the chapter we relate risk and vulnerability to the environmental, social, economic and political dimensions of sustainability. While hazards are occurring due to natural causes, their impacts on societies, communities and individuals vary because of societal forces and processes. Human induced environmental degradation and climate change directly affect risk, while poverty, distribution of political power and a variety of other societal factors render some people and groups more vulnerable than others.

Risk is described as a function of hazard and vulnerability (Wisner et al. 2004). Hazards are natural phenomena, occurring since time immemorial. With the advancement in scientific and technological know-how, natural hazards can be systematically studied. The world community now better knows hotspots of different types of hazards – such as seismically active areas, cyclone prone areas, flood prone areas, and so on. It is also recognized that some hazard types are very difficult or almost impossible to predict (such as earthquakes); however hydro-meteorological hazards such as floods and cyclones can be predicted and tracked almost on a real-time basis. Thus, preparedness, early warning and mitigation receives high attention in order to minimize consequences of these hazards on human settlements and the ecosystem they depend on. Vulnerability is much more dynamic and difficult to comprehend and hence emerged as a major area of concern to protect humankind from environmental and climate related hazards.

This chapter will revisit risks in general and those from natural disasters in particular. Various dimensions of vulnerability are highlighted in order to demonstrate that efforts towards achieving sustainable development will remain incomplete without considering risk and vulnerability interaction. We conclude the chapter by discussing risk and vulnerability interactions and by introducing a sustainable development perspective to risk and vulnerability.

## 3.2 Revisiting and Framing Risk

Humans are by nature risk takers. It is very common to see people taking risk in their everyday life (Fischhoff and Kadvany 2011). Depending on the perception and acceptance level, risk taking behavior may vary among individuals even if they belong to the same family, society, nationality or geography. It is a human tendency to 'perform balancing act' by way of judging dangers and rewards associated with everyday risks (Adams 1995). Thanks to the efforts of the World

Health Organization (WHO) and a large number of other organizations that have been successful in clearly conveying that smoking tobacco is one of the top three causes of preventable death, in most countries it is mandatory to warn about the dangers of smoking on each pack of cigarettes. Government and non-governmental organizations also highlight this fact and run awareness campaigns to inform masses about the negative impacts of smoking on human health. Despite this, WHO estimated that the tobacco epidemic is responsible for killing up to six million people annually, from direct tobacco use and second-hand smoke (WHO 2014). This example illustrates that information is not enough to change risk behavior. There is room for more innovative awareness raising communication and preventive techniques to prevent tobacco use and there is a long way to go to achieve success. While risks may emanate from a variety of sources, "they arise from our own acts and are imposed on us" (Fischhoff and Kadvany 2011). Beck (1992) noted that in the modern times "social production of wealth is systematically accompanied by the social production of risks". This assertion still holds true even in the twenty-first century. However, climate change related risks are introducing a new dimension where unsustainable lifestyles and consumption patterns in one part of the world are bringing environmental calamities in other parts.

This section will view risk from the lenses of natural hazards, climate change, environmental degradation and psychological perspectives.

### *3.2.1 Disaster Risk*

Traditionally, disasters have often been categorized as natural disasters and man-made disasters. However, in recent times, this perspective is changing. A growing body of knowledge acknowledges that disasters are not 'natural', rather they are products of a built-environment created through specific human decisions without due consideration to natural hazards or considering the impact of decisions on the environment at large. Accurate quantification and estimation of losses and damages associated with disaster risks and benefits from risk reduction efforts are difficult, lengthy, data intensive and complex. "The combination of severe consequences, rarity, and human as well as physical determinants makes disasters difficult to study" (IPCC 2012). However, there is no doubt that risk reduction measures significantly reduce loss of lives and assets in disaster prone areas. Therefore, there is a strong emphasis on investing more in risk reduction measures instead of waiting for more accurate and convincing estimates through scientific research. The World Risk Report (UNU 2014) studied over 170 nations and developed an index to map risk. As per this report, global risk hotspots are Oceania, South East Asia, Central America and the southern Sahel region, "where high exposure to natural hazards and the effects of climate change, such as sea level rise, coincides with very highly vulnerable societies" (ibid.).

While extreme events and large-scale disasters often receive international attention, frequent small-scale disasters remain unnoticed. "Accumulated losses from small-scale, highly frequent and localized disasters are similar in magnitude to those of major catastrophes. These losses contribute to declines in social welfare, economic growth and ecosystems. Analysis of new national disaster loss data confirms this regularly ignored truth. Communities, particularly in low and middle-income countries, experience regular small disasters that undermine local development as well as national competitiveness" (UNISDR 2013). Small (local) disasters are expected to be managed by sub-national and local governments, there remains a huge capacity gap at this level. Local level higher education institutions, community based organizations and businesses need to work hand in hand with local government to address small-scale disasters effectively.

### *3.2.2 Risk from Climate Change*

Climate change is often highlighted in terms of growing concentration of greenhouse gases (GHGs) resulting in global temperature rise. Climate change is considered as a main cause of increase in frequency and intensity of hydro-meteorological hazards. Extreme climate related hazards are now commonly noticed in every part of the world. Societies, businesses and governments in small islands and least developing countries are facing numerous challenges to recover from intense climate change induced hazards. The Intergovernmental Panel on Climate Change (IPCC) came out with a special report titled 'Managing the risks of extreme events and disasters to advance climate change adaptation (SREX)' which assesses extreme climate events in the wake of climate change. The report highlighted that climate change is affecting climate extremes, although to different degrees in different parts of the world. Heat waves, record high temperatures and heavy precipitation are some of the examples through which climate extremes are manifested (IPCC 2012). The SREX report noted that "Disaster risk emerges from the interaction of weather or climate events, the physical contributors to disaster risk, with exposure and vulnerability, the contributors to risk from the human side" (IPCC 2012). Climate change impacts are likely to be observed differently in different geographies. For example, coastal zones, especially low elevation coastal zones, are likely to be facing constantly increasing threat of sea level rise and storm surges. Inland areas are subject to increases in prevailing temperatures and rainfall anomalies, droughts and heatwaves (Wisner et al. 2012). Climate and weather related health hazards, such as water and vector borne diseases and heat-stress related disorders are expected to worsen health concerns.

### 3.2.3 Risk from Environmental Degradation

Environment and disaster interactions are not yet explored fully in the highly urbanizing world. In 2007, urban population surpassed rural population. Urban areas are growing both horizontally as well as vertically. Urban expansion is eating up land from adjoining peri-urban and rural areas, including agricultural fields, reclaimed land, natural drainage channels, low lying areas, hills and slopes, even forest lands. Urban expansion also results in an increase in paved/hard surfaces in the form of roads, parking areas, built forms, other infrastructure, etc. Hard surfaces do not allow rainwater to percolate affecting the ground water recharge process. Recent increases in flash floods, and prolonged water logging in human settlements are also attributed to the ill effects of urbanization. Deforestation and forest degradation, altering the land use without considering ecological sensitivity, dumping waste in open fields without scientific treatment, polluting surface and ground by discharging human and industrial waste, air-water-noise pollution – all fall in the category of those activities which cause environmental degradation. Environmental degradation either leads to disasters or worsens the effects of natural hazards (Surjan et al. 2011). One of the major problems noticed especially in urban areas is that poor quality urban environmental management leads to deepening the impact of small-scale disasters. In the cities like Mumbai, poor civic sense and weaknesses in solid waste management bring localized flooding even with not-so-high amounts of rainfall (Surjan and Shaw 2009). Promoting community based good practices "which are boon to the local environment can also invariably reduce disaster risk" (Surjan and Shaw 2008).

### 3.2.4 Risk from Psychological Perspective

Given the nature of psychological perspectives of risk, one has to bear in mind that a psychological state never causes any hazard. Whatever the internal state of individual or collective social (group) dynamics is, it has no effect on the increase or decline of extreme weather, floods, volcanic eruptions, mega earthquakes, landslides, or tsunami. In fact, psychological research on disasters usually does not deal with the hazard. Rather, disaster risk is considered in terms of vulnerability and aftermath damage, resilience, and adaptation.

However, there are few psychological factors that indirectly affect the occurrence of hazard, such as in a way human activity interacts with global warming and sea level rise. One of them is ignorance, indifference or avoidance of people towards phenomena relating to a hazard. Today climate change and its effects is one of most concerning and challenging global issues. Notwithstanding its importance, the priority afforded to it is low amongst the population in England (Spence et al. 2010), as it used to be in Germany (Stoll-Kleemann et al. 2001). Stoll-Kleeman's research makes it clear that people would not take actions that would change their lifestyles; rather, they engage in rationalization of inaction, put blame on others to bear responsibility, and devalue actions. Randall (2013), from the psychoanalytic (psychology of unconscious) perspective, stated that these attitudes are underpinned by psychological defense mechanisms such as denial, repression, and projection, which are utilized against guilty feelings of consuming resources, loss and grief of comfortable lifestyles. In short, they represent resistance against pain of the fact. In a community based workshop on climate change held by her in London, one woman rejected to turn unnecessary lights off in her house and said that she needed to make the house warm and bright as a good neighbor and mother. Painful choice is likely to be avoided as such, in conflicts between variant economical, cultural, social, and political interests. Consequently, this contributes to the occurrence of hazard indirectly.

In another case, the Fukushima Nuclear Power Stations (FNPS) accident on March 11, 2011, caused a maximum level of seven hazard on the International Nuclear Event Scale, and forced the evacuation of 130,000 people still in March 2014. The Investigation Committee on the Accident at FNPS of Tokyo Electric Power Company (2012) insisted in their final report that such a huge disaster was built up on the organizational culture of the company that could not consider risks of accidents. Beliefs that an accident would not occur and even if it occurred it would not become a severe one prevailed. This "myth of safety", and the belief in the omnipotence and infallibility, was concluded to be an important background factor of hazard. Per contra, Nishiyama and Imada (2012) argue that the infallibility was expected by the people in the region, and even by all Japanese, reflecting their hope of zero accident risk, now called "zero risk illusion" or "zero risk bias". And it was this pressure put on the organization that made it impossible to think about and discuss risk. The hazard can thus be seen as caused indirectly by the interaction and co-construction by stakeholders on both sides. Though these tendencies may have cultural, social and political roots, the myth and illusion are considered as psychological defense mechanisms against anxiety and pain of facing a fact.

While psychological state is indirectly related to the occurrence of hazard, the type and nature of hazard is discussed in direct connection with psychological state. In reviewing and analyzing 160 empirical researches, Norris et al. (2002) found that different disaster types cause different severity levels of psychological or psychopathological impairment. According to their report, mass violence is the disaster that causes highest rates of severe impairment: that is, 67 % of victims of mass violence in their analysis were severely or very severely impaired. It was significantly higher than the corresponding rates for natural and technological disaster (34 % and 39 % respectively). Moreover, Norris et al. insisted that there was no minimal or fleeting effects in mass violence, which means all population exposed to mass violence were victimized to a certain intensity. The tendency that psychological damage caused by a natural disaster is relatively small is observed in research conducted by Galea et al. (2005). They compared human-made disasters, such as gun shooting, terrorist bombing, and technological disasters with natural disasters using a collection of research data, and showed that the prevalence of PTSD (Post Traumatic Stress Disorder), which is used as one of representative indices of psychological damage after disaster, marked 30–60 % 1 year after the human-made or technological disaster, as compared with 5–60 % for natural disasters. Furthermore, in the latter case most existed in the lower half of the rate. Not surprisingly, the rates are dependent on the intensity and duration of the hazard itself, and people's exposure to it. However, in general, a man-made disaster would cause more adverse psychological reactions, characterized by anger, hate and hostility and likelihood of mental health problems.

When one considers risk, damage caused by a hazard is usually treated as a dependent variable affected by the vulnerability of individuals and communities. However, in case of psychological risk, the physical damaged caused by a hazard is counted as the independent variable. This is because a psychological reaction is brought on not by the hazard itself but by meanings and experiences of the hazard. Norris et al. (2002) listed multiple stressors in hazard which bring psychological problems: bereavement, injury to self or family member, life threat, panic during the disaster, peri-traumatic responses, horror, separation from family, property damage or financial loss, relocation, and collective exposure. The most notable stressors in the list are injury and threat to life. These are said to cause a long lasting and intensified psychological reaction, and frequently contribute to the development of psychopathology. In fact, DSM (Diagnostic and Statistical Manual of Mental Disorders) version 5 (American Psychiatric Association 2013) includes direct or witnessing or other indirect form of exposure to (threatened) death, injury, and sexual violence against self or others as its first criteria of PTSD. While loss of something important is the results of a hazard, the sense of loss is considered a risk of psychological damage.

## 3.3 Dimensions of Human Vulnerability

Human vulnerability has a number of dimensions that are closely interlinked. At its simplest, 'vulnerability' refers to the degree to which communities or societies are "susceptible to the damaging effects of a hazard" (UNISDR 2009: 30). Yet, it has been pointed out that the term is used in so many ways by different (scientific) disciplines involved that it lacks a common understanding (Taubenböck and Geiss 2014).

Here we use the definition of vulnerability as the characteristics of a person or group that affect their capacity to anticipate, cope with, resist and recover from the impact of a (natural) disaster (Blaikie et al. 1994: 9).

Disaster risk can be expressed as a function of the exposure to hazard, people's vulnerability (including their settlements and livelihoods) and the degree to which society engages in disaster mitigation (Wisner and Uitto 2009). In this formulation, vulnerability takes center place in analyzing disaster risk.

Human vulnerability can be defined as the combination of physical (i.e. vulnerability in the built environment) and social (vulnerability experienced by people and their social, economic and political systems) vulnerability (Pelling 2003: 5).

### 3.3.1 Physical Vulnerability

Traditionally both theory and practice have tended to focus on physical vulnerability. Physical vulnerability is fundamentally dependent on exposure to a hazard. This in turn is closely tied with location and the spatial distribution of the hazard. Consequently, it seems that it would be feasible to address vulnerability by addressing exposure through means such as land-use planning, zoning and structural engineering.

Such solutions appear attractive to politicians and authorities because they are concrete and seemingly apolitical. There is no doubt that engineering measures are important to reduce vulnerability. Well constructed buildings and other structures, for example, can withstand shaking caused by earthquakes or heavy winds during storms. This was unequivocally demonstrated during the 1995 Great Hanshin Earthquake that struck western Japan causing around US$ 100 billion worth of damage and killing more than 6,000 people. In the city of Kobe, buildings built after the 1984 building code changes performed quite well during the earthquake (RMS 1995).

It is, however, an illusion to assume that reducing physical vulnerability through managing exposure would not be political in nature. First of all, politicians and municipal officials use them for political gain often showing they are addressing public safety. Secondly, engineering solutions, such as building seawalls or embankments around rivers, involve major investments that tend to benefit engineering and construction companies. At times, such companies have been known to have close links to authorities in the cities and municipalities they work in. Consequently, they present opportunities for corruption and kickbacks, but even if this is not the case, they may concentrate resources and power to fewer hands.

Thirdly, although legislative tools such as land-use planning and zoning are very important, they may also reinforce existing differences in wealth and vulnerability between different geographical areas. For instance in many cities, especially rapidly growing urban centers in the developing countries, poor people and migrants from the countryside often congregate in informal settlements that are out of reach of urban planning and frequently vulnerable. Rigorous application of land-use planning and zoning may also affect then negatively by denying the settlement areas municipal services or by resulting in eviction. Taken together, focusing only on physical vulnerability has the tendency of reinforcing traditional power structures and perpetuating wealth differences.

From an environmental point of view, major engineering works transform nature and may have significant unintended and often negative consequences. In Japan, during the years of rapid economic growth after World War II, the government emphasized engineering works that would channel rivers and construct dams for electricity generation and flood control. The power was firmly in the hands of technocrats who believed strongly that this was the right approach to take. However, construction of dams often resulted in extensive inundation of large tracts of land and entire communities, especially in politically and economically weak mountainous areas were submerged. River ecosystems were destroyed and with them wildlife and fauna disappeared. The aesthetic and recreational value of rivers was also destroyed, which was particularly evident in cities where the river had for centuries been the center of life. Yet, floods continued to wreak havoc and at times of typhoons, such as when a major typhoon hit Ise Bay in 1959, disruption and damage were extensive. Since the 1960s, citizen opposition to dam construction started to rise, resulting in a gradual but relatively rapid change in national policy. As a consequence, the present policies in Japan give due consideration to the value of rivers from an ecological and environmental point of view (Takahasi and Uitto 2004).

Removal or altering natural vegetation often increases physical vulnerability by eliminating protective barriers. This was demonstrated drastically during the Indian Ocean Tsunami on Boxing Day of 2006. Coastal areas where mangrove forests had been cleared to give way to aquaculture or other coastal developments where left open to the advancing tidal wave to penetrate deep inland causing widespread death and destruction (Wun'Gaeo 2009).

Small Island Developing States (SIDS) can be seen as particularly vulnerable from a physical point of view. Often low-lying, they are exposed to a range of natural hazards ranging from high winds and storm surges to earthquakes and tsunamis. They are particularly vulnerable to the impacts of intensifying storms and sea-level rise that may be associated with global climate change. Saltwater intrusion may pose a direct threat to their freshwater supply. Yet, it would be erroneous to focus only on the physical vulnerability without consideration to the social side. Island economies are vulnerable due to internal constraints (including limited natural and human resources), as well as external factors, such as trade relations. Climate change itself is a global phenomenon with its main causes far away in the rich industrialized world but with SIDS bearing the brunt of the negative consequences (Pelling and Uitto 2001).

The initial critique of the focus on physical vulnerability calling attention to the social and political dimensions of vulnerability is often attributed to Hewitt (1983). His work criticized the prevalent view of disasters as "originating in dangerous agents outside society, such as flood and fire, or as unscheduled events or Acts of God" (Hewitt 2013). He made a strong argument for understanding disasters in their social, political and structural contexts.

### 3.3.2 *Social Vulnerability*

Social vulnerability is here meant to encompass dimensions, such as economic status, and social characteristics and political power of individual people, groups and communities. The degree of social vulnerability is defined by a variety of related factors.

Social vulnerability is linked to physical vulnerability and has a geographical dimension. Disadvantaged people and groups are often consigned to vulnerable locations. In cities, poor people live in neighborhoods that are exposed to natural and manmade hazards (poor neighborhoods even in wealthy cities tend to be located close to industrial facilities, including chemical plants that increase the risk of industrial disasters). Immigrants, especially in rapidly urbanizing developing countries, often settle in shanties and informal settlements on the outskirts of areas where employment opportunities exist. Often, these areas are on hillsides that are susceptible to landslides during heavy rains or earthquakes. Vulnerability thus varies spatially because of natural environments as well as housing and social structures.

Vulnerability has a distinct economic dimension. There is a strong correlation between poverty and vulnerability, but the relationship is not direct, nor is the causality unidirectional. Not all poor people are equally at risk from hazards; and not all rich people are impervious to them. Poverty and vulnerability are therefore not synonyms, even if they are closely related. Poor people tend to be more vulnerable to disasters through a variety of avenues, the first of which is again closely linked to physical vulnerability. The fact that poor people are often forced to live in geographical locations that are exposed to hazards, whether natural (e.g., coastal flats or hillsides exposed to storms) or technological (e.g., close to hazardous industries), contributes to their vulnerability. They also lack the financial and other resources to move to a safer location, improve their dwellings or make them more resilient towards a hazard. When exposed to a hazard, like a typhoon or a tropical storm, their dwellings are likely to be damaged resulting in further loss of assets. A disaster may also lead to a loss of livelihood opportunities if, for example, the place of work or tools of trade are damaged or disruption of transportation prevents access to places of earning. The lack of financial resources, assets and savings also poses challenges to recovery from and coping after a disaster. Therefore, not only does poverty breed vulnerability; disasters often also perpetuate poverty thus leading to a vicious cycle.

It is important to understand vulnerability in context. The root causes that render certain communities, groups or individuals more vulnerable than others are often structural and have deep roots in history. While exposure to natural hazards is primarily determined by geography and physical characteristics of the location, vulnerability is largely determined by the structural and societal situation, including the actors' access to resources, assets and rights (Pelling 2003). Social processes, including power relations in a society, generate unequal exposure to risk that renders some people more vulnerable to disasters than others. In many cases, such structures and processes have been in the making for centuries (Hilhorst and Bankoff 2004). A recent study by Simon (2014) makes the case of historical roots to the 1991 Oakland Hills Firestorm in California, where land development strategies starting from the mid-1800s contributed to the construction of vulnerable conditions. These were further exacerbated by homeowner politics and state tax policies from the 1950s through 1980s.

Blaikie et al. (1994) developed an influential analytical model that describes the chain of explanation of disasters. This pressure and release model, which is highly applicable to any type of disaster (both natural and manmade) enables us to trace a progression connecting a hazard to its impact on people through an analysis of the social factors that render them vulnerable. This progression starts with the root causes of vulnerability that include political and economic structures, such as access to power and resources and the overall political and economic systems. These lead to dynamic pressures that manifest themselves in (lack of) local institutions, skills, investments, freedoms etc., as well as macro forces, such as population growth, urbanization and environmental degradation. Taken together, these lead to unsafe conditions: fragile physical environment, fragile local economy, vulnerable society and lack of public actions to address these. The root causes of vulnerability can be rolled into a triangle with political, social and economic structures at the apexes (Wisner et al. 2012).

Apart from the more generic societal relations determining vulnerability, there are specific characteristics that may render specific individuals or groups particularly vulnerable. These include people with disabilities (Alexander et al. 2012); children and youth (Babugura 2012); and the elderly (Ngo 2012). All of these groups may lack in their ability to respond to, cope with and recover from disasters and will, thus, merit specific attention. Similarly, gender and sexual orientation (Fordham 2012), caste, ethnicity or religious affiliation (Gaillard 2012) may be causes of discrimination that in turn increases their vulnerability.

A study of four megacities – Los Angeles, Manila, Mexico City and Tokyo – around the Pacific Rim (Wisner and Uitto 2009) identified groups that were perceived as particularly vulnerable by disaster managers in each of the cities. In Manila and Mexico, characterized as being poor and less developed, squatters were overwhelmingly identified as the most vulnerable, with children following as the second group (especially in Manila, street children were identified as being particularly vulnerable). In the wealthier cities of Tokyo and Los Angeles, disaster managers suggested the elderly and disabled people were the most vulnerable. In these cities, foreigners were also seen as vulnerable, largely due to language issues and access to information. In all four cities, there was significant overlap between the categories of the most vulnerable groups. To counter bias or discrimination against specific groups (such as immigrants, minorities, the disabled), it is important that policies are designed and implemented in an inclusive manner.

Vulnerability is not a static concept and people do not need to be condemned to a state of vulnerability throughout their lives. At the same time, while some people's vulnerability may be reduced, others fall into increasingly vulnerable positions. Vulnerability varies temporally because of people moving through different life stages with varying mixes of resources and liabilities (Uitto 1998: 9). Vulnerability should not be regarded as a property, but as a result of social relations (Hilhorst and Bankoff 2004: 2).

The opposite of vulnerability is resilience, which IPCC defines as "the ability of a social or ecological system to absorb disturbances while retaining the same basic structure and ways of functioning, the capacity for self-organization, and the capacity to adapt to stress and change" (quoted in Pelling 2011: 55). It is important to note that resilience can be applied either to physical (for instance, an ecosystem or earthquake resistant buildings) or to social systems. The IPCC definition emphasizes capacity, which can be at various societal levels. Important aspects pertain for example to utilizing local knowledge and building capacity at the local level.

The four megacity project (Wisner and Uitto 2009) identified several common problems related to local capacity, which included limited or ritualistic involvement of community or neighborhood groups, political hostility towards NGOs and the low capacity of NGOs, all of which posed constraints on utilizing local capacity for disaster risk reduction.

Yet, an excessive focus on local capacity and local-level solutions may also be problematic. Given that many of the root causes of vulnerability are historical and structural, or stemming from global processes (like international trade or climate change), there is a risk of focusing on local capacity and resilience, thus placing the onus on the vulnerable people without addressing the root causes.

### *3.3.3 Psychological Vulnerability*

Psychological after-effects in people who experience damage of natural and man-made disaster and/or war have been described in many literatures. Academic research and surveys started to deploy around era of the First World War. Mental health problems, named war neurosis, shell shock, etc., under pressure of battles turned to be called PTSD, which became one of the important indices of psychologically detrimental responses after disaster. Research showed a number of problems other than the PTSD, such as strong anxious and depressive emotional states, difficulties in thinking and cognition, like lack of perspective, withdrawal and other interpersonal problems, and behavioral problems of increase in substance and alcohol abuse and suicide (Ursano et al. 1994; Galea et al. 2005; Masten and Narayan 2012; Milan et al. 2012; North and Pfefferbaum 2013).

However, as mentioned above, these damages are not directly caused by the hazard itself. As an earthquake collapses houses, for instance, shaking of ground does not destroy people's minds. There is mediation of fear that a person may die or someone important for the person may die in the event, and to discuss vulnerability from the psychological perspective one must think about what such subjective experiences (=vulnerability) consist of. Here, one also has to know that people generally suffer from psychological injuries when they are exposed to a disaster. Raphael (1992) proposed a model of trajectories of common reactions as is in Fig. 3.1, in which most people unfold maladaptation especially in the immediate aftermath of disaster. Other researchers presented similar models too (e.g., Masten and Narayan 2012). When a disaster hits people, they relatively soon begin to take actions needed for survival, and sometimes they show hyper-adjustment and generate happy feeling of having survived (called the Honey moon phase); but in the course of time they find what they lost and their adjustment levels decline (called the Disillusion phase). Then people start the real recovery process gradually. Some cannot go along this path and have to return to their own daily lives while keeping maladaptive conditions; and yet others cannot go back to their own lives at all (dotted line). This is the time when the incidence of PTSD occurs, and persons unable to fully recover are those who are vulnerable in psychological and psychiatric terms. In other wards, vulnerability coincides with the resilience process from a psychological perspective (Paton et al. 2000), though each has different definitions and factors from one

**Fig. 3.1** Model of trajectories of common reactions

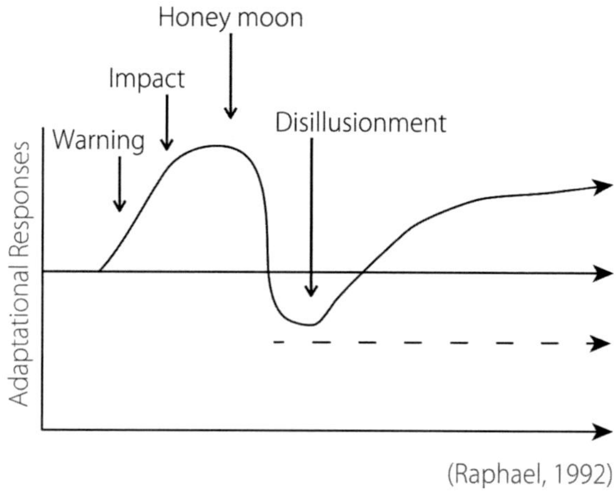


another. Therefore, factors consisting of vulnerability are those that prevent recovery and being resilient.

These factors vary from biological responsiveness to individual and communal levels. According to Ursano et al. (1994) factors include individual developmental history, pre-existing (mental) illness, previous exposure to stressful events, social supports, meaning, appraisal, and attribution, and can be summarized as stress mediators. As seen in the name of PTSD, psychological research deals with hazard caused disaster as "stressor", and its physical and mental result is called "stress reaction". Lazarus and Folkman (1984) were the ones who established this field proposing a cognitive model of stress that explains how an external event affects individuals' somatic and psychic conditions. Parameters are cognitive appraisal of the event and of coping mechanisms one can take. The more severe one's appraisal of loss and damage caused by the event is and the less capable his/her appraisal of coping options is, the more influenced the individual is by the event (i.e., vulnerable). This means the existence of a transactional and interactive model again in which resilience related factors, how and what coping is possible, plays an important role in construction of vulnerability. Thus, Paton et al. (2000) involve variables regarding coping at the individual level of stress mediator, that is, avoidance of threatening situations (avoidance of problem solving), social skills deficits, and inadequate problem-solving behavior. Characteristics of a vulnerable population, often include social and demographic variables like sex, minority status, poverty, younger and older population, and can be understood as they weaken stress mediators. The quality of attachment relationship (Bowlby 1969) is also one of stress mediator, specifically for children (Masten and Narayan 2012). Good quality of attachment with his/her caregivers provides a secure base and matrix of healthy development of personality for a child. The importance of social capital or social connectedness can be explained from this perspective.

Among those factors researchers refer to are pre-existence of psychiatric illness or unresolved precedingly experienced trauma (Raphael 1992; North and

Pfefferbaum 2013). For example, after the Oklahoma City bombing in 1995, 63 % of those who were diagnosed having a psychiatric disorder had had a psychiatric disorder prior to the disaster (North et al. 2008), Likewise, after Hurricane Katrina, the rate of people who received therapy for post-disaster disorder at a mental health clinic was 40 % amongst the population with a pre-existing mental illness, as opposed to 24 % those with no previous mental illness (North et al. 2008). The same association of preceding conditions and incidence of PTSD is observed in population placed in undesirable family environments (loss of caregivers, abuse and maltreatment) in their early lives (Stovall-McClough and Cloitre 2006). Again, given that people in a low socio-economic status are more likely to show prevalence of mental illness and unstable family relationships (e.g., inability to learn warm communication due to material deficiencies, high exposure to violence and crime, unlearnability of adequate stress coping, difficulty in receiving necessary treatments, unsupported by family or friends, etc. contribute to this problem), one can recognize how social layers are intertwined with vulnerability.

With regard to community level vulnerability, there are two different perspectives. One is what community factors heighten psychological risks of individuals; and another is what factors heighten communal vulnerability. The former is usually discussed in terms of social, cultural and economic factors. One representative factor is the existence of a social network (Schwarzer et al. 1994). A few studies have been conducted regarding the latter, introducing the concept of group dynamics. As part of army psychiatric services for rehabilitation in military hospitals during the Second World War, Bion (1961) and his colleagues adopted a group approach called therapeutic communities. When a group operated functionally, it was named 'working group'; but when a group was not organized and did not augment freedom because of, for instance, absence of a leader, the group fell into a category named 'basic assumption group' in which strong anxiety and suspicion dominated. This form of group raised conflict within and between groups and drove its members to being neurotic; therefore the group could not carry out functional work. This finding contributed to the development of group psychology and psychotherapy. In consideration of the role of community in a disaster, group dynamics characterized by distrust, complaining, deep dependence on a leader, and elation might be expected to boost disaster risk and communal level vulnerability.

## 3.4 Risk and Vulnerability Interaction

IPCC's SREX report keeps disaster risk in the center and dwells significantly on vulnerability as well as exposure. Figure 3.2 illustrates the core concept of SREX, which "evaluates the influence of natural climate variability and anthropogenic climate change on climate extremes and other weather and climate events that can contribute to disasters, as well as the exposure and vulnerability of human society and natural ecosystems" (IPCC 2012). Socio-economic development also translates into physical development. Climate change has forced communities to adapt

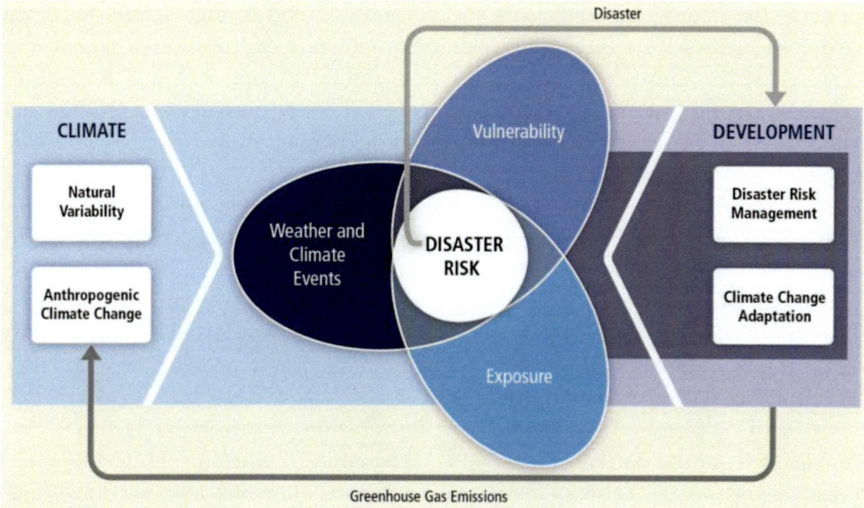


**Fig. 3.2** Core concept of IPCC SREX report

sometimes in a knee-jerk manner, which is often referred to as maladaptation. Maladaptation increases both exposure and vulnerability, and thus disaster risk itself. The World Risk Index which calculates and compares risk values for over 170 countries worldwide, ranked countries "based on four key components that take both natural hazards and social factors into account: exposure (to natural hazards and potential risks), susceptibility (likeliness of suffering harm, susceptibility as a function of public infrastructure), coping capacities (governance and capacity to reduce negative consequences of hazards) and adaptive capacities (capacity for long-term social change)" (UNU 2014). Risk can be reduced by building capacities. Vulnerabilities are the product of processes, which prevail in the society due to a number of reasons and may include poor governance, weak social structures and unstable economic systems. As opposed to vulnerability, resilience is considered to be instrumental in reducing risks.

## 3.5 Role of Sustainable Development Approach in Reducing Risk and Addressing Vulnerability

Our aim in this chapter has been to demonstrate how risk and vulnerability are intimately related to the concept of sustainability. Risk is defined as a function of hazard and vulnerability. Although hazards may be natural phenomena, their occurrence and intensity is influenced by human actions. Global climate change and the associated increases in climatic hazards is itself largely human induced. Disasters occur when people and communities exposed to a hazard are vulnerable.

Vulnerability in turn is dependent on social, economic and political forces that often have deep historical roots. Furthermore, environmental degradation often leads to increased vulnerability and reduces people's ability to cope with and respond to disasters. All of the above has strong psychological dimensions at the individual and group levels. It is evident that addressing risk and vulnerability requires a holistic perspective of sustainable development.

The year 2015 is a milestone year in international development. The global community is awaiting the post-2015 framework for disaster risk reduction, which will be successor of the Hyogo Framework of Action adopted in 2005 for 10 years. The Conference of Parties to the UNFCCC (COP-21) is expected to conclude ongoing negotiations and come up with a successor agreement to the Kyoto Protocol which may includes internationally binding emission reduction targets for the industrialized countries with specific timeframes.

UNESCO's Decade of Education for Sustainable Development (DESD) also completed its target period (2005–2014). Education for sustainable development spread across social, economic and environmental domains also encompassing basic tenets of disaster risk reduction. In the recent past, disaster education both at school level as well as at university level received ample attention. A number of new courses at bachelor, master and doctoral levels are launched, with significant emphasis of risk reduction, in different parts of the world.

There is tremendous hope and emphasis placed on various international processes and United Nations agencies to ensure coherence and mutual reinforcement between disaster risk reduction, sustainable development and climate change. It is essential that these outcomes should be complementary instead of contradictory. Sustainable development and climate change adaptation essentially should encompass well-understood disaster risk reduction and disaster risk management policies. Sustainable development is an overarching framework aimed at achieving well being and enhanced and inclusive quality of life for every human without jeopardizing the same for future generations. Sustainable Development Goals (SDGs) for post-2015 world are expected to place enhanced emphasis on reducing risk by addressing the root causes of vulnerability and building resilience.

**Acknowledgements** The first author would like to acknowledge knowledge gained and experience received from a multi-country research project titled "Advancing Locally Based Green Practices to realize establishment of Sound Material Cycle Society in Asian Cities" supported by the Asia-Pacific Network for Global Change Research (APN). This project was especially helpful in developing thorough understanding of small scale disaster, vulnerability of marginalized populations and disaster-environment interlinkages.

## References

Adams J (1995) Risk. Routledge. http://www.routledge.com/books/details/9781857280685/

Alexander D, Gaillard JC, Wisner B (2012) Disability and disaster. In: Wisner B, Gaillard JC, Kelman I (eds) The Routledge handbook of hazards and disaster risk reduction. Routledge, London/New York, pp 413–423

American Psychiatric Association (2013) Diagnostic and statistical manual of mental disorders: DSM-5, 5th edn. American Psychiatric Publishing, Virginia
Babugura AA (2012) Children, youth and disaster. In: Wisner B, Gaillard JC, Kelman I (eds) The Routledge handbook of hazards and disaster risk reduction. Routledge, London/New York, pp 436–446
Beck U (1992) Risk society: towards a new modernity. Sage, Thousand Oaks
Bion WR (1961) Experiences in groups. Tavistock, London
Blaikie P, Cannon T, Davis I, Wisner B (1994) At risk: natural hazards, people's vulnerability, and disasters. Routledge, London
Bowlby J (1969) Attachment and loss, vol. 1: attachment. New York: Basic Books http://www.routledge.com/books/details/9781857280685/, http://www.e-elgar.com/shop/handbook-of-sustainable-development
Fischhoff B, Kadvany J (2011) Risk: a very short introduction. Oxford University Press, Oxford
Fordham M (2012) Gender, sexuality and disaster. In: Wisner B, Gaillard JC, Kelman I (eds) The Routledge handbook of hazards and disaster risk reduction. Routledge, London/New York, pp 424–435
Gaillard JC (2012) Caste, ethnicity, religious affiliation and disaster. In: Wisner B, Gaillard JC, Kelman I (eds) The Routledge handbook of hazards and disaster risk reduction. Routledge, London/New York, pp 459–469
Galea S, Nandi A, Vlahov D (2005) The epidemiology of post-traumatic stress disorder after disasters. Epidemiol Rev 27:78–91
Hewitt K (ed) (1983) Interpretations of calamity from the viewpoint of human ecology. Allen and Unwin, London
Hewitt K (2013) Disasters in 'development' contexts: contradictions and options for a preventive approach. Jàmbá – J Disaster Risk Stud 5(2). Open Access journal: http://www.jamba.org.za/index.php/jamba/article/view/91/196
Hilhorst D, Bankoff G (2004) Introduction: mapping vulnerability. In: Bankoff G, Frerks G, Hilhorst D (eds) Mapping vulnerability: disasters, development and people. Earthscan, London, pp 1–9
Investigation Committee on the Accident at Fukushima Nuclear Power Stations of Tokyo Electric Power Company (2012) Executive summary of the final report
IPCC (2012) Managing the risks of extreme events and disasters to advance climate change adaptation. A Special Report of Working Groups I and II of the Intergovernmental Panel on Climate Change. In: Field CB, Barros V, Stocker TF, Qin D, Dokken DJ, Ebi KL, Mastrandrea MD, Mach KJ, Plattner G-K, Allen SK, Tignor M, Midgley PM (eds) Managing the risks of extreme events and disasters to advance climate change adaptation. A special report of working groups I and II of the intergovernmental panel on climate change. Cambridge University Press, Cambridge
Lazarus R, Folkman S (1984) Stress, appraisal, and coping. Springer Publishing Company, New York
Masten AS, Narayan AJ (2012) Child development in the context of disaster, war, and terrorism: pathways of risk and resilience. Annu Rev Psychol 63:227–257
Milan S, Zona K, Acker J, Turcios-Cotto V (2012) Prospective risk factors for adolescent PTSD: sources of differential exposure and differential vulnerability. J Abnorm Child Psychol 41:339–353
Ngo EB (2012) Elderly people and disaster. In: Wisner B, Gaillard JC, Kelman I (eds) The Routledge handbook of hazards and disaster risk reduction. Routledge, London/New York, pp 447–458
Nishiyama N, Imada T (2012) Waver of zero risk illusion and myth of safety: change in risk conscicousness in Japanese through the Great East Japan Earthquake and Fukushima nuclear power station accident. View Vis 34:57–64
Norris FH, Friedman MJ, Watson PJ, Byrne CM, Diaz E, Kaniasty K (2002) 60,000 disaster victims speak: part I. An empirical review of the empirical literature, 1981–2001. Psychiatry 65(3):207–239

North CS, Pfefferbaum B (2013) Mental health response to community disasters: a systematic review. JAMA 310(5):507–518

North CS, Nixon SJ, Shariat S, Mallonee S, McMillen JC, Spitznagel EL, Smith EM (1999) Psychiatric disorders among survivors of the Oklahoma City bombing. J Am Med Assoc 282(8):755–762

North CS, King RV, Fowler RL, Polatin P, Smith RP, LaGrone HA, Tyler D, Larkin GL, Pepe PE (2008) Psychiatric disorders among transported hurricane evacuees: acute-phase findings in a large receiving shelter site. Psychiatry Ann 38(2):104–113

Paton D, Smith L, Violanti J (2000) Disaster response: risk, vulnerability and resilience. Disaster Prev Manage 9(3):173–179

Pelling M (2003) The vulnerability of cities: natural disasters and social resilience. Earthscan, London

Pelling M (2011) Adaptation to climate change: from resilience to transformation. Routledge, Oxon

Pelling M, Uitto JI (2001) Small island developing states: natural disaster vulnerability and global change. Environ Hazards 3:49–62

Randall R (2013) Great expectations: the psychodynamics of ecological debt. In: Weintrobe S (ed) Engaging with climate change: psychoanalytic and interdisciplinary perspectives. Routledge, London, pp 87–102

Raphael B (1992) Disasters management. The National Health and Medical Research Council

RMS (1995) Japan – the Great Hanshin earthquake. Event report. Risk Management Solutions, Inc./Failure Analysis Associates, Inc., Menlo Park

Schwarzer R, Hahn A, Schroder H (1994) Social integration and social support in a life crisis: effects of macrosocial change in East Germany. Am J Community Psychol 22(5):685–706

Simon GL (2014) Vulnerability-in-production: a spatial history of nature, affluence, and fire in Oakland, California. Ann Assoc Am Geogr 104(6):1199–1221

Slovic P (2010) The feeling of risk: new perspectives on risk perception. Place and Publisher? http://www.routledge.com/books/details/9781857280685/

Spence A, Venables D, Pidgeon N, Poortinga W, Demski C (2010) Public perceptions of climate change and energy futures in Britain: summary findings of a survey conducted in January-March 2010. Technical report (Understanding risk working paper 10–01). School of Psychology, Cardiff

Stoll-Kleemann S, O'Riordan T, Jaeger CC (2001) The psychology of denial concerning climate mitigation measures: evidence from Swiss focus groups. Glob Environ Chang 11:107–117

Stovall-McClough KC, Cloitre M (2006) Unresolved attachment, PTSD, and dissociation in women with childhood abuse histories. J Consult Clin Psychol 74(2):219–228

Surjan A, Shaw R (2008) Eco-city to disaster resilient eco-community: a concerted approach in the coastal city of Puri, India. Sustain Sci 3(2):249–265, Springer Publication

Surjan A, Shaw R (2009) Enhancing disaster resiliency through local environment management: case of Mumbai, India. Disaster Prev Manag 18(4):418–423, Emerald Publication, UK

Surjan A, Takeuchi Y, Shaw R (2011) From disaster and climate risk to urban resilience: approaching through community based environmental improvement. Research Publishing Services, Singapore

Takahasi Y, Uitto JI (2004) Evolution of river management in Japan: from focus on economic benefits to a comprehensive view. Glob Environ Chang 14:63–70

Taubenböck H, Geiss C (2014) Vulnerability and resilience research: a critical perspective. Int J Disaster Risk Sci 5:86–87

Uitto JI (1998) The geography of disaster vulnerability in megacities: a theoretical framework. Appl Geogr 18(1):7–16

UNISDR (2009) Terminology: disaster risk reduction. United Nations Office for Disaster Risk Reduction, Geneva

UNISDR (2013) Global assessment report on disaster risk reduction 2013. United Nations Office for Disaster Risk Reduction, Geneva. Fact Sheets in English available at: http://www.

preventionweb.net/english/hyogo/gar/2013/en/home/media.php#press-kit. Accessed on 12 Jan 2015

UNU (2014) World risk report 2014, Fact Sheet. Available at: http://www.ehs.unu.edu/file/get/11896.pdf. Accessed on 10 Jan 2015

Ursano RJ, Fullerton CS, McCaughey BG (1994) Trauma and disaster. In: Ursano RJ, McCaughey BG, Fullerton CS (eds) Individual and community responses to trauma and disaster: the structure of human Chaos. Cambridge University Press, Cambridge, pp 3–27

WHO (2014) WHO Fact Sheet No. 339, updated May 2014. Retrieved from http://www.who.int/mediacentre/factsheets/fs339/en/. Accessed on 18 Dec 2014

Wisner B, Uitto JI (2009) Life on the edge: urban social vulnerability and decentralized, citizen-based disaster risk reduction in four large cities of the Pacific Rim. In: Brauch HG, Kameri-Mbote P, Oswald Spring U, Behera NC, Grin J, Chourou B, Mesjasz C, Krummenacher H (eds) Facing global environmental change: environmental, human, energy, food, health and water security concepts. Springer, Berlin/Heidelberg, pp 215–231

Wisner B, Blaikie P, Cannon T, Davis I (2004) At risk: natural hazards, people's vulnerability, and disasters, 2nd edn. Routledge, London

Wisner B, Gaillard JC, Kelman I (2012) Framing disaster: theories and stories seeking to understand hazards, vulnerability and risk. In: Wisner B, Gaillard JC, Kelman I (eds) The Routledge handbook of hazards and disaster risk reduction. Routledge, London/New York, pp 18–33

Wun'Gaeo S (2009) Environment as an element of human security in Southeast Asia: case study on the Thai tsunami. In: Brauch HG, Kameri-Mbote P, Oswald Spring U, Behera NC, Grin J, Chourou B, Mesjasz C, Krummenacher H (eds) Facing global environmental change: environmental, human, energy, food, health and water security concepts. Springer, Berlin/Heidelberg, pp 1131–1142. http://www.springer.com/us/book/9783540684879

# Chapter 4
# Building Urban Climate Resilience: Experiences from Vulnerability Assessment in Hue City, Viet Nam

**Phong Tran, Richard Friend, Kenneth MacClune, and Justin Henceroth**

**Abstract** Urban areas are particularly vulnerable to the impacts of climate change. This is due to the dependence on infrastructure, and the density of population, services and economic activity. Most cities have been developed, designed and built to cope with the historic climate, not the future climate trends and extremes. Moreover, cities' increasingly sophisticated and interdependent supply chains and transportation logistics, for water, energy, workforce, food and consumables make it harder to assess the cities' vulnerabilities. These characteristics further compound climate risks and create greater susceptibility to disruption. There is no agreed method for assessing urban vulnerability to climate change. This chapter argues that to overcome this constraint and to make the concept of vulnerability operational one should use a resilience approach that allows for consideration of complex systems, and from this, look at interactions between key systems of the cities. This chapter present a process and key findings from Hue city climate vulnerability assessment, in which climate change was not used as the starting point for the vulnerability assessment, instead, we have started with urbanization, and then factored in climate change into the process. We also looked more at urbanization as a process of change, rather than just the city as an administrative unit.

**Keywords** Urban • Climate resilience • Disaster risk reduction • Resilience approach • Vulnerability assessment

P. Tran (✉)
Institute for Social and Environmental Transition (ISET), Hanoi, Vietnam
e-mail: phongtran@i-s-e-t.org

R. Friend
Institute for Social and Environmental Transition (ISET), Bangkok, Thailand

K. MacClune • J. Henceroth
Institute for Social and Environmental Transition (ISET), Boulder, CO, USA



J.I. Uitto, R. Shaw (eds.), *Sustainable Development and Disaster Risk Reduction*, Disaster Risk Reduction, DOI 10.1007/978-4-431-55078-5_4

## 4.1 Introduction

Urban areas are particularly vulnerable to the impacts of climate change. This is due to the density of infrastructure, population, services and economic activity. Most cities have been developed, designed and built to cope with the historic climate record, not the future climate trends and extremes. Moreover, cities' increasingly sophisticated and interdependent supply chains and transportation logistics, for water, energy, workforce, food and consumables make it harder to assess the cities' vulnerabilities. These characteristics further compound climate risks and create greater susceptibility to disruption. There is no agreed method for quantifying urban vulnerability to climate change. This chapter argues that to overcome this constraint and to make the concept of vulnerability operational is to use a resilience approach that allows for consideration of complex systems, and from this, to look at interactions between key systems of the cities. However, putting a resilience framework into practice has faced the challenges. Thus, experience from practical application is important step in the feedback loops to improve the framework so that it not only captures the complexities of cities and but also can be applied by city-level planners, professionals, and practitioners in a wide range of circumstances. This chapter will present the process of urban climate vulnerability assessment using resilience approach, the key findings and lessons learned from the case study of urban climate vulnerability assessment in Hue city under the MBRACE project.

## 4.2 Why Resilience Approach

Traditionally, disaster risk management refers to the systematic management of administrative decisions, organization, operational skills and abilities to implement policies, strategies and coping capacities of the society or individuals to lessen the disaster impacts. Disaster risk management planning involves understanding natural hazards, vulnerabilities and potential losses and developing appropriate preparedness and mitigation strategies to mitigate such losses. However, there are two main challenges in existing disaster risk management work: (1) the problem of disaster risk assessment and disaster risk management planning in the context of emerging risks caused by climate change and the dynamic of vulnerability, and (2) the common practice of technocratic and traditional top-down disaster risk management approaches.

First, disaster risk assessment often works at the intersection of two domains: hazard and vulnerability, see Fig. 4.1. This intersection represents the proportion of current risks needing to be addressed by current disaster risk management planning. The Fig. 4.2 shows a conceptualization of disaster risk planning that highlights how current disaster risk management planning does not address the complete suite of vulnerabilities at any given point in time due to various limitations with the risk assessment process. Figure 4.2 shows the progression of risk over the time and successive disaster risk management planning and revisions as bars.

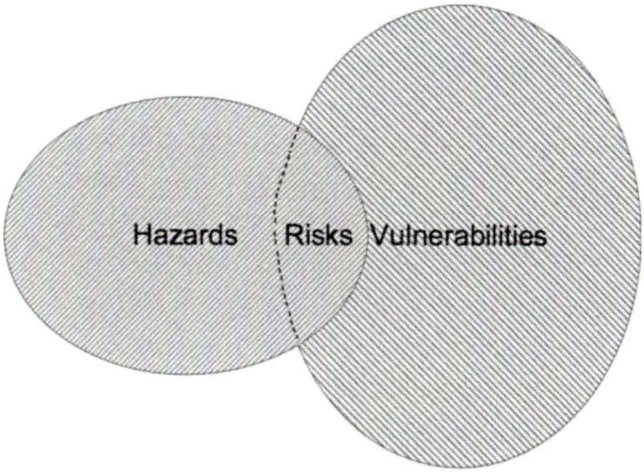


**Fig. 4.1** Domain of disaster risk management

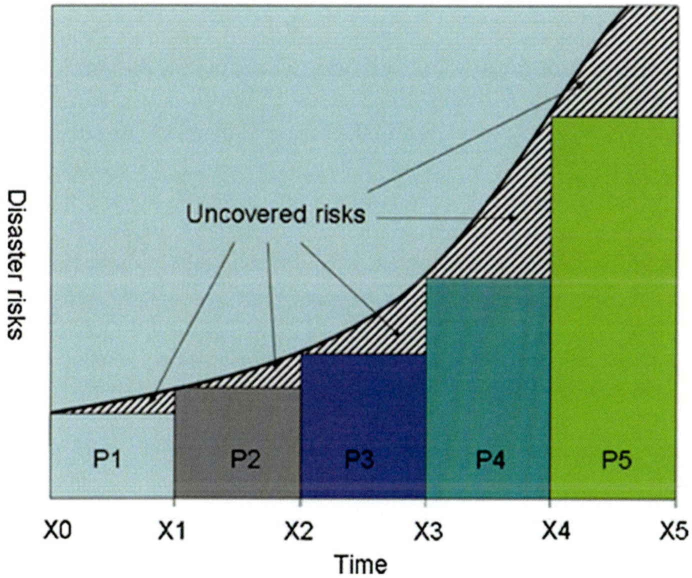


**Fig. 4.2** Uncovered risk in disaster risk planning under growing risk conditions. Planning processes produce static policies and procedures that do not change with the growing risk profile (Source: Prabhakar et al. 2008)

The figure depicts that the vulnerabilities are dynamic, applicable universally, and ever increasing in some parts of the world as it depends on the effectiveness of human interventions as a response. As vulnerabilities grow continuously in some locations, either in linear or exponential manner as there are no clear assessments available on risk progression in the available literature, as a result of climate change and many other dynamic pressures, a disaster risk management plan prepared based on a risk assessment done at X1 point of time remains static during X1–X2 while disaster risks could continue to grow, particularly in urban areas in developing countries due to development in hazardous sites, degradation of infrastructures, population growth in flood prone areas. This is because of the fact that the vulnerabilities are difficult to assess and assessments do not often consider the growing risks in future. This indicates that the Plan 2 (P2) will soon be outdated relative to a case of the growing vulnerabilities and risks until it is revised by Plan 3 (P3) and so on (shaded area in Fig. 4.2). A regular reassessment of hazards and vulnerabilities is advised due to the dynamic nature of the risk. Some have suggested a regular review of once in a year while others have suggested to revise the plan after a major disaster. However, regular revision of disaster risk management plans is far from reality as hazard and vulnerability assessments are done when funds are made available. In many contexts this funding is usually made available by an external agency and any revision is not possible after completion of the assessment. So, almost inevitably, disaster risk management plans lag actual disaster risks. The magnitude of the lag depends on the diligence and frequency of plans being updated, and on the pace of other changes.

In Viet Nam, the more technocratic and traditional top-down disaster risk management approach is still commonly practiced. In general, the more technocratic approach generally leads to engineering intervention responses to risks and disasters (Tran and Shaw 2010). However, in the face of climate change and rapid urbanization, conditions will become increasingly variable, dynamic and uncertain. With these impacts local planners face difficulty using historical climate data to address risk. Additionally, future projections of climate and urbanization are uncertain enough to make identify engineering specifications to address these impacts impossible (Tyler and Moench 2012; IFRC 2012). Furthermore, traditional disaster management tends to underemphasize the role of local communities and institutional factors that can reduce or increase disaster risks. In short, 'predict and prevent' approaches have been criticized for their limited ability to deal with uncertainty in the context of urban and climate change.

According to Tyler and Moench (2012), instead of focusing on discrete measures to reduce the specific perceived future risks, it may be more effective for cities to consider the problem as one of building resilience. In the context of rapid urbanization and impacts of climate change, we might know only the trend, but it is difficult to quantify the future risks. However, the traditional 'predict and prevent' strategies only work against known problems. Thus, the concept of urban resilience, where there is enhanced capacity to face various impacts, is increasingly relevant as we face future uncertainties associated with climate change and rapid urbanization.

The urban climate assessment described in this chapter applied the Urban Resilience Framework developed and tested by ISET (Tyler and Moench 2012). There are three key components in the framework: (i) urban infrastructures/ecosystems, (ii) agents which are usually but not always people, and (iii) institutions that regulate the interactions between them. The key elements of URF also link closely with concept of sustainable development that considers the balance amongst environment, economy and society in the process of development. The framework is presented in Fig. 4.3. It should be noted, this framework assumes each component will affect, and will be affected by every other component. This complexity is necessary to capture interdependencies within the systems.

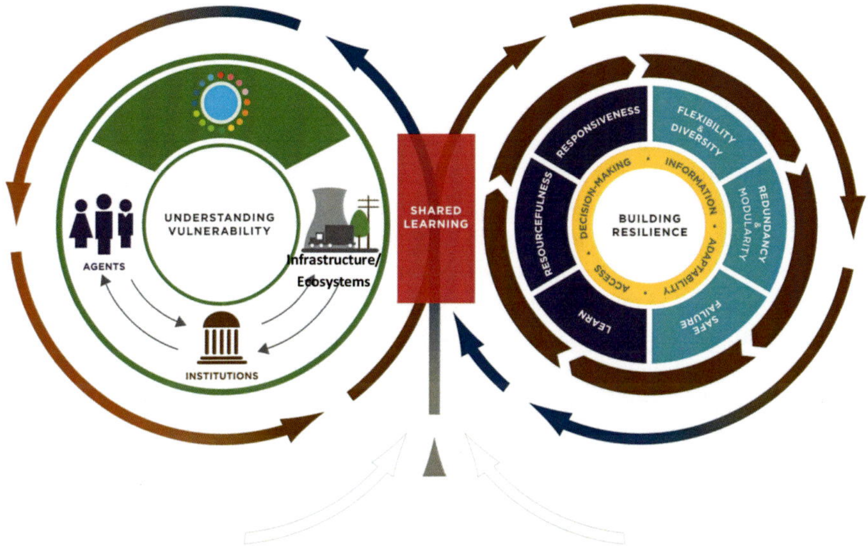


**Fig. 4.3** The urban climate resilience framework (Modified from Tyler and Moench 2012)

The Urban Resilience Framework (URF) process begins with a Shared Learning and Dialogue (SLD) to understand what it means to be a resilient city and to define principles that will guide the city's vision and process into the future (entry arrows at bottom of Fig. 4.3 are the local and scientific knowledge to understand the resilience of the city). In an SLD, a process whose goal is to maximize open, broad stakeholder participation and knowledge sharing, city partners first assess current and future vulnerability through identifying existing factors that contribute to vulnerability of city residents (left-hand loop in Fig. 4.3). They then develop strategies and interventions to address those vulnerabilities (right-hand loop in Fig. 4.3). Core to this approach is an iterative and deliberative assessment of vulnerability that takes into account not just currently vulnerable systems but the reasons for those vulnerabilities, including exposure to climate hazards, low capacity of organizations, households and individuals in responding to climate shocks, fragile supporting infrastructure and ecosystems, and the governance, social conventions and behavioral norms that act to constrain or enhance access by vulnerable users to supporting urban systems. The key elements of the URF are urban infrastructure/ecosystems, social agents, and institutions, and, for each, the degree to which it is exposed to climate change hazards.

**Urban infrastructures/ecosystems** in the urban setting include both infrastructure (e.g. water supply and wastewater treatment systems, roads, power lines) and ecosystems (e.g. agricultural land, parks, wetlands, fishing grounds). Physical environments are designed and managed through deliberate human intervention, but their performance depends on a multitude of factors that are difficult to control, including human behavior and institutional context, which often lead to unintended side effects (e.g. pollution, congestion). Fragile environments are easily disrupted or broken under stress, though their basic functioning may appear stable.

**Agents** are individuals, households, and communities. They are human beings functioning either alone or in groups. Agents, unlike systems, are capable of careful thought, independent analysis, voluntary interaction, and strategic choice in the face of new information. This makes agent behavior more difficult to predict than system behavior. Agent thinking, analysis, interaction and choice will generally, but not always, reflect the agent's location and structure within society (i.e. government entities will likely act very differently from individuals acting on their own behalf), their preferences, and the opportunities and constraints they perceive.

**Institutions** are the rules, laws, customs, social norms and conventions that guide, enable, and constrain agent behavior and exchange in social and economic transactions. They are the "rules of the game" that define the range of perceived possible responses or actions in a given situation. Institutions are created to reduce uncertainty, to maintain continuity of social patterns and social order, and to stabilize forms of agent interaction in more predictable ways. In so doing institutions can also create or maintain disparities in agent access to systems and system outputs and services, particularly for outputs and services that are limited or otherwise highly prized.

These three core elements in the URF provide distinct lenses through which urban climate vulnerability is assessed in order to build urban climate change resilience. Each aligns with specific interests and backgrounds associated with key practitioners and decision makers responsible for planning in urban areas. They are analyzed and assessed using different tools and methodologies. Separation of these major components provides a practical basis for engaging with key actors in urban areas about climate resilience (Moench et al. 2011). Collectively they provide a holistic view of urban climate vulnerability: urban physical environments relate to what are vulnerable (infrastructure, ecosystems, etc.); agents relate to who are vulnerable and how; institutions relate to the drivers of vulnerabilities (legal or regulatory frameworks and processes, laws, authority, agreements, etc.).

## 4.3 Vulnerability Assessment Methodologies

### *4.3.1 Engage Stakeholders and Define the Context*

The starting point of the URF is a process of shared learning dialogues (SLD) that fosters knowledge exchange, dialogue, and deliberation to mutual learning, innovation and flexibility. This is the core process through which vulnerabilities are assessed and stakeholders are able to reflect and review in a participatory, iterative manner. The SLD process is the keystone of the vulnerability assessment bringing together different stakeholders and different types of knowledge both scientific and local to construct and define the objectives of the assessment (ISET, NISTPASS and TEI 2012), see Fig. 4.4. In Hue, the stakeholders are members from technical departments at provincial level, researchers from Hue University, local NGOs, local

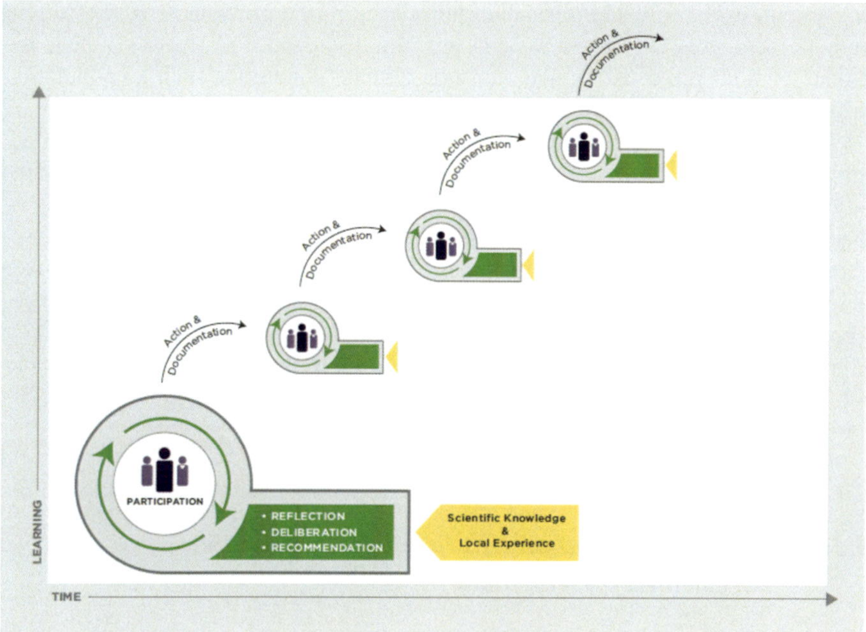


**Fig. 4.4** Share learning and dialogue process

governments from ward to provincial levels, and researchers from national and international organizations. The assessment team was also established with members from those stakeholders to implement the vulnerability assessment. The first SLDs are designed to define the urbanization and climate change context based on the discussions of the following topics:

- the historical and future climate trend, and climate impacts;
- who and what are the most vulnerable groups, areas, physical systems, and how they may be affected;
- institutional capacities to adapt to climate change and urbanization;
- the range of factors that systematically combine to make them vulnerable, including both direct (e.g. exposure to hazards) and indirect (e.g. supply chain, urbanization) factors.

### 4.3.2 *Define Key Urban Infrastructure/Ecosystem and Vulnerability Assessment Objectives*

This step is to identify key urban infrastructure/ecosystem in which local government or community manages, plans, or makes policy affecting those systems' services for enhancing the city resilience. After identifying key urban infrastructure/

ecosystem, using climate information introduced in the previous step, the current and expected stresses to those systems are then discussed. The stresses include the effects of both projected climate and non-climate conditions such as population growth, economic development and other major trends.

The key urban systems often interlink with other systems. However, at local levels, climate change is seen as a number of discreet changes (temperature, precipitation) that lead to specific types of impacts on certain places, people and activities. Therefore, in order to define key urban system to conduct the vulnerability assessment, a series of SLDs were organized to enable stakeholders to present his/her perceptions to explore the inter-linkages of systems in the context of climate change. These discussions provide the basis for stakeholders to move on to identifying key urban systems. By working together on identifying future vulnerabilities related to urbanization and climate change, stakeholders frame the scope and terms of reference for a process of vulnerability assessments.

### *4.3.3 Conduct Current and Future Urban Climate Vulnerability Assessment*

There are four components in the current and future climate vulnerability analysis: (i) Climate trend analysis using historical data from Thua Thien Hue Province's hydro-meteorology center and Ministry of Natural Resources and Environment (MONRE) climate change scenarios; (ii) Institutional capacity assessment using the Local Government Self-Assessment Tool (LGSAT) developed by UNISDR, in-depth interview, group discussion; (iii) Physical systems assessment using spatial data analysis, hydrological models and climate scenarios; and (iv) community and local residential adaptive capacity assessment using PRA tools. The overall process is presented in Fig. 4.5.

## 4.4 Results

Based on the process described above, this section presents the conclusions that the local team drew and included in the vulnerability assessment report conducted by local stakeholders (Hue Climate Change Working Group, ISET and NISTPASS 2013).

Climate change is not only a global issue of the future, but it is now happening in Hue city and affecting people's daily lives. Precipitation levels shown in monitoring documents of Hue Monitoring Station are increasing consistently in recent years, with some very intense rain events that led to unusual flooding situations. Impacts of climate change in Hue city are diverse, including: increased flooding in rainy seasons, droughts and water shortages in dry seasons, impacts on the city's ecological environment, etc. On the social side, climate change has negative impacts on infrastructure, tourism services and the life of a majority of Hue residents.

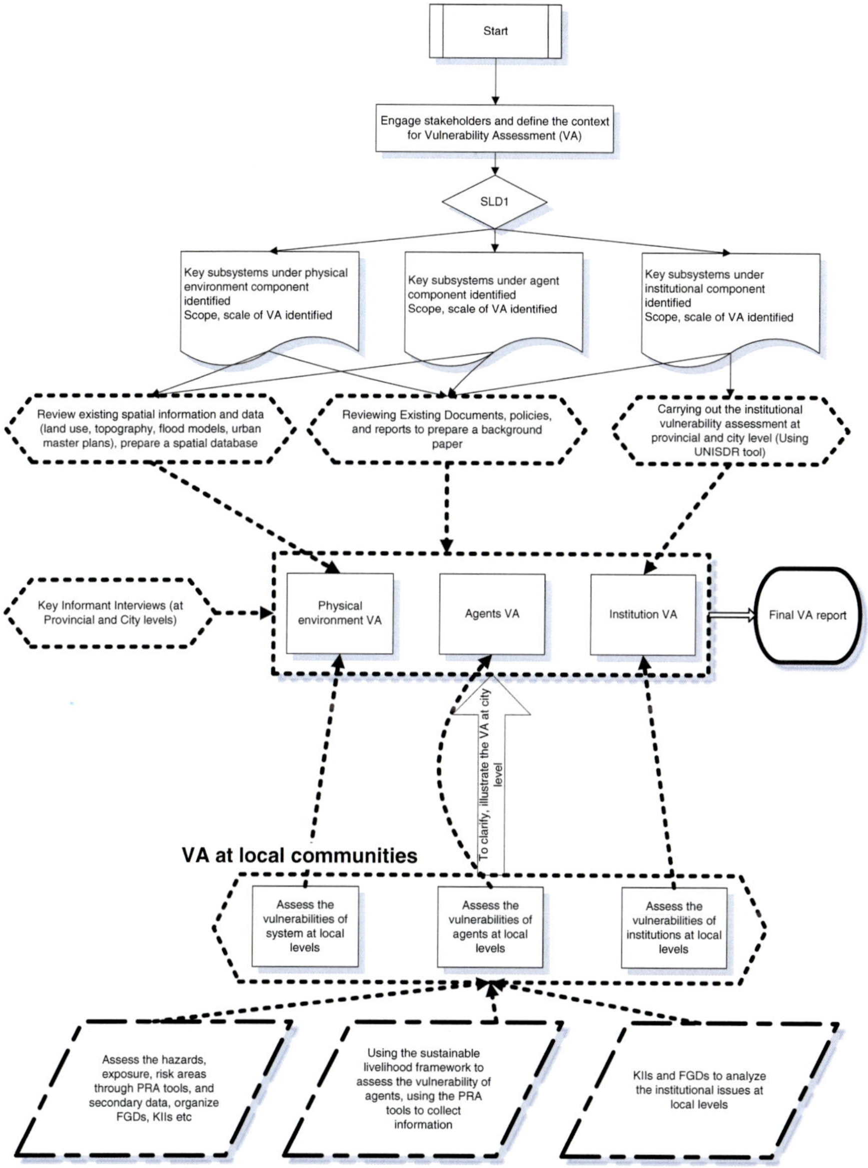


**Fig. 4.5** Vulnerability assessment methods and process

However, the increase in both the level of precipitation and in rain intensity is the most obvious impact at present with projections for the future showing this trend continuing. Compounding increased rainfall amount and intensity are the roles of rapid urbanization of Hue and the gaps in planning and implementation in leading to frequent and extreme flooding.

Flooding is a natural phenomenon that often occurs in Hue city; however, it has become more complicated in terms of duration, intensity and frequency due to rapid urbanization and impacts of climate change. Hue people are familiar with annual flooding, with roughly half of its population residing in low-lying areas; they have high awareness and understanding of ways to adapt to flood situations. However, flood levels and duration are increasing in some areas, especially the new low-lying urban areas. Floods no longer develop the way they did before, so many of the old adaptive approaches no longer work. In Hue, precipitation of some days of heavy rain could be equivalent to half a year's rainfall. According to the most recent Research project by Thua-Thien Hue Province's Department of Transportation, rains with frequency of 50 % (i.e. happening every 2 years, which is 72 mm of rainfall in 3 h) will cause 70 % of the city's roads to lie under 0.1–0.5 m of water. Land raising and filling, and road construction for build new urban areas in low land area and floodways is leading to local water logging and prolonged duration of floods due to slow drainage.

The assessment team was required to use the climate models output based on the IPCC SRES "B2 scenario" that are issued by Ministry of Natural Resources and Environment (MONRE) when developing their vulnerable assessment scenarios of Hue. This is because all provincial documents apply this scenario. However, Climate change projections from "B2 scenario" do not reflect reality of Hue City. Results of analysis show that precipitation data of recent years have already surpassed projections of the B2 scenario. In addition to climate change impact issues, flood risks will also increase in the coming decades due to challenges from the management of upstream reservoirs. Therefore, flood drainage planning and design of water supply and drainage construction in Hue city will have to take these extreme factors into account. Compounding the risk from rainfall and reservoir operations is the influence of sea level on floods in the city. Clearly the context of climate change and urbanization is a major challenge of the future.

Assuming that in the future, with regulation of large upstream reservoirs, downstream flood levels will be lowered (if reservoirs are operated properly to mitigate floods) scenarios project that flooding will still occur in the city. This is attributable mainly to the increase of rainfall in the city and its poor drainage system. In addition, plans to expand the city to the East and Southeast areas with low elevation also expand the area of its high-risk zones. Though climate change awareness of Hue city Government and its people are fairly high, the city has not had a climate change resilience plan. Sector plans (socioeconomic development plans, construction plans, land use plans, etc.) have no climate change impacts integration. In fact, when using historical flood data as the basis for construction of infrastructure and construction in floodway areas, solutions to respond to potential extreme weather developments in the future have not been considered.

The coordination between city departments and agencies in urban planning, urban management and urban flood management is limited, which lessens flood resilience of the city. In addition, guidelines and regulations on integration of climate change and extreme weather developments into development projects and plans are either absent or not specific enough. Limitations in finance, technology and expertise in climate change are major challenges in building resilience for the city. Development of new urban areas such as An Van Duong reduces drainage capacity of the city to the east, as this is the main drainage direction of the city. On the other hand, discrepancies in project planning and implementation of urban clusters in An Van Duong area caused local water logging. Moreover, spatial planning and socioeconomic development orientations still tend to point towards development in low-lying and floodway areas of the city.

## 4.5 Conclusions

In Hue, while floods had been a part of human life for centuries, and local communities have evolved ways of coping with flood's annual cycle, they completely lack experience to deal with the new risks induced by urbanization and climate change. In addition, in the past floodwater quickly drained away through the network of rivers and canals built in the flood prone area. In recent years, the increasing degradation of the natural environment through deforestation and the conversion of agricultural land to urban areas has made the impact of floods more serious and longer lasting in the lowland areas. For example, because of the expansion of the road network and other barriers to drainage, the flood level has increased beyond people's anticipation and the drainage and flood way capacity are no longer able to cope with the extreme rainfall. Fatefully, although most interviewed households agreed that floods are becoming worse both in severity and frequency, they were still preparing for floods at a level commensurate with the worst disaster they experienced, and not for the greater floods that are likely to strike in the future.

The flood management methods still follow 'old customs' mostly dependent on large-scale infrastructure once implemented by the government, but too costly to maintain nowadays. However, while these old practices are under pressure, the new methods necessary to deal with the modern risks have yet to be materialized. This phenomenon is visible in the cities where most households and industries are not well prepared to cope with the increasing level of floodwaters.

Climate change is real and already being felt at local levels Already the impact is more severe than IPCC B2 scenario suggests. However, the city's development plan does not factor in enough the impacts of climate change in planning for this urban expansion. The urban expansion of Hue rests on an assumption that key flood infrastructure will work well.

Ultimately, the worsening flooding situation in Hue appears to be the outcome of many different interacting factors including the natural setting, traditional practice, and institutional structure. These factors provide a backdrop for the more recent rapid environmental change brought about by the expansion of the market economy, population pressure, urbanization and industrialization, inappropriate or poorly implemented urban management policies, and lack of adequate knowledge and understanding of the changing environment. Hence while traditional risks remain high, as the city now enters a critical phase in development, new risks are rapidly appearing, and traditional autonomous adaptation measures are becoming inadequate. Without initiating new and more innovative measures, the future Hue city will be subjected to severe environmental deterioration and increased vulnerability to climate risks.

## 4.6 Lessons Learned

Responding to climate change requires strategies that address both the physical dynamics of systems and the social and institutional context of the city level. As a result, analytical and other strategies need to combine technical as well as social science-based approaches. Specialized technical studies as well as more "people-centered" forms of engagement are essential. Strategies that overemphasize one dimension to the exclusion of the other are likely to be ineffective.

Quality climate information is difficult to access, particularly at a scale useful to adaptation planners. Local-scale historical climate information and future projections are not always easy to find and often do not exist, even appropriate historical data can never tell us exactly what to expect in the future. Resilience planning, however, cannot wait for the ideal information.

Effective engagement within cities depends at a minimum on active commitment to resilience planning on the part of a number of individuals who are well connected with diverse local groups. Because for many urban areas, climate change is a "new" and poorly understood issue, and because effective responses must involve interaction among diverse groups of actors, identifying a few charismatic and articulate individuals who can serve as champions can greatly facilitate the growth of awareness and action.

While basic principles and broad process elements do apply across regions, results from Hue vulnerability assessment demonstrate that variations in local contexts can be a significant challenge for resilience planning, so strategies must be tailored to localities. Because cultures, bureaucratic structures, physical characteristics of urban areas, and a myriad of other factors affect how climate change impacts urban areas and what practically can be done, strategies must be locally grounded.

Planning for urban climate resilience involves integrating many new concepts and tools into already complex local planning processes, and under conditions in which local government resources are already strained. Time constraints are a fact of life, but short time horizons are the enemy of quality engagement and learning. Even using iterative processes, it may be difficult to anticipate how much time is needed for introduction of basic concepts, collection of relevant climate and planning information, sharing and digestion of new information, and building consensus on action. Resilience is unlikely to be achieved without carefully acquired, shared understanding about the interdependencies of systems and people. Attempts to shortcut this process even with skilled external support run the risk of yielding ineffective or even maladaptive results. Working with local partners also involves being flexible: scheduling conflicts, shifting priorities, staff changes, political and bureaucratic procedures are inherent to this work.

## References

Hue Climate Change Working Group, ISET and NISTPASS (2013) Hue climate vulnerability assessment report, MBRACE project

IFRC (2012) Understanding community resilience and program factors that strengthen them: a comprehensive study of Red Cross Red Crescent Societies tsunami operation. International Federation of Red Cross and Red Crescent Societies, Geneva

ISET-International, NISTPASS, & TEI (2012) Changing cities and changing climate: insights from shared learning dialogues in Thailand and Vietnam. Institute for Social and Environmental Transition- International, Boulder

Moench M, Tyler S, Lage J (eds) (2011) Catalyzing urban climate resilience: applying resilience concepts to planning practice in the ACCCRN program (2009–2011). Institute for Social and Environmental Transition-International, Boulder

Prabhakar S, Srinivasan A, Shaw R (2008) Climate change and local level disaster risk reduction planning: need, opportunities and challenges. Mitigation Adapt Strategy Global Change. doi:10.1007/s11027- 008-9147-4

Tran P, Shaw R (2010) Integrated river basin management for effective flood risk reduction. Research Publishing, Singapore

Tyler S, Moench M (2012) A framework for urban climate resilience. Clim Dev 4(4):311–326

# Chapter 5
# Climate Change Risks – Methodological Framework and Case Study of Damages from Extreme Events in Cambodia

**Kirsten Halsnæs, Per S. Kaspersen, and Sara Trærup**

**Abstract** Climate change imposes some special risks on Least Developed Countries, and the chapter presents a methodological framework, which can be used to assess the impacts of key assumptions related to damage costs, risks and equity implications on current and future generations. The methodological framework is applied to a case study of severe storms in Cambodia based on statistical information on past storm events including information about buildings damaged and victims. Despite there is limited data available on the probability of severe storm events under climate change as well on the actual damage costs associated with the events in the case of Cambodia, we are using the past storm events as proxy data in a sensitivity analysis. It is here demonstrated how key assumptions on future climate change, income levels of victims, and income distribution over time, reflected in discount rates, affect damage estimates and thereby the economic recommendations for climate change adaptation decision making. The conclusion is that taken vulnerabilities and equity concerns into consideration in adaptation planning for Least Developed Countries really makes a strong case for allocating economic resources to the protection of these countries.

**Keywords** Climate change risks • Least developed countries • Cambodia • Damage costs • Equity

## 5.1 Introduction

Climate change is most often presented as a long term phenomenon characterized by changes in mean parameters like temperature, precipitation, wind storms, etc., and this information plays an important role in climate policy target setting and in

K. Halsnæs (✉) • P.S. Kaspersen
Management Engineering, Technical University of Denmark, Roskilde, Denmark
e-mail: khal@dtu.dk

S. Trærup
UNEP (United Nations Environment Programme) DTU (Technical University of Denmark) Partnership, Copenhagen, Denmark

 
J.I. Uitto, R. Shaw (eds.), *Sustainable Development and Disaster Risk Reduction*, Disaster Risk Reduction, DOI 10.1007/978-4-431-55078-5_5

the development of long term adaptation and mitigation strategies. Little attention, however, is given to assessments of how extreme weather events already today cause serious losses particularly in least developed countries. There is limited information available about the consequences of these extreme events, and there are large uncertainties about the attribution of specific extremes to climate change due to large noise from natural variability in the short term (Hawkins and Sutton 2009, 2011). However, statistical records of past extreme weather events like severe storms already today, for some regions, show that the frequency and the intensity of events have increased over the last decades, and serious damages have been reported.

Least developed countries (LDCs) are particularly vulnerable due to low incomes, weak infrastructures and institutions, and a low capacity for coping with climate change (IPCC 2014), and this implies that damages associated with extreme events often are very serious. Despite uncertainties and data limitations, it is therefore important to study the consequences of extreme events in LDCs. The aim of this chapter is to develop a methodological framework for damage cost assessment, which reflects key assumptions reflecting the specific vulnerabilities in a developing country context. The framework is applied to the assessment of the consequences of severe storms in Cambodia based on 18 years statistical records of events. Given limited data availability both in terms of climate scenarios and damage costs, we are focusing on a sensitivity analysis, where a few general assumptions about trends in severe storms and associated damages are used to illustrate, how assumptions related to equity, in terms of impacts on low income households, and risk aversion, reflecting uncertainties about the probability of extreme events, can influence damage cost estimates significantly. Finally these findings are discussed in the context of the economics of climate change adaptation policies in LDCs.

## 5.2 Methodological Framework

It is well documented that climate change is likely to influence the frequency and severity of some extreme weather and climate events regionally[1] (IPCC 2012), even though extreme events by definition are events with low probabilities relatively to mean climate change variables. This opens some particular methodological issues related to how the risks of high consequence/low probability events are to be assessed. Assessing such risks involves specific methodological challenges related to key uncertainties and to economic assumptions, and these are again related to the multiple elements involved in climate change impact studies, frequently visualized as a "cascade" of uncertainties (Wilby and Dessai 2010). Methodologically, in the cascading picture uncertainty propagates through the different interlinked steps in a "top-down" assessment of climate risks, going from socio-economic scenarios through emission scenarios, global and regional climate model projections, and

[1] "Extreme events" are here defined as specific outcomes of individual or combinations of climate variables belonging to the tails of a given probability distribution.

impact models to local impacts and possibly adaptation responses. The uncertainties involved are however of a different nature dependent on disciplines, modelling tools, and approaches applied (IPCC 2005; Refsgaard et al. 2013). Thus some of the uncertainties reflect parameter uncertainties, while others are of a more structural character such as uncertainties related to economic valuation, risk perceptions, equity, and preferences (Weitzman 2011). All together this plethora of uncertainties provides a basis for a wide range of climate change risk estimates. In our case study of Cambodia we will highlight how substitute specific with key assumptions applied to specific development aspects influence the outcome of risk assessments.

A systematic assessment is carried out on how risk estimates and uncertainties are related to climate scenario and impact uncertainties and, in particular, to economic assumptions seen in the context of LDCs. The methodological framework we use for linking physical and economic models is inspired by a paper by Weitzman (2011) describing the role of "Fat-tailed Uncertainty in the Economics of Catastrophic Climate Change". Weitzman argues that there are large uncertainties associated with the probability of extreme events as projected by climate models as well as deep structural uncertainties related to economic risk evaluations, including damage cost estimates, discounting, and risk aversion. The latter are key issues in terms of real-life decision making, i.e. how much society should be willing to pay for adaptation taking vulnerabilities and equity into consideration, which is often overlooked or severely simplified in many real-life climate change impact assessments.

Seen from the perspective of a climate change adaptation decision maker, society should be willing to pay adaptation costs, which are at least equal to the avoided costs of climate change impacts. Adaptation costs should here be adjusted for residual damages, to the extent where adaptation costs are exceeding these. According to this, residual damages are associated with climate change impacts which either have very low damage costs or where adaptation is very expensive or not feasible. The avoided costs of climate change in terms of risks depend on damages as well as on the probability of a given event.[2] Adhering to conventional usage of the term we define climate change risks as the probability times the consequence of a climate change event. For high consequence events with low probabilities, the estimated risks will depend on a sort of "race" between how fast the probabilities of climate events decline, compared with how fast damage costs increase, when we are moving further away from the mean (median) of a climate probability density function.

Climate change impact assessments generally involve integrated climate modelling and impact assessment. Specifically, future climate events such as temperature and precipitation extremes, wind storms, droughts, or combinations of these are used as drivers for impact assessments. Subsequently, economic consequence studies address damages to specific sectors, ecosystems, geographical locations, and human assets.

[2]A climate event should here be understood as a broad terminology covering particular weather events like hot spells, intensive precipitation, wind storms etc., which are associated with societal risks.

In the current study damage costs are based on a bottom up assessment, where cost parameters are assigned to different assets, which are expected to be at risk from severe storms in Cambodia. Our study is based on data reports on damages on houses and people, and we are using proxy values reflecting the number of incidences as an indicator of damage costs (see the more detailed explanations about assumptions subsequently). Damage costs are transformed to a measure of "willingness-to-pay" (WTP) reflecting welfare costs and taking risk aversion and income equality of victims into account. The Cambodian study is a simplified application of a more general methodological framework for risk assessment, which is illustrated in Fig. 5.1.

Figure 5.1 illustrates the different logical steps of an impact assessment for climate change risks. Generally, in terms of describing the risks associated with specific climate variables like temperature, precipitation, wind or sea level, the probability of a specific event is derived from climate projections. The probability may be expressed in the form of a probability density function (pdf), which is typically constructed on the basis of an ensemble of climate models. Principally, the pdf provides a comprehensive description with respect to both frequency and intensity of the occurrence of any given climate variable. In practical terms it is far from trivial though to construct such a pdf, and tail events representing extremes are particularly uncertain i.e. because these are very rare. Hence, in some cases more stylized shapes of pdfs are therefore used in sensitivity analysis, e.g. in order to explore the tails of the distribution given certain assumptions about uncertainties (Weitzman 2011).

To exemplify consider the two stylized pdf's illustrated in Fig. 5.2. The x-axis shows the value of some climate variable, e.g. daily mean temperature/wind speed,

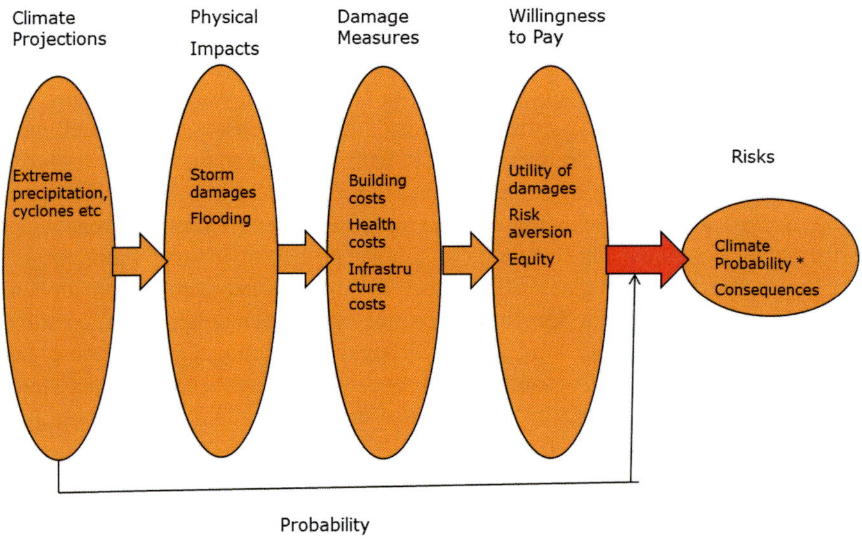


**Fig. 5.1** General structure of climate change impact assessment and risk analysis

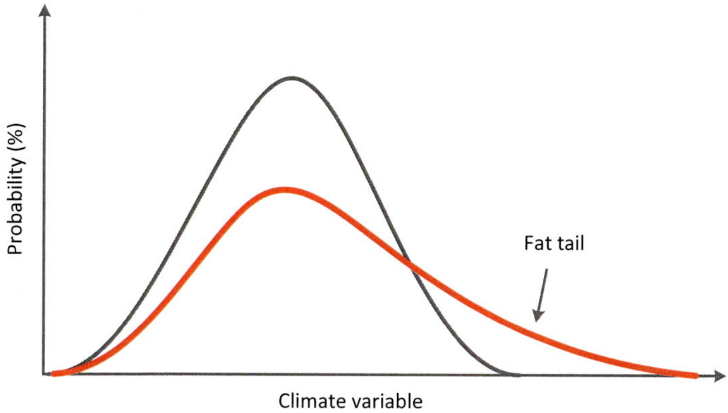


**Fig. 5.2** Stylized probability density functions

for a given time period while the y-axis expresses the probability of this value, e.g. as inferred from single model simulations or an ensemble of climate model projections. The mean value of the distribution is here expected to increase from present day conditions to future climate change, and the upper tail of the distribution is also fatter than in the pdf illustrating present day climate conditions. If we for example were considering wind speed, it would mean, that the mean expected wind speed would be expected to increase with climate change, and the probability of events that are very strong relatively to the mean value would also increase. The impact assessment of an extreme event will be directly influenced by the shape of these pdfs, which of course introduces a significant uncertainty in terms of determining and quantifying risks.

The perspective of the damage cost assessment in our approach is social welfare,[3] where the total damage cost is an aggregate measure of the costs to all individuals of damages to given assets, and total damages are calculated as the sum of damages in all sub categories.

In terms of climate change, the uncertainty surrounding future events and the specific character of extreme events with low probabilities and high consequences suggests that the social welfare function applied to damage cost evaluation is adjusted to reflect society's perspective on uncertain future risks (Heal and Kriström 2002; Weitzmann 2011). One way to include this type of uncertainty in economic analysis is to apply a risk aversion factor. Risk aversion by definition is the reluctance of a person to accept a bargain with an uncertain payoff rather than a bargain with a certain payoff, and as already pointed out, extreme consequences of e.g. high end climate scenarios are by their very nature uncertain.

Damages from climate extremes in LDCs will in many cases particularly harm people with low incomes, and we are therefore suggesting applying an income equality factor to the damage estimates to reflect that damages to low income house-

[3] Social welfare reflects societies perspectives as for example in relation to climate change impacts.

holds in terms of welfare counts more than to high income households (Markandya 1998).

We assume a social welfare function V, where $V_t$ is the contribution to the social welfare function of generation t consuming $c_t$. Since $c_t$ is uncertain, we consider the expected value $Eu\ c_t$ of consumption in our social welfare function. The concavity of the function u combines inequality aversion reflected in discount rate and risk aversion to reflect uncertainty[4]

$$\text{V-}\sum_{t=0}^{\infty} Eu(c_t)d(t)$$

where u $c_t$ is the contribution to the social welfare of consumption $c_t$ at time t. The factor $d$ is a discount factor, which reflects our preferences to equity among present and towards future generations (IPCC 2005).

We assume a constant risk aversion factor as defined by Arrow (1965):

$$A(w) = -U''(w) / U'(w)$$

where $A(w)$ is the risk aversion associated with a given social welfare change and $U(w)$ the utility of the social welfare change.

In the case of a utility function, which is a polynomial of order $n$, the risk aversion factor takes the form:

$$A(w) = nx^{n-1}$$

There are to the authors' knowledge no specific climate change risk attitude studies suggesting, what the level of risk aversion should be, so instead we consider two different risk aversion factors (i.e. high risk aversion and risk neutrality) based on an approach developed by Heal and Kriström (2002). Heal and Kriström (2002) suggests using a risk aversion values between 1 and 6 based on risk preferences revealed among investors.

Based on Markandya (1998) we apply an inequality aversion parameter (epsilon) to reflect the difference between the average income and the income of the persons suffering from the damages. The inequality factor can then be defined as:

$$\frac{\partial W}{\partial Y_i} = [\frac{\overline{Y}}{Y_i}]^{\varepsilon}$$

where $W$ is the social welfare, $Y_i$ is the income of an individual, and $\overline{Y}$ is a numeraire given the value of 1 representing the average income. The literature has estimates of

[4] Different assumptions can be applied to the choice of discount rate, whether the perspective is from a global societal point of view or whether the perspective is from a national point of view, where it can be expected that LDC would apply relatively high discount rates to welfare cost evaluations (IPCC, AR3 Costing Methodologies).

the inequality parameter of between 1 and 2 (Markandya 1998), where a factor of 1 implies that damages on all individuals are valued associated with the average income, and we are in the case study applying a high and a low parameter.

The risks of climate change impacts may now be calculated from:

$$\text{Risks} = \text{WTP for avoiding the event} * \text{probability of the event}$$

and the WTP for avoiding the events is:

$$\text{WTP} = \text{damage costs} * \text{risk aversion factor} * \text{the inequality weight to income classes}$$

To exemplify how uncertainties and economic assumptions individually and combined influence the levels of risk, we apply the methodological framework to assess the WTP for avoiding damages from severe storms in Cambodia.

## 5.3 Climate Change Risks from Severe Storms in Cambodia

Cambodia is located on the mainland of Southeast Asia covering an area of 181,035 $km^2$ with a population of 14.7 million in 2013 (UNFPA 2013) whereof the rural population accounts for 79 %. Cambodia is an LDC, with a GDP per capita of US$ 709 in 2013, whereof agriculture accounts for 36 % of GDP (World Bank 2014). In 2010, 20 % of the population lived below the national poverty line (World Bank 2014).

The climate is characterised by a dry season from mid-November to mid-May and a rainy season from mid-May to mid-November. The annual average temperature is 27 °C, and rises to a maximum of 38 °C in April or May and falls to a minimum of 14 °C in January or December (MoE 2006). Cambodia is highly vulnerable to the impacts of climate change and has been ranked as the country second most affected by extreme-weather events in 2011, with an annual GDP loss of 3.1 % caused by extreme events (Harmeling and Eckstein 2012).

A number of national studies have analysed potential vulnerability to climate change in Cambodia (MoE 2002, 2005, 2006, 2014), and they have identified particularly vulnerable sectors and areas including:

- Agricultural production which is dependent on the annual flooding and recession of the Tonle Sap Lake and the Mekong River.
- Sea level rise affecting the 435 km long coastline, which is already suffering from storm surges, high tide, beach erosion and seawater intrusion. Low-lying areas, including settlements, beach resorts, seaports, coastal fisheries, and mangrove forests, may become submerged with rising sea levels.
- Vector-borne diseases, in particular malaria, which may become more widespread under changing climatic conditions.

Studies on the anticipated climate change effects for Cambodia are limited, but national assessments suggest rainfall increases by 0–15 % over baseline levels in

2025 and potentially more for 2050 and 2100 (MoE 2002). IPCC results for the Southeast Asian region are more muted with little expected deviation in seasonal precipitation totals, but with more significant temperature and rainfall anomalies in the latter part of the century (Cruz et al. 2007). McSweeney et al. (undated) find a gradual warming trend through the 2090s, Johnston et al. (2010) predict no significant change in annual rainfall for the country as a whole. They also project decadal temperature changes of about 2.3 °C. This is significantly greater than Eastham et al. (2008) who initially tested output from 24 GCMs and then selected the 11 models best able to recreate historical temperature and precipitation data. They forecast a change of +0.7–0.8 °C in mean temperature in 2030 for most of Cambodia, with slight reductions in dry season precipitation and larger gains in wet season precipitation. In addition, DanChurch Aid/Christian Aid (2011) finds, in a literature review which summarizes the natural disasters affecting Cambodia since 1994, that floods and droughts are responsible for the greatest damages. This is in accordance with the findings in the National Adaptation Plan and in other government documents (MoE 2002).

## 5.4 Damages from Storms and Cyclones

We are in the following focussing on severe storms including cyclones and the number of victims and damaged houses in Cambodia, and we are here not distinguishing between cyclones and severe storms. Data has been drawn from the UNDP Disaster Information System (UNDP 2015), and it is important to recognize that reports about extreme events in Cambodia in international and national databases are surrounded by large uncertainties, related to the quality of the reported data and the inclusion of relevant events.

The UNDP Disaster Information System reports the development over time in severe storms and other extreme weather events, and as shown in Table 5.1, the number of severe storms reported for Cambodia has been increasing in the reporting period from 1996 to 2014.

The number of houses and victims affected by the storms and cyclones were reported in the UNDP database for the period from 2000 to 2014, and it can be seen from Fig. 5.3, that in particular the number of victims increased strongly during the period, and there were a lot of victims suffering from events in 2008, 2009, 2010, and in 2013.

### *5.4.1 Assumptions Applied to Damage Assessments*

We are in our case study working with two alternative scenarios for the frequency of severe storms in Cambodia associated with damages on houses and people namely:

**Table 5.1** Severe storms in Cambodia 1996–2014. http://camdi.ncdm.gov.kh/DesInventar/profiletab.jsp

| Year | Number of events |
|---|---|
| 1996 | 1 |
| 2000 | 16 |
| 2001 | 6 |
| 2002 | 10 |
| 2003 | 4 |
| 2004 | 22 |
| 2005 | 32 |
| 2006 | 43 |
| 2007 | 71 |
| 2008 | 55 |
| 2009 | 171 |
| 2010 | 125 |
| 2011 | 116 |
| 2012 | 155 |
| 2013 | 315 |
| 2014 | 71 |
| Total | 1,213 |

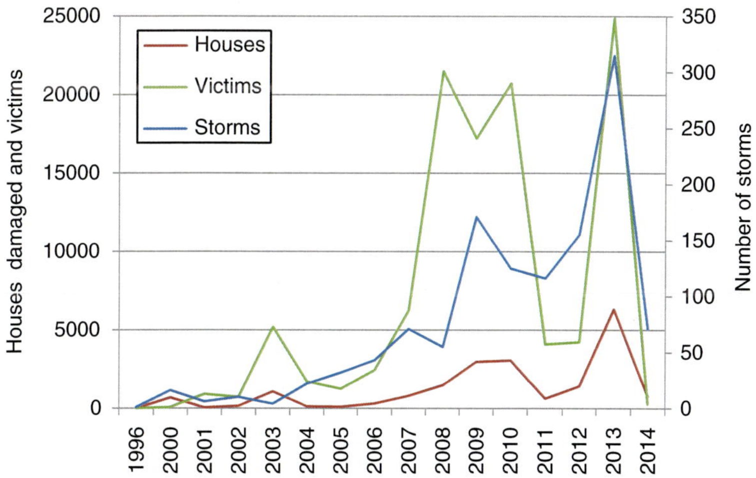


**Fig. 5.3** Development in number of houses damaged and victims of storms in Cambodia 1996–2014 http://camdi.ncdm.gov.kh/DesInventar/profiletab.jsp

- Climate Scenario 1 which represents a continuation of our current climate and assumes constant frequency of the severe storms corresponding to the mean value of the events for the period 1996–2014. The mean value is 76 storms/year (Fig. 5.4).

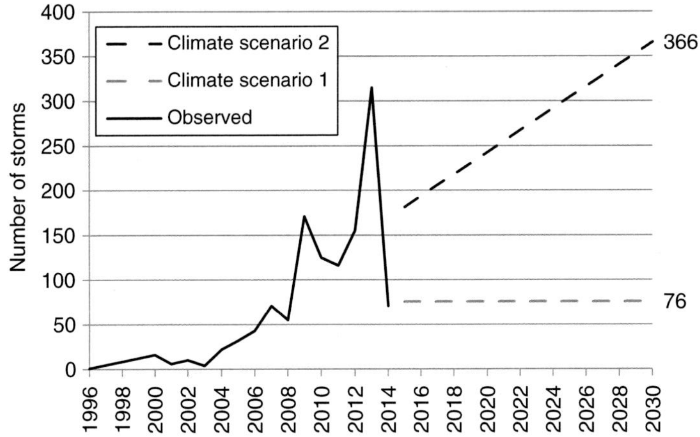


**Fig. 5.4** Observed number of severe storms in Cambodia in the period 1996–2014 and as projected in climate scenario 1 and 2 for the period 2015–2030

- Climate Scenario 2 which represents climate change with an increasing frequency of severe storms corresponding to a linear interpolation of the trends observed in the period 1996–2014. The trend value is 366 in 2030, and an average of 273 for the entire period of 2015–2030 (Fig. 5.4).

The two alternative climate scenarios are used to compare the index of damages, which given the alternatives could happen in the period 2015–2030 according to Climate Scenario 1 and Climate Scenario 2. The simple climate scenario assumptions on the future frequency of severe storms is of course not a good representation of climate change and related events in Cambodia, but due to the limited data availability, we have chosen to use these alternative scenarios as an illustrative sensitivity case of the implications of future increases in the frequency of extreme events.

Different to the definition of risk as the probability of the event multiplied with the WTP for avoiding damages as outlined in the methodological framework section, we are in the case of Cambodia not able to assign probabilities to the two climate scenarios. Our risk calculation is therefore in the current case study for Cambodia only assessing the WTP for avoiding damages associated with severe storms.

Damage cost reports for cyclones in Cambodia have not been available, and we are therefore focusing on the development over time in the number of houses and victims affected by severe storms, and are using these numbers as proxy values for damages. Using such proxies rather than damage cost data implicitly assumes that the magnitude of damages on individual houses and people are constant over time, and that the number of incidents thereby represents damages. An implication of these assumptions is also, that the damages associated with each severe storm is similar in Climate scenario 1 and 2, but the number of damages in terms of houses affected and victims are higher in Climate Scenario 2 than in Climate Scenario 1.

We are using the number of victims as the representative indicator of damages[5] associated with severe storms.

As suggested previously in the methodological framework section, the WTP for avoided damages is defined as:

WTP = damage costs * risk aversion factor * inequality weight to income classes

We are in the following combining the damage cost proxy given the climate scenarios with alternative assumptions for risk aversion and inequality weights to income classes.

Damage cost proxy

Using the development in victims reported for Cambodia in the period 1996–2014 as proxy index value,[6] now provides us with the following values for the period 2015–2030:

- Climate Scenario 1: Damage cost proxy = 1 (no change in number of victims)
- Climate Scenario 2: Damage cost proxy = 2.48 (change according to 1996–2014 trend)

Risk aversion

In order to reflect risk aversion, we are applying a risk aversion factor 3 and an alternative factor of 1, which represents risk neutrality.

Inequality weight to income classes

In order to reflect Inequality aversion towards damages imposed on low income groups, we are using income distribution weights. We are using an income distribution weight of 1 and 1.75 to reflect two different assumptions respectively in the damage assessment. A weight of 1 reflects that all individuals are valued with the weight corresponding to the average income in Cambodia, and the weight of 1.75 reflects that a higher value is given to low income groups. We are in our case study assuming that 30 % of the victims of the cyclones have an income which is only 25 % of the average income, 40 % have an average income, and 30 % have an income which is 30 % above the average.

The inequality factor then becomes:

- With income distribution weight 1: $0.30 * (1/0.25)^{1} + 0.40 * (1/1)^{1} + 0.30 * (1/1.3)^{1}$
- With income distribution weight 1.75: $0.30 * (1/0.25)^{1.75} + 0.40 * (1/1)^{1.75} + 0.30 * (1/1.3)^{1.75}$

[5] Victims are chosen as the only indicator since it would not be possible without any cost values to aggregate over houses destroyed, damaged and victims.

[6] The development trends in houses damaged are very similar to the trend in victims, so we have just chosen one of these damage categories.

**Table 5.2** Combinations of damage cost proxies in the climate scenarios, risk aversion, and income inequality factors

| | Climate Scenario 1 | Climate Scenario 2 |
|---|---|---|
| Damage cost proxy | 1 | 2.48 |
| Risk aversion High | 3 | 3 |
| Risk aversion (neutral) | 1 | 1 |
| Inequality factor High | 1.75 | 1.75 |
| Inequality factor Low | 1 | 1 |

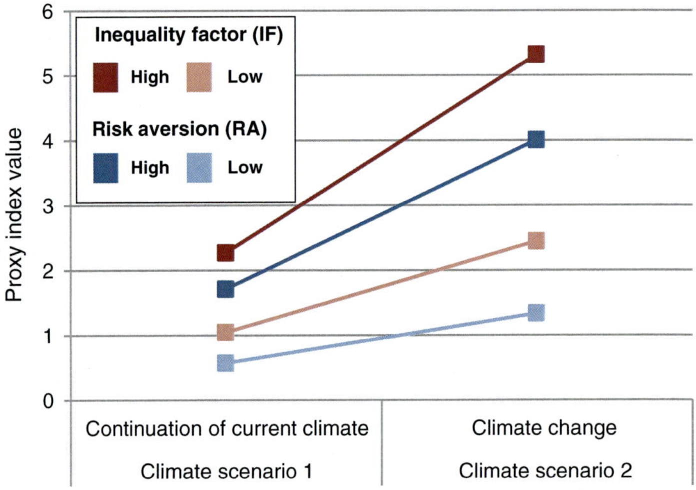


**Fig. 5.5** Damage proxy values in terms of NPV of WTP for avoided damages using different combinations of inequality factor, risk aversion and climate scenario (as shown in Table 5.2), using a discount rate of 7 %

We are then in our sensitivity analysis for Climate Scenarios 1 and 2 combining the risk and inequality aversion factors in order to reflect, how damage estimates could be affected by priorities of avoiding damages affecting low income groups in developing countries.

Table 5.2 shows the combinations of WTP assumptions in the sensitivity cases.

The WTP estimates are conducted for a 15 year timeframe, which means that we are calculating a proxy value of damages based on climate scenario assumptions for the period 2015–2030. We are then subsequently calculating the 2015 Net Present Value, NPV of the damages using a high discount rate of 7 % and a lower discount rate of 5 %.

Figure 5.5 shows the calculated NPV damage proxy values for the two climate scenarios for the different WTP assumptions represented in Table 5.2. A high discount rate of 7 % has here been used in the NPV calculation. We are here comparing the damage proxy for Climate Scenario 1, where the current climate is continued until 2030, and the proxy for a change in the frequency of severe storms up to 2030 as assumed in Climate Scenario 2.

As illustrated in Fig. 5.5 the damage proxy for severe storms increases considerably with the application of income inequality factors, as compared to the case with risk neutrality and no inequality factor (the green line in Fig. 5.6). This is both the case for the high and low factor, and it can here be seen that adding a dimension, where vulnerabilities of low income groups are valued as a particularly high welfare loss can make a big difference to economic estimates of, what society should be willing to pay for avoiding extreme climate events. Adding a risk aversion factor of three as illustrated by the blue line in Fig. 5.6 also increase the damage proxies.

The importance of the choice of discount rate in the NPV calculations is illustrated in Fig. 5.6, where combinations of the climate scenarios and the WTP assumptions are calculated as NPV's with a high 7 % discount rate and a low 5 % discount rate.

It can here be seen, that the application of high factors of income inequality and risk aversion factors make a much larger difference to the damage cost proxies, than the use of alternative discount rates of 7 % versus 5 %.

In conclusion it can be said that the sensitivity cases demonstrate that assessments of welfare implications of extreme events like severe storms in LDCs like Cambodia strongly depend on how particularly vulnerable people like low income households are addressed in the assessments. It must be considered that low income households most often will live in houses which are poorly protected against storms,

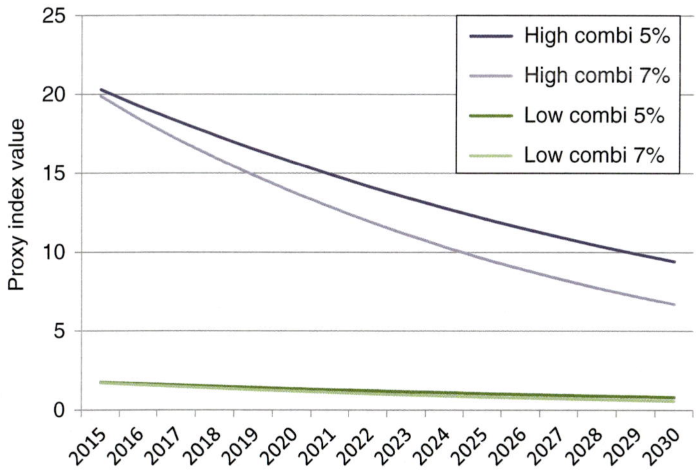


**Fig. 5.6** Index of the NPV's of damage cost proxies for assumptions with high inequality factor and high risk aversion assuming Climate Scenario 2 (RA=3, IF=1.75, Climate change) and for assumptions about low inequality factor and a risk aversion factor of 1 in Climate Scenario 1. Alternative cases applying a 7 % and 5 % discount rates

and that they also could be located in geographical areas suffering from extremes. Applying an evaluation framework for assessing the WTP for avoiding damages as suggested in this chapter will, following these arguments, suggest that seen from a welfare economic point of view society should give a high priority to adaptation investments in vulnerable areas with a high density of low income households in LDCs.

## 5.5 Conclusions

Least developed countries are particularly vulnerable to climate change due to low incomes, weak infrastructure, and limited institutional capacity for coping with climate change. Extreme events, which have happened over the last decades point to a threat of an increasing frequency of incidents and damages, and despite uncertainties about climate change attribution of the events, it is therefore important to further strengthen data and methodological frameworks for assessing risks in vulnerable countries.

Assessing climate risks in LDCs can be integrated in more general methodological frameworks for the assessment of climate risks based on probability density functions for climate events and welfare economic concepts for damage cost assessment. We have in this chapter suggested to apply specific assumptions to WTP estimates for avoided damages for LDCs reflecting risk aversion and an inequality factor, which gives relatively high weight to damages and thereby income losses of poor households. It is demonstrated that the application of these specific factors for LDCs strongly influence WTP estimates for avoided damages, and the assumptions therefore are very important seen in the context of economic arguments in adaptation investments for countries like Cambodia.

The current study has focused on damage costs related to risk from severe storms in Cambodia. However, the approach could be easily generalized to other types of high impact events. The implication of applying these relatively simple assumptions is that the Cambodian case study cannot be used as a basis for directly arguing about the magnitude of climate risks and adaptation costs in LDCs. However, critical assumptions are highlighted, and a case is made for the application of specific assumptions in adaptation cost studies for LDCs in order to reflect inequalities and vulnerability.

## References

Arrow KJ (1965) The theory of risk aversion. In: Arrow KJ (eds) Aspects of the theory of risk bearing. Yrjö Jahnssonin Säätiö, Helsinki. Reprinted in: Arrow KJ (1971) Essays in the theory of risk bearing, Markham, Chicago, pp 90–109

Cruz RV, Harasawa H, Lal M, Wu S, Anokhin Y, Punsalmaa B, Honda Y, Jafari M, Li C, Huu Ninh N (2007) Asia. Climate change 2007: impacts, adaptation and vulnerability. Contribution of Working Group II to the Fourth Assessment Report of the Intergovernmental Panel on Climate

Change (Parry ML, Canziani OF, Palutikof JP, van der Linden PJ, Hanson CE (eds)). Cambridge University Press, Cambridge, pp 469–506
DanChurchAid/Christian Aid (DCA/CA) (2011) Climate change and disaster management policy mapping and analysis in Cambodia, September 2011
Eastham J, Mpelasoka F, Mainuddin M, Ticehurst C, Dyce P, Hodgson G, Ali R, Kirby M (2008) Mekong river basin water resources assessment: impacts of climate change. CSIRO, Canberra
Harmeling S, Eckstein D (2012) Global climate risk index 2013. German watch http://germanwatch.org/fr/download/7170.pdf
Hawkins E, Sutton R (2009) The potential to narrow uncertainty in projections in regional climate predictions. Bull Am Meteorol Soc 90:1095–1107
Hawkins E, Sutton R (2011) The potential to narrow uncertainty in projections of regional precipitation change. Clim Dyn 37:407–418
Heal G, Kriström B (2002) Uncertainty and climate change. Environ Resour Econ 22:3–39
IPCC (2005) Guidance notes for lead authors of the IPCC Fourth Assessment Report on addressing uncertainties www.ipcc-wg1.unibe.ch/publications/supportingmaterial/uncertainty-guidance-note.pdf
IPCC (2012) Managing the risks of extreme events and disasters to advance climate change adaptation. A special report of Working Groups I and II of the Intergovernmental Panel on Climate Change (Field CB, Barros V, Stocker TF, Qin D, Dokken DJ, Ebi KJ, Mastrandrea MD, Mach KJ, Plattner G-K, Allen SK, Tignor M, Midgley PM (eds)). Cambridge University Press, Cambridge/New York
IPCC (2014) Climate change 2014: synthesis report. Contribution of Working Groups I, II and III to the Fifth Assessment Report of the Intergovernmental Panel on Climate Change (Core Writing Team, Pachauri RK, Meyer LA (eds)). IPCC, Geneva, Switzerland, 151 pp
Johnston R, Lacombe G, Hoanh CT, Noble A, Smakhtin V, Suhardiman D, Kam SP, Choo PS (2010) Climate change, water and agriculture in the Greater Mekong Sub-Region. International Water Management Institute research report 136
Markandya A (1998) Economics of greenhouse gas limitations. The indirect costs and benefits of greenhouse gas limitations handbook reports. UNEP Collaborating Centre on Energy and Envrionment, Risø National Laboratory, Denmark
McSweeney C, New M, Lizcano G (Undated) Cambodia climate change country profile, United Nations Development Program (UNDP). Retrieved from http://country-profiles.geog.ox.ac.uk
Ministry of Environment, MoE (2002) National communication. Royal Government of Cambodia, Phnom Penh
Ministry of Environment, MoE (2005) Vulnerability and adaptation to climate hazards and to climate change: a survey of rural Cambodian households. Royal Government of Cambodia, Phnom Penh
Ministry of Environment, MoE (2006) National adaptation programme of action to climate change (NAPA). Royal Government of Cambodia, Phnom Penh
Ministry of Environment, MoE (2014) 3rd state of the coastal environment, climate change and socio-economy report. Coastal Coordination Unit, Royal Government of Cambodia, Phnom Penh
Refsgaard JC, Arnbjerg-Nielsen K, Drews M, Halsnæs K, Jeppesen E, Madsen H, Markandya A, Olesen JE, Porter JR, Christensen JH (2013) The role of uncertainty in climate change adaptation strategies – a Danish water management example. Mitig Adapt Strateg Glob Change 18(3):337–359
UNDP (2015) Disaster information system. http://camdi.ncdm.gov.kh/DesInventar/profiletab.jsp
UNFPA (2013) Cambodia inter-censal population survey 2013. National Institute of Statistics, Phnom Penh. Available at http://countryoffice.unfpa.org/cambodia/drive/CIPS_Report_English_Final.pdf
Weitzmann ML (2011) Fat tails and the economics of climate change. Rev Environ Econ Policy 5(2):275–292
Wilby RL, Dessai S (2010) Robust adaptation to climate change. Weather 65(7):180–185
World Bank (2014) World development indicators http://data.worldbank.org/country/cambodia

# Chapter 6
# Food Security, Climate Change Adaptation, and Disaster Risk

**Umma Habiba, Md. Anwarul Abedin, and Rajib Shaw**

**Abstract** Climate change impacts and natural disasters are the leading cause of hunger and affect all dimensions of food security including access to food, availability and stability of supplies, and nutrition across the world. The global food crisis is exposing existing and potential vulnerabilities of households, governments, and the international system to food and nutrition insecurity. Most food-insecure people live in areas prone to natural hazards, and they are the least able to cope with shocks. Due to their vulnerability and limited capacity to manage risks, poor households are often trapped in a downward spiral of food insecurity and poverty. On the other hand, Millennium Development Goals, the first goal is to eradicate hunger and poverty, and everybody around the world is fighting together for this. At present, there are 925 million undernourished people in the world. Therefore, this chapter focuses its attention on the current scenarios of hunger and concept and dimension of food security systems, in order to understand their vulnerability to environmental change, linkages among food security, climate change, and natural disasters and to identify solutions of food security coupled with climate change and disaster risk.

**Keywords** Food security • Climate change • Hunger and disaster risk

U. Habiba (✉)
Department of Agriculture Extension, Ministry of Agriculture, Dhaka, Bangladesh
e-mail: shimuagri@yahoo.com

Md. A. Abedin
Department of Soil Science, Bangladesh Agriculture University, Mymensingh, Bangladesh

R. Shaw
Graduate School of Global Environmental Studies, Kyoto University, Sakyo-ku, Kyoto, Japan
e-mail: shaw.rajib.5u@kyoto-u.0ac.jp



J.I. Uitto, R. Shaw (eds.), *Sustainable Development and Disaster Risk Reduction*, Disaster Risk Reduction, DOI 10.1007/978-4-431-55078-5_6

## 6.1 Introduction

Global climate change and disaster risk are fundamental threats to food security and the eradication of poverty. The negative impacts threaten to roll back decades of food security. Building resilient and sustainable food security means addressing both climate and disaster risks and integrating these risks, as well as potential opportunities, into proper access of food. The entire global food system is failing almost one billion people. To put it in perspective, this exceeds the combined number of people who live in the European Union, the USA, Japan, Canada, and Australia. Adding to this billion, there are another one billion people who suffer from "hidden hunger" and do not have enough vitamins and minerals in their diets to be assured normal physical and mental growth.

Again, potential impact of climate change will act as a multiplier of existing threats to food security: it will make natural disasters more frequent and intense, land and water more scarce and difficult to access, and increases in productivity even harder to achieve. The implications for people who are poor and already food insecure and malnourished are immense. Mainly in the least developed countries and small island developing states, it is the livelihoods and lives of the poorest and most vulnerable, including women, children, and marginal communities, which are also at greatest risk to suffer from to the potential impacts of climate change. This is due to their high exposure to natural hazards; their direct dependence on climate sensitive resources such as plants, trees, animals, water, and land; and their limited capacity to adapt to and cope with climate change impacts.

On the other hand, food security as a concept has evolved over a period of time. Before early 1970s, adequate availability of food grains at the national level was considered a measure of food security. Emphasis was placed on food self-sufficiency at the national level, principally through domestic production. In 2002, FAO gave a functioning definition of food security for all countries: food security exists when all people, at all times, have physical, social, and economic access to sufficient, safe, and nutritious food which meets their dietary needs and food preferences for an active and healthy life. Therefore, at minimum, to achieve food security, four components such as availability, accessibility, stability, and utilization are needed.

Again, the links among climate change, food security, and sustainable development are strong. While climate change will know no boundaries, poor and developing countries, particularly the least developed countries, will be among those most adversely affected and least able to cope with the anticipated shocks and disaster risk to their social, economic, and natural systems. Internationally agreed frameworks and goals have set an agenda for integrating climate change and sustainable development. For instance, Agenda 21, which addresses climate change under its Chap. 9 (Protection of the atmosphere), recognizes "that activities that may be undertaken in pursuit of the objectives defined therein should be coordinated with social and economic development in an integrated manner, with a view to avoiding adverse impacts on the latter, taking into full account the legitimate priority needs of developing countries for the achievement of sustained economic growth and the eradication of poverty." Finally, this will help in the process of food security and sustainable development in the affected countries.

It is obvious that poor communities and countries among others, natural disasters, and climate change are a leading cause of hunger and affect all dimensions of food security including economic and physical access to food, availability and stability of supplies, and nutrition. Climate-related disasters are by far the most frequent today: nine out of ten natural disasters are climate related, and every year, they affect more than 200 million people. Trends indicate that less predictable, extreme climate events are becoming the norm. At the global level, climate change will increase the risk of food and nutrition insecurity to unprecedented levels, undermining current efforts to eradicate hunger and under nutrition. The scale and complexity of some recent disasters, such as the Horn of Africa crisis and Pakistan floods, illustrate the type of challenges that local communities, national authorities, and global actors will increasingly have to face in the future. This only reinforces the need for a more strategic forward-looking approach to scale up efforts to reduce disaster risk, support climate adaptation, and build resilience in food insecure and at-risk communities and countries. Promising and innovative initiatives that tackle hunger while building long-term resilience and food security are now being developed throughout the world. Food security, climate change and disaster risk reduction research, and policy agendas are also increasingly centered on resilience and how to bring together ideas, innovations, and lessons from these three fields. In this contrast, although climate change will result in more extreme weather events, which are difficult to predict locally and could result in large-scale ecological shocks and surprises, it will also cause slower changes to established weather patterns. In this respect, climate change offers time and opportunities for mitigation and adaptation: risk reduction, risk management, and risk coping. Hence, this chapter tries to address facts and figures regarding hunger in the present world, the food security concept, and how climate change affects food security. Finally, the focus shifts prime entry points on food security for climate change and disaster risk scenarios.

## 6.2 Facts About Hunger

In broad spectrum, hunger is more than missing a meal. But, there is enough food in the world to go around, but almost a billion people go hungry every day and a further billion people are undernourished – not getting enough of the vitamins and minerals they need to live healthy and productive lives.

By 2050, the world will have another two billion mouths to feed. Changing consumption patterns, climate change, and growing numbers of shocks, such as drought, price rises, and conflict, are increasing the risk of hunger in many places in the world. Families struggling with chronic food insecurity, hunger, and malnutrition don't consistently have the food their minds and bodies need to function, which then prevents them from having the resources to improve their lives. It's a perilous cycle that passes hunger from one generation to the next. Recent facts and figure of hunger are summarized in the following bullet points:

- In the world, about 842 million people do not eat enough to be healthy. That means that one in every eight people on Earth goes to bed hungry each night (Source: FAO 2013).
- The number of people living with chronic hunger has fallen by 17 % since 1990–1992. If the trend continues, we will fall just short of the hunger target in the Millennium Development Goals (Source: FAO 2013).
- Most of the world's undernourished people are still to be found in Southern Asia, closely followed by sub-Saharan Africa and Eastern Asia (Source: FAO 2013).
- A third of all deaths in children under the age of 5 in developing countries are linked to undernutrition (Source: IGME 2011).
- In the developing world, one child in four is stunted, meaning that their physical and mental growth is impaired because of inadequate nutrition (Source: The Lancet 2013).
- The first 1,000 days of a child's life, from pregnancy through age 2, are critical. A proper diet in this period can protect children from the mental and physical stunting that can result from malnutrition (Source: IGME 2011).
- If women farmers had the same access to resources as men, the number of hungry in the world could be reduced by up to 150 million (Source: FAO 2011).
- It costs just US $0.25 per day to provide a child with all of the vitamins and nutrients he or she needs to grow up healthy (Source: WFP 2011).
- By 2050, climate change and erratic weather patterns could have pushed another 24 million children into hunger. Almost half of these children would be in sub-Saharan Africa (Source: IFPRI 2009).

## 6.3 Definition and Dimensions of Food Security

### *6.3.1 Defining Food Security*

In the year 1974, food security concept was originated, in the discussions of international food problems at a time of global food crisis. The initial focus was the volume and stability of food suppliers. During that period, food security was defined by World Food Summit (WFS) in the 1974 as "availability at all times of adequate world food supplies of basic food stuffs to sustain a steady expansion of food consumption and to offset fluctuations in production and prices" (United Nations 1975). The most recent careful redefinition of food security is that negotiated in the process of international consultation leading to the WFS in November 1996. According to WFS (1996) definition, "Food security exists when all people, at all times, have physical and economic access to sufficient, safe and nutritious food that meets their dietary needs and food preferences for an active and healthy life." This definition addresses four key components of food supplies and security, namely, availability, stability, access, and utilization (Schmidhuber and Tubiello 2007).

A large number of different definitions have been proposed passing over time. However, at the World Summit of Food Security in 2009, this definition was reconfirmed, and the concept was extended and specified by adding that the "four pillars of food security are availability, access, utilization, and stability" and stated that "the nutritional dimension is integral to the concept." The strength of this definition is its comprehensiveness and imperative for "concerted actions at all levels" (that are "individual, household, national, regional, and global levels") and "coordinated efforts and shared responsibilities" across institutions, societies, and economies to tackle food insecurity effectively (FAO 1996).

## 6.3.2 *Dimensions of Food Security*

Food security is the outcome of food system operating efficiently. Efficient food system continues positively to all dimensions of food security. The dimensions of food security are described under the following sub headings (Fig. 6.1).

### 6.3.2.1 Food Availability

The dimension addresses supply side of the food security and expects sufficient quantities of quality food from domestic agricultural production or imports. This is simple mathematical calculation whether the food available in certain territory/country is enough to feed the total population in that particular territory and calculated from the level of local agriculture population at that territory, stock levels, and net import/export.

The dimension of food security at different levels can be assessed by precipitation record, food balance sheet, food market survey, and agricultural production planet. Similarly, indicators of food security for this dimension at different levels are fertility rate, food production, population flows, harvesting time, staple food production, food shortage, consumption of wild foods, etc.

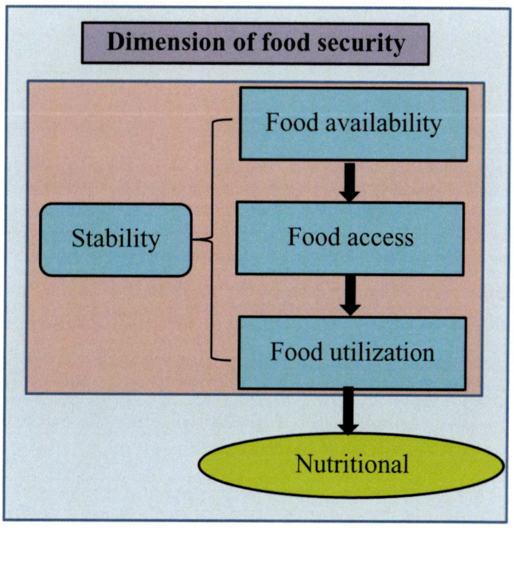


**Fig. 6.1** Dimensions of food security

#### 6.3.2.2 Food Access

Having sufficient food at national level or at certain territory cannot be taken as the proof that all the household or individuals in the country/territory have enough food to eat. Food access is another dimension of food security which encompasses income, expenditure, and buying capacity of households or individuals. Food access addresses whether the households or individuals have enough resources to acquire appropriate quantity of quality foods.

Some of the indicators of this dimension at different levels are food price, wage rate, per capita food consumption, meal frequency, employment rate, etc., and the dimension can be assessed by vulnerability assessment mapping (VAM), food access survey, food focus-group discussion, Intra-household food frequency questionnaire, etc. Interventions to improve this dimension of food security are inter alia on-farm, off-farm, and non farm employment creation, school-feeding program, breast-feeding campaign, etc.

#### 6.3.2.3 Food Utilization

Food Utilization is another dimension of food security which addresses not only how much food the people eat but also what and how they eat. It also covers the food production, intra-household food distribution, water and sanitation, and health-care practices. The nutritional outcome of the food eaten by an individual will be appropriate and optimum only when food is prepared/cooked properly, there is adequate diversity of the diet, and proper feeding and caring practices are practiced.

Stunting rate, wasting rate, prevention of diarrheal diseases, latrine usage, weight for age, goiter, anemia, night blinders, etc. are the indicators at different levels for this dimensions which can be assessed by demographic and health survey, immunization chart, etc.

#### 6.3.2.4 Stability

This dimension addresses the stability of the other three dimensions over time. People cannot be considered food secure until they feel so, and they do not feel food secure until there is stability of availability, accessibility, and proper utilization condition. Instability of market price of staple food and inadequate risk-bearing capacity of the people in the case of adverse condition (e.g., natural disaster, unexpected weather, etc.), political instability, and unemployment are the major factors affecting stability of the dimensions of food security.

This dimension of food security can be assessed by Global Information Early Warning System, anthropometric survey, weighing chart of pregnant women, etc. against certain indicators like food price fluctuation, women's BMI, pre harvest food practice, migration, etc. Interventions to address this dimension are saving and loan policy, inter-household food exchange, grain bank, food storage, etc.

## 6.4 Climate Change Affects Food Security

The direct implications of climate change for food security are through its impacts on food production worldwide. Furthermore, climate change is already affecting food security, and it is expected to have even greater impacts in the coming years. In this contrast, climate change is only one of many change factors that are either apparent or will emerge as driving forces of global food security in coming decades and will interact positively or negatively with each other. There are at least five channels by which climate change affects food security that will be briefly mentioned here:

- **Temperature increase**: Effect of higher temperatures leads to heat stress for plants, increasing sterility, and lowering overall productivity. Higher temperatures also increase evaporation from plants and soils, increasing water requirements while lowering water availability.
- **Changing patterns**: In many places in the globe, growing seasons are changing, ecological niches are shifting, and rainfall is becoming more unpredictable and unreliable both in its timing and its volume. This is leading to greater uncertainty and heightened risks for farmers and potentially eroding the value of traditional agricultural knowledge such as when to plant particular crops.
- **Rising sea levels:** The effect of rising seas contaminates coastal freshwater aquifers with saltwater. Several small island states are already having serious problems with water quality, which is affecting agricultural productivity. Higher seas also make communities more vulnerable to storm surges which can be 5–6 m high. The storm surge from cyclone Nargis traveled 35 km inland, killing 140,000 people and flooding around 14,400 km, an area one third the size of Switzerland.
- **Water:** The linkage between climate change, water scarcity, and declines in agricultural productivity could lead to regional tensions and even open conflict between states already struggling with inadequate water supplies due to rising populations and over pumping of groundwater. For instance, water scarcity currently affects some 1.7 billion people, and this number is projected to rise to about five billion by 2025. As with the food security impacts of climate change, "water insecurity" will affect people living in the subtropics most severely, due to more erratic rainfall, more frequent droughts, and increased evaporation.
- **Politics**: Climate change has impacts on global politics on food security. Hence, climate change could be "politically stratifying" at the global level. Losers from climate change could find themselves increasingly marginalized and eventually abandoned by wealthier "winners," which might abrogate responsibility for global poverty and withdraw behind bureaucratic fortresses designed to restrict in-migration from affected countries. Shortages of food and water "could potentially destabilise the geo-political environment [and] could contribute materially to an increasingly disorderly and potentially violent world" (Schwartz and Randall 2003). One epicenter of this instability could be chronically food-deficit and water-stressed regions such as the Sahel and Horn of Africa, where tensions over access to the Nile already contribute to the long-running conflict between north and south Sudan.

## 6.5 Climate Change, Food Security, and Disaster Risk Reduction Nexus

Climate change (CC)), food security (FS), and disaster risk reduction (DRR) are relatively young concepts that continue to evolve. Global platforms, for example, the climate dialogues under the United Nations Convention on Climate Change (UNFCCC), the FS dialogues driven by the Food and Agriculture Organization (FAO), and the milestone Hyogo Framework for Action (HFA) on DRR, are setting the course internationally. However, the dialogue is only now starting to move toward identifying and acting on the multiple inter linkages among DRR, FS, and CC. This is not easy because there are different sets of actors at global and regional level, who seldom find each other at shared platforms and remain ignorant of the nature and importance of this nexus.

The multifaceted dynamics that characterize the CC-FS-DRR nexus is simplified in Fig. 6.2. The linkages (intersections in the diagram) sometimes operate in both directions. For example, agriculture (implied within the FS circle) is affected by climate change, but it also contributes to global emissions of greenhouse gasses, which cause climate change. The diagram that shows the nexus between DRR, FS, and CC is not well understood or discussed and is poorly reflected in policies, planning, and programming, resulting in a lack of shared purpose and fragmented implementation. Climate change is changing the frequency, intensity, and duration of disasters and also the type (e.g., hailstorms where they never occurred before, changing and unpredictable rainfall patterns). Disasters impact directly (production) and indirectly (access to food) on food security. More gradual climate changes

**Fig. 6.2** Disaster risk reduction, climate change, and food security nexus nexaRR/FS/CC nexus

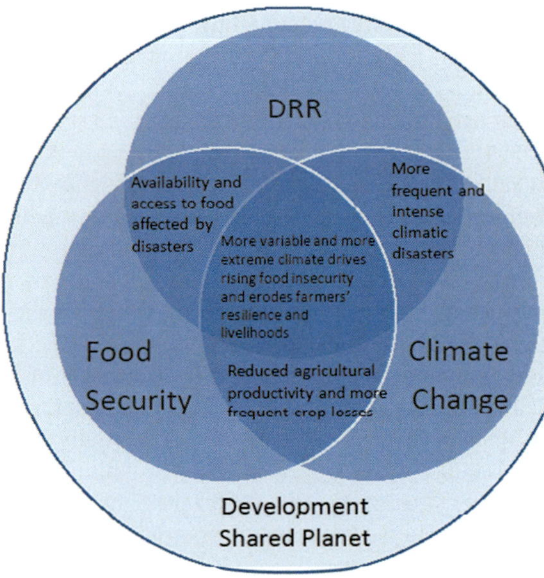

(e.g., warming) also impact directly on food production by reducing and destabilizing yields of sensitive crops. In the middle, where all three overlap, the overall impact on agriculture-based livelihoods becomes clear and shows the need for an integrated and long-term building of resilience, which could save millions of livelihoods.

Outstandingly, all three issues have component drivers that do not intersect with one another: food insecurity (especially issues of access to food) is driven by numerous socio economic factors and shocks to the food system; CC has causes and impacts unrelated to DRR and FS; and DRR also includes non-climatic disasters, such as earthquakes and tsunamis.

## 6.6 Role of Institutions for Adaptation and Disaster Risk Reduction to Ensure Food Security

Institution is the hub of mainstreaming adaptation; vulnerability reduction and risk into development activities such as food security are thus important policy goals for responding to climate change and disaster risk. But, according to O'Brien et al. (2006), implementing these changes often requires fairly radical shifts in thinking and new institutional architecture. However, typically with extreme weather events the focus, particularly in developing countries, is on the recovery from a disaster rather than vulnerability reduction before the event, and this system is reinforced by the investment policies of donors (Mirza 2003). This system is beginning to change with integrated disaster risk management and the acceptance that the time frame of focus for risk reduction needs to consider pre-event vulnerability reduction, as post-event response. For this to occur, different institutions need to be involved at various stages (UNISDR 2004).

Many countries are having to modify their disaster management institutions at the national level, to reflect this new paradigm, or design disaster management policies and setting up institutions to formalize and implement such policies. In the USA, disaster response has typically been coordinated by the Federal Emergency Management Association (FEMA), which was integrated into the Department of Homeland Security in 2001. Still, the institutional arrangements have been criticized (Gopalakrishnan and Okada 2007). According to Gopalakrishnan and Okada (2007), eight characteristics have been outlined that are necessary for institutions to implement the disaster risk management activities. These are:

- Awareness/access
- Autonomy (in that the institution must have the authority to act in the case of a disaster/state of emergency being declared)
- Affordability
- Accountability
- Adaptability (to take into account cultural norms as well as the nature of the risk)

- Efficiency (how well they do all of the above)
- Equity
- Sustainability

While the reorientation of existing disaster management frameworks can be problematic, even countries introducing disaster management policies and institutions from scratch can come up against barriers. The information available with regard to climate change and disasters is increasing through vulnerability and risk assessments, and having this information has been shown to be correlated with the number of lives saved and general quality of response after a disaster (NRC 2007). Information sharing is dependent on understanding of the importance of that information, knowing to whom to disseminate it and how. Individuals act as key hubs within a network and thus play a large gate-keeping role. In addition to this, administrative structures are also important to create institutional frameworks that allow for participation of other relevant stakeholders. For instance, NGOs have a long history of providing emergency humanitarian assistance after disasters and longer-term reconstruction. They are arguably suited and indeed often have a comparable advantage in this role, due to their location on the ground and understanding of local context and conditions.

## 6.7 Ensuring Food Security in a Changing Climate and Disaster Risk

Primarily, four entry points for climate change adaptation and risk reduction strategies are aiming at increased food security in view of climate change and natural disaster. Part of the solution is to increase food availability. Another lies in strategies that ensure that those who are at greatest risk of hunger can actually access and benefit from increased amounts of food and that protect the most vulnerable from the immediate impacts of climate change. This involves improving disaster risk management; enhancing social protection schemes, including the delivery of direct nutrition interventions; and strengthening resilient community-based development.

### *6.7.1 Increasing Agricultural Productivity, Resilience, and Sustainability*

Local people are the on-site land managers and real experts who play central roles in adapting agriculture and food systems to meet their needs under changing climate conditions. The concept of adapting to climate impacts is not new to them. Traditionally, coping mechanisms for adapting to seasonal and annual climate variability have included sharing local knowledge on varieties, farming systems, management technologies, etc., but the need to increase production, coupled with the speed and magnitude of the expected changes in climate, poses new challenges.

In reality, traditional coping mechanisms and wisdom will not be sufficient to ensure food security and prevent effects on nutritional status. They must be complemented by the introduction of technical innovations and enabling frameworks. More research is needed on the breeding of new and adapted as well as the preservation of traditional, locally adapted varieties that can tolerate climate variability and are suitable for changed climatic conditions; the development of innovative but practical technologies such as alternative cropping systems, conservation and precision agriculture, and sustainable forest management; and the application and improvement of technologies for more efficient use of inputs such as energy, fertilizer, water, and seeds. For all technological innovations in agriculture, it is crucial that they will be easily accessible and affordable for the communities in need.

Institutions and enabling policy and legal frameworks are interconnected with adaptation strategies. Incentives and services for rural producers that can stimulate and guide adaptation processes and link producers to markets are also important supporting mechanisms. Adaptation to climate change can incorporate a range of successfully tested methods and technologies derived from sustainable agriculture and natural resource management and equitable and inclusive rural development approaches, building on the "noregrets" principle. However, adaptation often involves substantial investments and changes in practices that may take a long time to implement or show benefits. It must therefore be complemented by other responses that address the immediate effects of climate change and protect those who cannot adapt.

### 6.7.2 *Strengthening Disaster Risk Management*

It is stated that the number of disaster-affected people has more than tripled since the 1990s. With this contrast, over 74 million people were victims of humanitarian crisis in 2007, as climate change leads progressively toward increased extremes – storms, droughts, and high temperatures – the challenge to the humanitarian community is not only to respond to the crises but also to be better prepared and to be able to manage the risks more effectively.

Recently developed approaches that integrate relief and response in long-term risk management have begun to influence the way disaster management programs are planned and financed. In order to enhance community safety and resilience, the complex interactions between long-term risk reduction and short-term response need to be better understood. At the same time, the most vulnerable to food insecurity must be protected from the immediate impacts of climate change now. In this connection, appropriate risk reduction planning and response requires an understanding of risks and vulnerabilities in terms of who are the vulnerable, where they are, and why they are vulnerable. There is a need for improved monitoring, information systems and forward-looking risk analysis. Particular efforts are needed to target the poorest and food-insecure people without assets and entitlements in risk reduction or response interventions. In addition to a rural focus, attention has to be given to urban and peri-urban areas.

Ideally, both vulnerable people and communities themselves should always be the primary owners and drivers of any actions aimed at increasing their resilience to disasters; it is crucial to directly involve them in planning and implementation of disaster risk reduction. At the same time, in order to achieve greater effectiveness, disaster risk management and climate change adaptation management should also be linked and better integrated into national development plans and strategies, starting from poverty reduction strategies, food security strategies, and sustainable development. Furthermore, the gaps between sectoral organizations must be bridged in order to share timely and relevant information concerning risks and their management. Climate information must be made accessible to affected communities and decision-makers. Last but not least, sufficient financial resources are a prerequisite for effective disaster risk reduction. Current practice indicates that less financial resources are being made available for disaster risk reduction than for adaptation.

### *6.7.3 Enhancing Social Protection Schemes*

Adverse consequences of climate change further widen the existing inequities in food security, food safety, and nutrition. Adapting food production systems has the potential to significantly increase the resilience of poor farmers to changing climate conditions. However, the vast majority of the one billion undernourished people do not have sufficient capacities and resources in order to adapt to or cope with the risks posed by climate change. They are in urgent need of public support in the form of social protection schemes, safety nets, and other supportive measures. Such public actions have large potential to increase resilience to climate change by contributing to breaking vicious cycles that lead into chronic poverty traps. Droughts, for example, frequently force poor families to sell off productive assets such as livestock; other shocks often lead to families taking children out of school and to reduction in households' food intake, number of meals, restriction of portion sizes, and purchase of less expensive but less nutritious foods – each with immediate and long-term physical and mental consequences for children. Eventual recovery becomes much more difficult as a result of such emergency "coping" measures. Environmental risks are among the most frequent, costly, and impactful causes of such shocks – a problem that will grow immensely with climate change.

Social protection relevant to food insecurity, climate change, and resource scarcity includes cash and in-kind transfers, such as Ethiopia's Productive Safety Net Programme, which transfers cash (and food) during seasonal food insecurity through employment on public works; employment guarantee schemes, which can be used to invest further in climate resilience, for example, strengthening embankments or planting trees; mother and child health and nutrition and school-feeding programs; weather-indexed crop insurance; micro-finance services; as well as emergency food assistance interventions. It also encompasses essential nutrition interventions, such as the distribution of micronutrient supplements for mothers and young children as well as fortification of foods as appropriate and the treatment of severe acute malnutrition.

Still developing countries have very limited access to formal social protection systems. Currently, only 20 % of the world's people have access to formal social protection systems. Financing social protection support is complicated by the fact that safety nets need to be financed in a counter-cyclical manner, given that needs are greatest when economic performance is weakest. Effective targeting of the poorest and most vulnerable people is also critical, which fundamentally depends on policymakers understanding the vulnerabilities of these people. Apart from financial resources, formulating social protection policies hence demands significant institutional capacity, which international actors can help to build.

### 6.7.4 Strengthening Resilient Community-Based Development

Life saving interventions are crucial to protect the food-insecure people and their livelihoods from rapid-onset emergencies caused by climatic events. It is equally important, however, to create enabling conditions to ensure that communities affected by disasters are able to build back systems, which are better adapted to changing climate conditions. Supporting a transition toward "climate-smart" relief, rehabilitation, and development that improves the livelihoods of low-income farmers and rural people and thereby increases their overall resilience must be considered the basis of adaptation.

Farmer's farm soil is often degraded in two thirds of developing country. They form the majority of the food insecure and are most vulnerable to climate change. Yet experience shows the right strategies that can transform their lives and create climate-resilient communities. Agro ecological paths show enormous potential, if combined with equally crucial – and often neglected – strategies to empower farmers to influence policy formulation and implementation. In many cases of degraded farms, doubling production within a few years after adopting agro ecological practices shows that such farms can "jump" to a higher threshold and enter a "virtuous circle" of environmental restoration, renewed productivity, and greater resilience to current seasonal climatic stresses. Agro ecological measures for delivering food security, climate change adaptation, or mitigation typically deliver the other two objectives as well, delivering a "win-win-win" outcome. Achieving resilient communities, which involves people achieving increased material welfare and reduced risk, is bound up with people attaining greater capacity to determine their own destiny. Three factors are crucial: (a) the prospect of major new investment flows focused on previously neglected lands, (b) incentivizing farmers through investments in agro ecological practices and in providing environmental services, and (c) communities influencing policy making and implementation.

## 6.8 Conclusion

Addressing climate change and disaster risk in food security goals will help ensure that these goals will be maintained and achieved in the face of changing climatic conditions and disaster events and prove to be truly sustainable. The negative impact of climate change and natural disasters will have profound implications for food security across the globe, but these implications are far from clear, and the causal pathways from changes in climate to changes in food security outcomes are complex and likely to vary from region to region. The examination of food security needs to consider the broader range of sectors and activities contributing to food production, including agriculture, fisheries, and forests. It also requires increasing attention to urban and peri-urban areas rather than only a rural focus, as these areas become increasingly important areas for markets, storage, and production as well as consumption.

Again, global climate change and disaster risk pose an unprecedented challenge to the aim of eradicating hunger and poverty. In order to meet the growing demand for food security and nutrition under increasingly difficult climatic conditions and in a situation of diminishing resources, the world must urgently move toward embracing a twofold approach: First, we must invest in and support the development of more efficient, sustainable, and resilient food production systems. Second, we must improve access to adequate food and nutrition by the most vulnerable and at risk populations and communities and enhance social protection systems and safety nets as part of the adaptation agenda. Protecting the most vulnerable also requires enhancing our capacities to manage weather-related disaster risks and accelerating community development. Only if we succeed in making significant advances on all fronts – increasing food availability, enhancing access to food, and strengthening resilience and development – will we reduce the risk of dramatic increases in the number of hungry people among the poorest countries in the most vulnerable regions of the world.

Attaining food security under a changing climate requires substantial increases in food production on the one hand, as well as improved supply and access to adequate and nutritious food and capacities to cope with the risks posed by climate change on the other hand. Governments must be assisted in enhancing food production, supply, and access, scaling up social protection systems; and improving their ability to prepare for and respond to disasters. Community-based development processes need to be fostered in order to enable the poorest and most vulnerable to build sustainable and climate-resilient livelihoods and move out of chronic poverty and food insecurity. On the contrary, it is true that a changing climatic system is expected to impact the availability of basic necessities like freshwater, food security, and energy, while efforts to redress climate change, both through adaptation and mitigation, will similarly inform and shape the global development agenda which will at the end enhance sustainable development outcomes.

## References

FAO (Food and Agriculture Organisation) (2002) The state of food insecurity in the World 2001. FAO, Rome

FAO (2011) The state of food and agriculture. Food and Agriculture Organization of the United Nations, Rome, Italy. http://www.fao.org/docrep/013/i2050e/i2050e.pdf. Accessed Nov 2014

FAO (2013) The state of food insecurity in the world. Food and Agriculture Organization of the United Nations, Rome, Italy. http://www.fao.org/docrep/018/i3458e/i3458e.pdfFAO. Accessed Nov 2014

FAO (Food and Agriculture Organization) (1996) Rome declaration on world food security and world food summit plan of action. http://www.fao.org/DOCREP/003/W3613E/W3613E00.HTM. Accessed on 23 Sept 2014

Gopalakrishnan C, Okada N (2007) Designing new institutions for implementing integrated disaster risk management: key elements and future direction. Disasters 31(4):353–372

IFPRI (2009) Climate change impact on agriculture and costs of adaptation. International Food Policy Research Institute, Washington, DC. http://www.ifpri.org/sites/default/files/publications/pr21.pdf. Accessed Nov 2014

IGME (2011) Levels & trends in child mortality. United Nations Children's Fund, New York, USA. http://www.unicef.org/media/files/Child_Mortality_Report_2011_Final.pdf. Accessed Nov 2014

Lancet (2013) http://www.thelancet.com/journals/lancet/article/PIIS0140-6736(13)60937-X/abstract. Accessed Nov 2014

Mirza MMQ (2003) Climate change and extreme weather events: can developing countries adapt? Clim Pol 3(3):233–248

NRC (2007) Committee on the effective use of data, methodologies and technologies to estimate subnational populations at risk. Tools and methods for estimating populations at risk from natural disasters and complex humanitarian crises. National Academies press, Washington, DC, 264p

O'Brien G, O'Keefe P, Rose J, Wisner B (2006) Climate change and disaster management. Disasters 30(1):64–80

Schmidhuber J, Tubiello FN (2007) Global food security under climate change. Proc Natl Acad Sci U S A 104(50):19703–19708

Schwartz P., Randall D. (2003) An abrupt climate change scenario and its implications for United States national security. Report commissioned by the US Defense Department, October, Emeryville. Global Business Network, San Francisco

United Nations (1975) Report of the world food conference, New York, Rome, 5–16 November 1974

UN ISDR (International Strategy for Disaster Reduction) (2004) Living with risk. UN-ISDR Secretariat, Geneva

WFS (World Food Summit) (1996) http://www.foodsecurityatlas.org/bgd/country/food-security-at-a-glance, Accessed on Dec 2014

WFP (2011) Two minutes to learn about: school meals. http://documents.wfp.org/stellent/groups/public/documents/communications/wfp220221.pdf. Accessed Nov 2014

# Chapter 7
# Human Health as Precondition for Achieving Sustainable Development

**Minako Jen Yoshikawa and Akhilesh Surjan**

*"History teaches that the improvement of human health at the population level is largely determined by good policies that protect the environment and people...."*
*(Peralta and Hung 2003)*

**Abstract** This chapter revisits sustainable development from perspectives of human health. The authors unfold interlinkages between human health and sustainable development and then identify gaps, which need to be narrowed.

Factors known to influence human health include climate change, environmental degradation, urbanization, societal issues, and disasters. The recent effects of climate change like extreme weather patterns and events have demonstrated negative implication for health, e.g., impacts of heat waves on older persons. Environmental degradation like changing ecosystems due in part to urbanization and microbial adaptation saw emergence of *Nipah Virus* infection. Societal issues such as changing human demographics and behavior, poverty and social inequality, and international travel may have contributed to India's excruciating economic damage after the episode of plague in Surat. Post-disaster disrupted living conditions might give rise to epidemics, as in the recent case of outbreaks of infectious diseases after the Tohoku disaster.

Insufficient understanding of the role of human health in sustainable development generates a gap between health and development policies. Weak health policies could keep healthcare systems underdeveloped, which could undermine human health. A popular belief remains that only developed economies like Japan can afford to build good health systems, but her GNP rose sharply in the early 1960s only after roundworm prevalence was controlled. Examining public health in relation to disaster and development, the authors show that promotion of population

---

M.J. Yoshikawa (✉)
Inter-Graduate School Unit for Sustainable Development and Survivable Societies,
Kyoto University, Kyoto, Japan
e-mail: mjenyjp@gmail.com

A. Surjan
Faculty of Engineering, Health, Science and the Environment, Charles Darwin University,
Darwin, Australia



J.I. Uitto, R. Shaw (eds.), *Sustainable Development and Disaster Risk Reduction*,
Disaster Risk Reduction, DOI 10.1007/978-4-431-55078-5_7

health is an essential precondition to achieve development and nutrition to sustain development. This chapter reveals that human health is deeply interconnected with social, economic, and environmental aspects, which form vital pillars of sustainable development.

**Keywords** Human health • Sustainable development • Post-disaster health issues

## 7.1 Introduction

This chapter will revisit sustainable development by focusing on human health, especially health at population level or public health. Referring to post-disaster public health issues, especially infectious diseases, elements of human health threatening sustainable development will be elucidated. In addition, an example of public health improvement, which could have contributed to economic development in a Southeast Asian country, will be discussed. Then, interlinkages of human health and development will be presented to show that gaps between the two must be narrowed.

The recent effects of climate change, such as extreme weather patterns and events, have mostly shown negative implication to human health, psychology, and survival. Respiratory failure, myocardial infarction, strokes, and heat stroke can be caused by heat waves. For example, an estimated 40,000–50,000 people with lung disorders and cardiovascular diseases died in Western Europe in August 2003 (Robine et al. 2008; Fouillet et al. 2006).

Disasters could instantly and often in a prolonged way devastate social functions including health system. Death is the most serious damage that disasters can cause to humans. Health issues of humans owing to disasters are not limited to only mortality and injuries. Fractures, lacerations, blunt trauma, crush, projectile, burn injuries, as well as drowning could occur during or following natural disasters. To enrich well-being, health, and survival, human societies seek to obtain security, services, wealth growth, "skill sharing," and sound social relationship (McMichael and Lindgren 2011). If injuries result in long-term disabilities, quality of life of humans deteriorates severely.

If the disaster generates substantial displacement of population and/or environmental changes, risk of post-disaster health issues, especially outbreak of infectious diseases, must be considered, as in the case of the 2004 Andaman Nicobar earthquake and tsunami in Indonesia as well as the 2011 Great East Japan Earthquake. Natural disasters such as floods, tropical cyclones, earthquakes, tsunamis, and tornadoes could lead to incidence of infectious diseases including diarrheal diseases, acute respiratory infections, influenza, malaria, leptospirosis, measles, dengue fever, viral hepatitis, typhoid fever, meningitis, tetanus, and cutaneous mucormycosis (Kouadio et al. 2012). Furthermore, the environmental changes may result in increasing breeding sites of disease-carrying vectors like mosquitoes; higher exposure of human being to rodents and/or pathogens; overcrowded living conditions (in shelters or refugee settings); inadequate air, water, and sanitation conditions; compromised nutrition, personal hygiene, and personal protection; low levels of immunity or vaccine coverage; reduction in health education; less access to healthcare facilities; and less resources for case management of diseases at healthcare settings.

Disaster-related impacts on human health, especially mortality, diseases with long-term disabilities, and serious spread of infectious diseases, can threaten sustainable development of any community or nation. Both slow onset and fast onset disasters affect human health adversely. Slow onset disasters include drought, saline intrusion, sea-level rise, and so on. Water scarcity, water pollution, etc., also cause serious health damage. Collapses of healthcare settings and discontinued healthcare services might be inevitable after total destruction of facilities, blackouts, water outage, and fuel shortage. The three affected prefectures by the Great East Japan Earthquake saw approximately 80 % of hospitals and one third of medical/dental clinics having suffered damage (Saito and Kunimitsu 2011). Furthermore, interruption of disease surveillance, health programs (e.g., vaccine and health education), and treatment including medication could occur.

Although some experts caution that risk of post-disaster outbreak of infectious diseases may have been overemphasized (Howard et al. 1996), it is important to pay attention to potential health risks. This is because significant and rapid spread of infectious diseases among a family or a community could force each affected individual to be unable to earn a living or even provide support for those who require care, such as small children and older persons, leading to serious economic damage and heavy social burden. On the national level, significant loss of workforce means loss of productivity, which threatens sustainable development. Indeed, human health is thus interconnected with sustainable development.

This chapter also acknowledges that disasters have both direct and indirect health effects. While direct health losses have been dealt in this chapter, indirect health losses such as stress, psychosocial disasters, etc., are also important area of study but are not exclusively discussed in this chapter.

## 7.2 Revisiting Sustainable Development

Much of the discussion of sustainable development has focused on social, economic, and environmental aspects, which are considered as three main pillars of sustainable development. However, less attention has been paid to human health as a necessary component of sustainable development. This section will delve into some of the historical milestones, which have helped to deepen the idea of sustainable development.

The concept of sustainable development was initially propagated with the view that the planet earth's resources are not infinite. The idea received prominence when Club of Rome commissioned a study to a group of researchers based in the Massachusetts Institute of Technology (MIT). Researchers at the MIT developed a global computer model "World3." This model simulated resulting outcomes of the interaction of earth and human systems. The study outcome led to the publication of a milestone book in the year 1972 with the title, "*The Limits to Growth*." Although this book received some criticism from the 1970s onwards, it remains one of the most pioneer works laying the foundation stone toward understanding sustainability. Over 40 years ago, the book interestingly used five variables, *viz.*, global population, industrialization, pollution, food production, and resource depletion, which are very much relevant in the present times as well.

Both the growing population and the concentration of the majority of the population in urban areas are an issue of great concern for scholars in public health. Population densities in certain pockets and global movements of people have increased manifold and might carry potential risk of turning easily preventable diseases into a major epidemic. Industrialization and pollution are often closely correlated and are creating new forms of health risks. New Delhi, India, is a city with one of the highest ratios of the road fatalities as well as respiratory diseases. Smog, which is a toxic air pollution, has been continuously horrifying residents of major cities in the People's Republic of China. Food production has seen enormous growth in the past decades; however, unruly usage of fertilizers and chemicals have degraded huge chunk of productive land globally. New techniques including genetically modified food could offer promises to produce enough food for the growing population but have already been subject of enormous concern. Resource depletion is another serious concern alarming time and again through ideas such as the peak oil.

1972 was a remarkable year because 113 countries gathered under the umbrella of the United Nations to convene the 1972 Stockholm Conference, "United Nations Conference on the Human Environment." The outcome of this meeting, popularly known as "Stockholm Declaration," was comprised of 26 principles, which were directly related to environment and development. This was one of the first global conferences having heightened awareness about development while equally paying attention to global environmental concerns. Nongovernmental organizations also joined this conference as active partners in the conference diplomacy although they had no roles to play in decision making (Kanninen 2013).

The International Union for the Conservation of Nature (IUCN) with its presence in 160 countries and broad base of membership is the world's oldest and largest global environment network. In 1980, the IUCN announced "World Conservation Strategy," which helped further strengthening the idea of sustainable development.

The Stockholm Conference and the IUCN's strategy paved the way for much larger global intervention. In this backdrop, the World Commission of Environment and Development, also known as the Brundtland Commission, was formed in 1983. The Brundtland Report (WCED 1987), also known as "Our Common Future," was one of the milestone outcome having highlighted clearly the use of integrated approaches and collaborative global efforts by all nations to achieve sustainable development. Sustainability was formally placed on political agenda of the world through this report (Dovers 2004). This report not only helped to define deeper meaning of the concept of sustainable development but also clearly illustrated that economic growth must not happen at the cost of serious negative impacts on the environment. The report was a defining work influencing global policies in many ways; it also highlighted that poverty reduction and gender equality including equity is crucial to ensure conservation of the environment.

In the year 1992, the UN Conference of Environment and Development (UNCED) was held in Rio de Janeiro, popularly known as the Rio Earth Summit. The Rio Summit was one of the largest meetings of that time on the issues concerning sustainable development. One hundred seventy-six countries attended this summit of which two-thirds were heads of the state and government. In addition, 1100 NGOs joined the meeting, bringing 30,000 people to participate. The Rio Summit's outcome includes the adoption of the Agenda 21, which is a nonbinding and voluntarily implemented action plan prioritizing actions toward sustainability. These actions were identified at global, regional, national, and local levels. This was a major shift to engage cities and towns including their local governments to implement actions of the Agenda 21. Conventions on climate, biodiversity, and forests were also born out of the Rio Summit. The Agenda 21 additionally mentioned about promoting health, control of pollution, managing radioactive waste, etc.

While human health or public health were not given clear positions nor prioritized mentioning in these international events discussed above, it can be deduced that improving the environmental conditions or reducing environmental degradation was intended to ensure good living conditions and decent quality of life of future generations to come. Because of the less explicit recognition as a part of these discourses, human health might not have greatly influenced policies toward public health at national and local levels in some parts of the world.

## 7.3 Human Health in the Twenty-First Century

The first Human Development Report (HDR) published by the United Nations Development Program (UNDP) in 1990 used empirical data for the first time to help understand and measure development. Mahbub ul Haq (1934–1998), the founder of the HDR, mentioned that the "objective of development is to create an enabling environment for people to enjoy long, healthy and creative lives" (HDR 2014). The report informs of greater access to knowledge, better nutrition and health services, more secure livelihoods, security against crime and physical violence, satisfying leisure hours, political and cultural freedoms, and sense of participation in community activities. Clearly, human well-being is at the core of HDR and human health is intrinsic part of it.

The Millennium Summit of the United Nations held in the year 2000 was instrumental in establishing eight international development goals of which three goals are directly related to human health. All 189 United Nations member states of the UN at that time shared the common commitment to achieve the Millennium Development Goals (MDGs) by 2015. The entire world's leading development institutions also agreed to contribute to meeting these goals. The MDGs included several important targets such as: to halve the number of undernourished people; to achieve universal primary education; to promote gender equality and empower women; to reduce child mortality; to improve maternal health; to combat HIV/AIDS, malaria, and other diseases; to ensure environmental sustainability; and to develop a global partnership for development.

> During the past two decades, the likelihood of a child dying before the age of five has been nearly cut in half, which means about 17,000 children have been saved every day. The maternal mortality ratio dropped by 45 per cent. Antiretroviral therapy for HIV-infected people has saved 6.6 million lives. An estimated 3.3 million deaths from malaria were averted due to a major expansion of simple preventions, such as bed nets, and treatments. Efforts to fight tuberculosis have saved an estimated 22 million lives. (UN 2014)

While the global community is gearing toward the preparation of new global development goals also known as the Post 2015 development process, to be followed after 2015, it is clear that there is a long way to go toward ensuring health for all. It is expected that the new set of goals will be much more ambitious with targets and indicators attached to them with active monitoring and evaluation mechanism in order to meet the goals set during the time frame.

## 7.4 Sustainable Development and Human Health: Inter-linkages and Gaps

"Consumption, economic growth and environmental degradation impact sustainable development in complex and often apparently contradictory ways" (Atkinson et al. 2007). There are numerous factors undermining human health, including environmental degradation, prevalence of lifestyle diseases and infectious diseases, and ineffective (health) policies. In the rest of this section, we will focus on post-disaster health issues in relation to climate change, disaster risk, environmental degradation, societal issues, and weak policies.

### *7.4.1 Climate Change*

The recent effects of climate change, such as extreme weather patterns and events, have demonstrated adverse implication for public health, such as impacts of heat waves on elderly people resulting in deaths and hospitalization due to myocardial infarction, stroke, respiratory disorders, and other diseases. In tandem with various environmental changes, it affects food yields and nutritional quality, water system, and disease-carrying vector populations, such as mosquitoes. There are two prominent approaches to deal with climate change including adaptation and mitigation. There is certainly a need for much dense and wider research to explicitly understand what consequences will be there on human health with adaptation and mitigation.

The warmer temperature with other variables like precipitation is thought to affect spatial and temporal transmission of mosquito-borne infectious diseases like dengue infection. Pei-Chih Wu and colleagues show that cases of dengue fever at “township level” were affected by urbanization and the monthly number of days with average temperature exceeding 18 °C (Wu et al. 2009). Having conducted a statistical analysis on the weekly reported cases of dengue infection and climate variables in Singapore from 2004 to 2007, Y. L. Hii et al. reveal a linear increase of the disease incidence in the 5–16 weeks following higher weekly mean temperature and 5–20 weeks after high weekly cumulative precipitation (>75 mm). Since these two climate parameters recorded higher levels for the investigated period when compared to previous years, this study postulates that the surged reported dengue cases were induced by increased temperature and rainfall in the country (Hii et al. 2009).

Climate change is also expected to increase food insecurity. Relationship between malnutrition and diseases is well established and both can be cause and result of each other. Threat to food security will have severe effect on poor and vulnerable people, especially children. Sustainable development will be compromised if malnutrition-related health impacts are not addressed.

### 7.4.2 *Disaster Risk*

Post-disaster disrupted living conditions give rise to disaster-related epidemics, mostly due to exposure to unhygienic environment and malnutrition resulting from power failure, insufficient food provision, and a lack of running water. The Tohoku Disaster alone saw spread of respiratory tract infection such as tsunami-related pneumonia (Daito et al. 2013), legionellosis (Takahashi et al. 2012), and influenza (Tohma et al. 2012). Additionally, outbreak of wound infection, tetanus, was reported in both Miyagi and Iwate prefectures (Takahashi et al. 2012) and in Ache, Indonesia (Ache Epidemiology Group 2006). An outbreak of norovirus gastroenteritis was reported among the elderly evacuees after the 2007 Noto Peninsula Earthquake (Okumura et al. 2008). Post-cyclone Nargis, diarrheal diseases, dysentery, and acute respiratory infections were reported (Myint et al. 2011). Other infectious diseases that follow natural disasters include malaria, leptospirosis, measles, dengue fever, viral hepatitis, typhoid fever, meningitis, and cutaneous muscormycosis (Kouadio et al. 2012).

Furthermore, disaster could directly contribute to degradation of surrounding environments of humans. For example, if sewage treatment facilities are damaged by disaster, water-borne infectious diseases such as cholera could occur as a result of unavailability of safe drinking water (Bhunia and Ghosh 2011). Cholera outbreak after the 2010 earthquake in Haiti was reported (Hendriksen et al. 2011) in relation to post-disaster human movements. In Thailand, norovirus, rotavirus, and enteric hepatitis A virus were confirmed in flood water following the 2011 Thai Flood (Ngaosuwankul et al. 2013).

Both water and vector-borne diseases play important roles following flood, waterlogging, and such incidences. Poor municipal services often pave the way for proliferation of water- and vector-borne diseases. In developing countries, local governments often struggle to serve basic services to population living in its jurisdiction. Availability of potable water, decent sanitation, and health facilities are considered as mere essentials for human survival. Incidents of water- and vector-borne diseases due to climate change are going to pose serious health risks to human dwindling sustainable development.

### 7.4.3 Environmental Degradation

The environmental degradation here refers to the conditions threatening air quality, water quality, food safety, healthy housing, waste/sanitation management, and vector control (Ratnapradipa et al. 2012). Environmental changes, such as air pollution, are known to cause respiratory diseases, and extreme cases may result in lung cancer (Loomis et al. 2014). An example of serious disease caused by water pollution in Japan is Minamata disease suffered by local residents who consumed fish and shellfish contaminated with methylmercury from industry waste (Tamashiro et al. 1986).

Environmental degradation, especially changing ecosystem and microbial adaptation due in part to urbanization, was seen in a painful example of *Nipah Virus* infection in Malaysia. Development activities of humans such as deforestation created new opportunities for bats, the natural reservoir hosts of the virus, to come in contact with pig farms, which caused infections to both animals and humans in 1998–1999 (Muniandy and Aziz 2004).

### 7.4.4 Societal Issues

Societal issues including changing human demographics and behavior, poverty and social inequality, and international travel may be blamed for India's excruciating economic damage after the episode of plague in the province of Gujarat, India, in September 1994 that caused panic and global overreactions such as banning flights and ships to and from India (Cash and Narasimhan 2000). On the other hand, India's Pulse Polio Campaign is an encouraging story of how polio vaccination is reaching out to the entire population (Rajput and Sharma 2010). The campaign involved organizing massive awareness generation programs from national, state, and district to individual level. Mass media engaged popular people such as cinema artists, famous sportspersons, and well-known personalities from various fields to spread the message about giving polio vaccine (oral drops) to all children. The campaign was organized periodically. A huge number of volunteers were trained to visit door to door in rich and poor neighborhoods alike and offer polio vaccination to children at the doorstep.

Aging has been recognized as an emerging risk factor of diseases, such as cancer, accompanied by the increasing demand of healthcare services and resources (Institute of Medicine 2013), often contributing to higher medical cost to individuals and/or governments. Lifestyle diseases like diabetes can be problems in developed and developing countries alike, and disruption of treatment after large-scale disasters has drawn attention as an issue to be overcome.

### 7.4.5 *Weak Policies*

Although policies may be in place, limitations of the existing law was revealed as the challenge in providing uninterrupted supply of pharmaceutical items for the elderly survivors of the 2007 Noto Earthquake (Okumura et al. 2008). In many places of the world, there seem to be serious gaps between sustainable development and health policies especially due to lack of political will. This weakness may be indeed blamed for underdeveloped healthcare system on the national levels that threaten human health of the population.

Is development a prerequisite of a good health system? It is popularly believed that Japan's economic progress contributed to the successful control of conventional infectious diseases, but her GNP actually rose sharply only after roundworm prevalence was controlled from 60 % to 5 % and schistosomiasis from nearly 15 % to 1–2 % in the early 1960s (Takeuchi 2004). There is yet another country, the Republic of Singapore, where we saw its political will at a nascent of developing nation in improving population health possibly contributed to development, which has been sustained over almost five decades until now. It is therefore possible for developing economies to build national health system.

To explain interconnectedness between human health and development, we will review the industrialization strategy of Singapore prior to the 1980s. Components including legal system, tax incentives, and wage management are studied well, but much less attention has been paid to the role of health policy along housing and education policies that were promoted during the process of industrialization of the country. Above all, the mechanisms of vigilant control of infectious diseases that resulted in the improvement of population health will be discussed in the next section.

## 7.5 A Case of Good Practice on Health Issue Linked to Development[1]

"Risk decisions do not occur in isolation, but in the context of a society's shared and conflicting norms and practices" (Fischhoff and Kadvany 2011). Whether or not healthy labor force is locally available is an important factor for multi national

[1] This section is heavily drawn from the first author's doctoral dissertation (Yoshikawa 2012b).

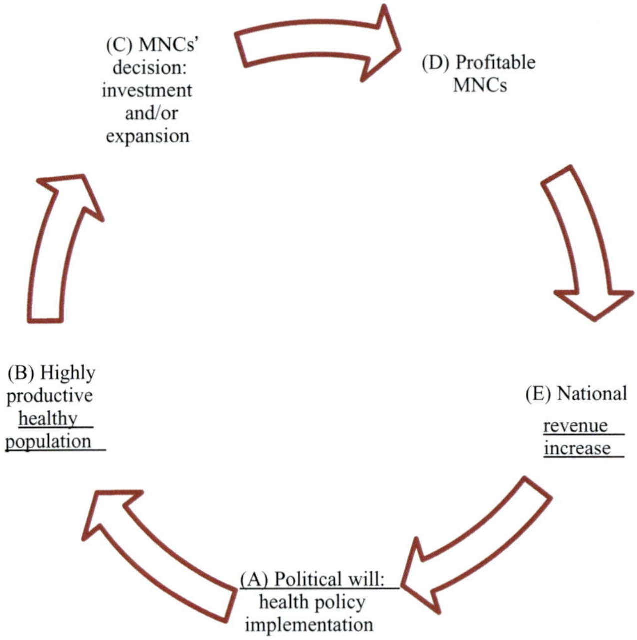


**Fig. 7.1** A circular model to explain the role of health improvement in attracting foreign direct investments (Source: Yoshikawa 2012b)

corporations (MNCs) when evaluating country risks for investment. While an acute "health shock" can impact short-term capital markets as seriously as a coup d'état or an assassination of an important politician, long-term foreign direct investments (FDIs) react more slowly. FDIs are likely to link unhealthy citizens to system failure or the incapability of policy makers in the country (Tandon 2005). In other words, a government wishing to industrialize itself must first address problems of public health nature.

Although investors evaluate a variety of conditions, for simplification, health condition of labor force as a factor is shown by a circular model (Fig. 7.1) to explain the role of health improvement in attracting FDIs. Implementing a health policy by a government (as shown in the point A) improves health of citizens at point B, and more healthy and productive labor force can then attract MNCs' investments (or expansions of existing investments) at point C. If the businesses succeed, the MNCs can generate profits at point D. The higher output can be reflected in a rise of GDP at point E. Increases in the income of businesses and/or individuals (e.g., employees) contribute to host country's national revenue through tax and other payments.

The final arrow from the point E to point A is the key. The additional national revenue (E) may not achieve the goal of healthy populations (B) unless governments form and implement effective health policies (A). A connection therefore can be established if increased revenue is allocated to the national health policy directly or indirectly. Apparently, any private enterprise can contribute to the improvement of population health by building healthcare institutions, but the circular model here deals with an explanation of a national FDI strategy of a government in achieving industrialization.

In view of scarce land, shortages in manpower could have been fatally disadvantageous for Singapore's budding development following the independence in 1965. Securing the healthy labor force became an utmost important prerequisite for the government, which needed to depend on FDIs to create jobs quickly and build up profitable industries to promote economic growth. Furnishing citizens with a solid structure in which they could "learn, work hard, be productive and be rewarded accordingly" (Lee 1998) became the responsibility of political leaders.

However, outbreaks of cholera and typhoid occurred almost annually in the early years. In the year of independence, less than 80 % of about 50,000 street hawkers involved in food services were licensed and trained to become knowledgeable of public hygiene. As a result, these hawkers attracted flies and rats (National Archives of Singapore 2008). Intensive investigations of food-borne infectious diseases often traced the route of infection to local hawker centers and schools. In response, the Ministry of Health occasionally issued warnings and introduced regulations to food handlers. Table 7.1 summarizes public health-related laws. For example, the Environment Public Health Act in 1969 made it illegal for premise occupiers failing to "dispose of refuse or filth within 48 hours" and for anyone to litter in public places.

In September 1972, the Ministry of the Environment was newly inaugurated, and the ministry assumed responsibilities of environment public health from the Ministry of Health including vector control, food hygiene, and sanitation. The establishment of the ministry occurred after Prime Minister Lee Kuan Yew set "an official national goal" to clean Singapore's waterways, especially the Singapore River. In October 1973, both ministries formed the Joint Coordinating Committee on Epidemic

**Table 7.1** Laws to control infectious diseases introduced/amended by the Singapore government in the first 11 years of independence

| Year | Description of laws |
|---|---|
| 1968 | Destruction of Disease-Bearing Insects Act |
| 1969 | Environmental Public Health Act |
| | Environmental Public Health (Markets) Regulations |
| | Environmental Public Health (Hawkers) Regulations |
| 1970 | Environmental Public Health (Public Cleansing) Regulations |
| 1973 | Environmental Public Health (Food Establishment) Regulations |
| | Environmental Public Health (Food Handlers) Regulations |
| 1974 | Environment Public Health (Manufacture and Sale of Ice Cream) Regulations |
| 1976 | Infectious Disease Act |

Source: Constructed from National Archives of Singapore (2008) and Goh (1983)

Diseases with the National University of Singapore, the defense ministry, and the Ministry of National Development. A series of public persuasion campaigns was used to change the collective and individual habits like littering, spitting, and smoking to improve the public environment. The anti-spitting campaigns in the 1960s cautioned the act would spread diseases like tuberculosis (Lee 2000).

Recognizing the interlinkage between loss of manpower and disease prevalence, painstaking efforts began to improve population health across the nation. The 5-year plan announced by the Ministry of Health in 1970 revealed "[g]overnment's effort of providing a better and healthier environment for the people to live in" (NUS, June 1957–1981). The country had neither sufficient healthcare providers nor facilities then. The government therefore encouraged the local medical education system to produce more Singaporean doctors while at the same time ensuring sufficient doctor numbers by recruiting foreign doctors. Five hospitals were built between 1960 and 1974; 13 outpatient dispensaries in 1959 was doubled to 26 in 1974 (Barr 2005). In 1976, the Ministry of Education supported the health ministry in promoting the National Health Campaign to fight against infectious diseases. In addition, a series of housing policies were implemented, which contributed to the improvement in living conditions of those who had previously lived in urban slum conditions. Infant mortality rate of 26.3 per 1000 live births and life expectancy at birth of 64.5 years in 1965 improved to 7.6 and 73.9, respectively, in 1985. The coordination across ministries and sectors in addition to campaign-based health education efforts characterizes Singapore's public health management even today (Yoshikawa 2012a).

It is well known that the city-state has attracted a number of MNCs in various promising industries such as oil refining, petrochemical, banking, tourism, semiconductor, telecommunication, electronics, and biomedicine. It can be said that the strong political leadership and governmental efforts at the nascence of the state established the good health system, which in part created more favorable condition for investments in the country with limited natural resources. Due to lack of data, it is difficult to prove that the health of Singaporean citizens, i.e., supply of healthy labor, has indeed attracted the FDIs. However, it is apparent that the nation today commands economic power that is disproportionate to the size of its territory or the total population of about just five million (Yoshikawa 2012a).

## 7.6 Conclusions and the Way Forward

Examining public health in relation to disaster and development, the authors hitherto showed that promotion of human health and nutrition is essential to sustainable development. This chapter also reaffirms that human health is not a standalone issue, but rather it is deeply interconnected with social-economical-environmental aspects, which form vital pillars of sustainable development. Policy makers especially in developing economies are strongly encouraged to consider human health in the planning of national development strategies.

## References

Aceh Epidemiology Group (2006) Outbreak of tetanus cases following the tsunami in Aceh Province, Indonesia. Glob Public Health 1(2):173–177

Atkinson G, Dietz S, Neumayer E (eds) (2007) Handbook of sustainable development. Edward Elgar Publishing Limited, London. http://www.e-elgar.com/shop/handbook-of-sustainable-development

Barr M (2005) Singapore. In: Gauld R (ed) Comparative health policy in the Asia-Pacific. Open University Press, Berkshire, pp 146–173

Bhunia R, Ghosh S (2011) Waterborne cholera outbreak following Cyclone Aila in Sundarban area of West Bengal, India, 2009. Trans R Soc Trop Med Hyg 105:214–219

Cash R, Narasimhan V (2000) Impediments to global surveillance of infectious diseases: consequences of open reporting in a global economy. Bull World Health Organ 78:1358–1367

Daito H, Suzuki M, Shinhara J et al (2013) Impact of the Tohoku earthquake and tsunami on pneumonia hospitalizations and mortality among adults in northern Miyagi, Japan: a multicenter observational study. Thorax 68(6):544–550. doi:10.1136/thoraxjnl-2012-202658

Dovers S (2004) Sustainability and disaster management. Aust J Emerg Manag 19(1):21–25

Fischhoff B, Kadvany J (2011) Risk: a very short introduction. Oxford University Press, Oxford. http://ukcatalogue.oup.com/product/9780199576203.do

Fouillet A, Rey G, Laurent F et al (2006) Excess mortality related to the August 2003 heat wave in France. Int Arch Occup Environ Health 8:16–24

Goh KT (1983) Epidemiological surveillance of communicable diseases in Singapore. Southeast Asian Medical Information Center, Tokyo

HDR. About human development, webpage accessed at http://hdr.undp.org/en/humandev. Accessed on 30 July 2014

Hendriksen RS, Price LB, Schupp JM, Gillece JD, Kaas RS, Engelthaler DM, Bortolaia V, Pearson T, Waters AE, Upadhyay BP, Shrestha SD, Adhikari S, Shakya G, Keim PS, Aarestrup MF (2011) Population genetics of Vibrio cholerae from Nepal in 2010: evidence on the origin of the Haitian outbreak. mBio 2(4):e00157–e001511. doi:10.1128/mBio.00157-11

Hii YL, Rocklőv J, Ng N, Tang CS, Pang FY, Sauerborn R (2009) Climate variability and increase in intensity and magnitude of dengue incidence in Singapore. Glob Health Action 11(2). doi:10.3402/gha.v2i0.2036

Howard MJ, Brillman JC, Burkle FM (1996) Infectious disease emergencies in disasters. Emerg Med Clin North Am 14:413–442

Institute of Medicine (2013) Delivering high-quality cancer care: charting a new course for a system in crisis. The National Academies Press, Washington, DC

Kanninen T (2013) Crisis of global sustainability. Routledge, London. http://www.amazon.com/Crisis-Global-Sustainability-Institutions/dp/0415694175

Kouadio I, Aljunid S, Kamigaki T, Hammad K, Oshitani H (2012) Infectious diseases following natural disasters: prevention and control measures. Expert Rev Anti Infect Ther 10(1):95–104

Lee KY (1998) The Singapore story. Times Editions, Singapore

Lee KY (2000) From third world to first: the Singapore story: 1965–2000. Time Media, Singapore

Loomis D, Huang W, Chen G (2014) The International Agency for Research on Cancer (IARC) evaluation of the carcinogenicity of outdoor air pollution: focus on China. Chin J Cancer. doi:10.5732/cjc.014.10028. Available at http://www.cjcsysu.com/ENpdf/2014/4/189.pdf. Accessed June 8 2014

McMichael AJ, Lindgren E (2011) Climate change: present and future risks to health, and necessary responses. J Intern Med 270:401–413

Muniandy N, Abdul Aziz J (2004) Effects of intensification of the traditional farming system on the environment and bio-safety of the human population: Nipah Virus outbreak in Malaysia. In:

Furukawa H, Nishibuchi M, Kono Y, Kaida Y (eds) Ecological destruction, health, and development: advancing Asian paradigms. Kyoto University Press, Kyoto, pp 303–317

Myint NW, Kaewkungwal J, Singhasivanon P, Chaisiri K, Panjapiyakul P, Siriwan P, Mallik AK, Nyein SL, Mu TT (2011) Are there any changes in burden and management of communicable diseases in areas affected by Cyclone Nargis? Confl Health 5:9. doi:10.1186/1752-1505-5-9

National Archives of Singapore (2008) 10 years that shaped a nation. National Archives of Singapore, Singapore

National University of Singapore (NUS). Medical and health services in Singapore *(Straits Times Press Cuttings)* (1957–1981). Central Library, National University of Singapore, Singapore, multiple years

Ngaosuwankul N, Thippornchai N, Yamashita A, Vargas REM, Tunyong W, Mahakunkijchareon Y, Ikuta K, Singhasivanon P, Okabayashi T, Leaungwutiwong P (2013) Detection and characterization of enteric viruses in flood water from the 2011 Thai flood. Jpn J Infect Dis 6:398–403

Okumura J, Nishita Y, Kimura K (2008) Pharmaceutical supply for disaster victims who need chronic disease management in aging region based on lessons learned from the Noto Peninsula Earthquake in 2007. Yakugaku Zasshi 128(9):1275–1283 (in Japanese)

Peralta GL, Hunt JM (2003) A primer of health impacts of development programs. Asian Development Bank, Manila

Rajput M, Sharma L (2010) The threat of vaccine associated poliomyelitis in India: medicolegal issues involved. Hum Vaccin 6:1071–1075, PMID: 21157179; http://dx.doi.org/10.4161/hv.6.12.13354. Accessed 8 June 2014

Ratnapradipa D, Conder J, Ruffing A, White V (2012) The 2011 Japanese earthquake: an overview of environmental health impacts. J Environ Health 74(6):42–50

Robine JM, Cheung SL, LeRoy S et al (2008) Death toll exceeded 70000 in Europe during the summer of 2003. C R Biol 331:171–178

Saito T, Kunimitsu A (2011) Public health response to the combined Great East Japan Earthquake, tsunami and nuclear power plant accident: perspective from the Ministry of Health, Labor and Welfare of Japan. Western Pac Surveill Response J 2(4):7–9

Takahashi T, Goto M, Yoshida H, Sumino H, Matsui H (2012) Infectious disease after the 2011 Great East Japan Earthquake. J Exp Clin Med 4(1):20–23

Takeuchi T (2004) Reporting Japan's experiences of parasite infection control to the world. In: Takeuchi T, Nakatani H (eds) Global jidai no kansensho [Infectious diseases in the era of globalization]. Keio University Press, Tokyo, pp 177–179

Tamashiro H, Arakaki M, Futatsuka M, Lee ES (1986) Methylmercury explore and mortality in southern Japan: a close look at causes of death. J Epidemiol Community Health 40:181–185

Tandon A (2005) Population health and foreign direct investment: does poor health signal poor government effectiveness? ERD Policy Brief 33:1–3

Tohma K, Suzuki A, Otani K, Okamoto M, Nukiwa N, Kamigaki T, Kawamura K, Nakagawa H, Oshitani H (2012) Monitoring of influenza viruses in the aftermath of the Great East Japan Earthquake. Jpn J Infect Dis 65:542–544

UN (2014) Press release at the launching of new report, MDG advocates outline challenges and opportunities for achieving the Millennium Development Goals by end of 2015. Released on 25 Sept 2014. http://www.un.org/millenniumgoals/advocates/pdf/MDG%20Leader%20Press%20Release%20-%20final.pdf. Accessed on 30 Sept 2014

WCED 1987: World Commission on Environment and Development (1987) Our common future. Oxford University Press, Oxford. http://ukcatalogue.oup.com/product/9780192820808.do

Wu PC, Lay JG, Guo HR, Lin CY, Lung SC, Su HJ (2009) Higher temperature and urbanization affect the spatial patters of dengue fever transmission in subtropical Taiwan. Sci Total Environ 407:2224–2233

Yoshikawa MJ (2012a) Singapore's prescription for successful control of transnational emerging infectious diseases. Southeast Asian Stud 1(2):301–331

Yoshikawa M (2012b) Research on emerging and re-emerging mosquito-borne and related infectious diseases in southeast Asia: prescriptions from the City-State of Singapore and a tourist destination of Bali, Indonesia. PhD dissertation, Kyoto University, Kyoto

# Chapter 8
# Ecosystem-Based Disaster Risk Reduction: Experiences, Challenges, and Opportunities in the Post-2015 Development Agenda

**Noralene Uy, Rafaela Jane P. Delfino, and Rajib Shaw**

**Abstract** Ecosystems, climate change, and disaster risk reduction are among the cross-cutting issues highlighted in the Rio+20 Conference. In view of the post-2015 development agenda, the chapter discusses the important role of ecosystem-based disaster risk reduction in sustaining ecosystems and building disaster-resilient communities. It describes ecosystem management strategies that link ecosystem protection and disaster risk reduction, elucidates the challenges in advancing the use of ecosystem-based disaster risk reduction and linking it to policy, and identifies opportunities for scaling up.

**Keywords** Ecosystem-based disaster risk reduction • Ecosystem management • Post-2015 framework for disaster risk reduction • Post-2015 development agenda

## 8.1 Introduction

For the past few decades, the linkage between poverty, ecosystem degradation, and disaster risk has already been widely discussed in the science and policy arena. However, it was only during the UN Conference on Sustainable Development 2012 (Rio+20) that this relationship was given greater political attention (Beck et al. 2012). Additionally, scientists have only recently begun to systematically establish the influence of ecosystems on disaster risk. According to UNDP (2007), the

N. Uy (✉)
Independent Researcher, San Mateo, Rizal, Philippines
e-mail: noralene@gmail.com

R.J.P. Delfino
The Oscar M. Lopez Center for Climate Change Adaptation and Disaster Risk Management, Foundation, Inc., Pasig City, Philippines

R. Shaw
Graduate School of Global Environmental Studies, Kyoto University, Sakyo-ku, Kyoto, Japan
e-mail: shaw.rajib.5u@kyoto-u.ac.jp



J.I. Uitto, R. Shaw (eds.), *Sustainable Development and Disaster Risk Reduction*, Disaster Risk Reduction, DOI 10.1007/978-4-431-55078-5_8

attention given to these issues at this time may be attributed to major disasters, an active civil society to promote democratic political change, the engagement of particularly dynamic individuals, and a well-educated and participative population.

In the outcome document adopted at Rio+20, member states acknowledge planet Earth and its ecosystems as mankind's home and the rights of nature in the context of sustainable development (UN 2012). They recognize that it is necessary to promote harmony with nature to achieve a just balance among the economic, social, and environmental needs of present and future generations. There is, thus, a need to seize and create opportunities to achieve sustainable development through economic growth and diversification, social development, and environmental protection. With this in mind, ecosystems, climate change, and disaster risk reduction are among the thematic areas highlighted at Rio+20 and in the discussions on the post-2015 agenda on sustainable development, disaster risk reduction (DRR), and climate change.

### 8.1.1 Linking Ecosystem Management, Disaster Risk Reduction, and Sustainable Development

Ecosystem management has grown in theory and application from its ecological beginnings. This evolution is largely due to the changing views and processes within various scientific and social disciplines, technology, decision making, and policy. As shown in Table 8.1, many events in the environment, DRR, and climate change arena contributed to the growing link between these issues.

Environment, one of the pillars of sustainable development, figures highly in all discussions on sustainable development. As mentioned in the Millennium Assessment Report, attaining environmental sustainability requires an end to the current unsustainable uses of ecosystem services (e.g., fisheries and freshwater) as well as an end to the degradation of other services (e.g., water purification, natural hazard regulation, disease regulation, climate regulation, and cultural amenities) (MA 2005). An examination of selected key documents that advance sustainable development shows that environment has always been a main concern (Table 8.2).

The Hyogo Framework for Action (HFA), the first international plan to substantially reduce disaster losses by 2015, outlines five priorities for action. Sustainable ecosystem and environmental management are listed under Priority 4, to "reduce the underlying risk factors." Among the activities identified to achieve this priority are (i) encourage the sustainable use and management of ecosystems, including through better land use planning and development activities to reduce risk and vulnerabilities, and (ii) implement integrated environmental and natural resource management approaches that incorporate disaster risk reduction, including structural and non structural measures, such as integrated flood management and appropriate management of fragile ecosystems (UNISDR 2005).

In the following, the chapter discusses the important role of ecosystem-based disaster risk reduction (Eco-DRR) in sustaining ecosystems and building

**Table 8.1** Timeline of selected events in environment, disaster risk reduction, and climate change

| Period | Environment | Disaster risk reduction | Climate change |
|---|---|---|---|
| 1960s–1970s | 1962: Rachel Carson's *Silent Spring* (published by Houghton Mifflin)<br>1970: 1st Earth Day<br>1971: Greenpeace<br>1972: Stockholm Conference; UNEP; Meadows et al.'s *Limits to Growth* (published by Universe Books)<br>1974: *Chlorofluoromethanes Destroy the Ozone Layer* (Rowland and Molina, Nature 249, 810–812, 28 June 1974)<br>1975: Convention on International Trade in Endangered Species (CITES)<br>1976: UN Conference on Human Settlements<br>1977: UN Conference on Desertification<br>1979: Three Mile Island Nuclear Accident; Convention on Long-Range Transboundary Air Pollution | 1971: UN Disaster Relief Office<br>1971–1985: Drought in Afghanistan, Ethiopia, and Other Countries in Africa | 1974: *Chlorofluoromethanes Destroy the Ozone Layer* (Rowland and Molina, Nature 249, 810–812, 28 June 1974) |
| 1980s–1990s | 1980: IUCN's World Conservation Strategy<br>1982: UN Convention on the Law of the Sea; UN World Charter for Nature<br>1984: Bhopal Toxic Leak<br>1985: WMO Climate Change Meeting<br>1986: Chernobyl Nuclear Accident<br>1987: Montreal Protocol on Substances that Deplete the Ozone Layer<br>1987: Brundtland Report<br>1988: Intergovernmental Panel on Climate Change (IPCC)<br>1991: Global Environment Facility (GEF)<br>1992: Rio Summit<br>1995: World Trade Organization<br>1995: World Summit for Social Development<br>1996: ISO 14001 | 1990: International Decade for Natural Disaster Reduction<br>1991: Bangladesh Cyclone<br>1994: World Conference on Disaster Reduction; Yokohama Strategy<br>1995: Great Hanshin Earthquake | 1985: Antarctic Ozone Hole Discovered; WMO Climate Change Meeting<br>1987: Montreal Protocol on Substances that Deplete the Ozone layer<br>1988: IPCC<br>1990: IPCC FAR<br>1992: UNFCCC<br>1995: IPCC SAR; COP1 |

(continued)

**Table 8.1** (continued)

| Period | Environment | Disaster risk reduction | Climate change |
|---|---|---|---|
| 2000s – | 2000: Millennium Development Goals (MDGs)<br>2002: Johannesburg Plan of Implementation<br>2005: Kyoto Protocol; Millennium Ecosystem Assessment<br>2006: Stern Review on the Economics of Climate Change<br>2007: Al Gore's "An Inconvenient Truth"<br>2008: Green Economy in the Mainstream<br>2010: The Economics of Ecosystems and Biodiversity (TEEB)<br>2012: Rio+20 Conference | 2000: UNISDR<br>2004: Indian Ocean Tsunami<br>2005: Kashmir Earthquake<br>2005: HFA<br>2006: Global Platform on Disaster Reduction<br>2007: First Session of the Global Platform on Disaster reduction<br>2008: Sichuan Earthquake<br>2010: Haiti Earthquake<br>2011: Great East Japan Earthquake and Tsunami<br>2013: Typhoon Haiyan | 2001: IPCC TAR<br>2005: Kyoto Protocol<br>2006: Nairobi Work Program<br>2006: Stern Review on the Economics of Climate Change<br>2007: AR4; Bali Action Plan<br>2009: Copenhagen Accord<br>2010: Cancun Agreements<br>2011: Durban Platform for enhanced action 2013/2014: AR5 |

Source: http://www.iisd.org/pdf/2012/sd_timeline_2012.pdf; http://www.unisdr.org/who-we-are/history; http://unfccc.int/essential_background/items/6031.php

**Table 8.2** Issues in focus in selected documents on sustainable development

| Document | Issues in focus | Summary |
|---|---|---|
| Stockholm declaration | Protection and improvement of 'Human Environment' | The declaration emphasizes that protection and improvement of natural and man-made environments is an urgent desire of people and a major duty of all governments and most environmental problems in developing countries were attributable to under development and natural disasters where poverty alleviation became an important step to improve environmental conditions |
| Rio declaration/ Agenda 21 | Social and economic development; *conservation* and *management of resources*; strengthening the role of major groups; means of implementation | Key outcomes: |
| | | The convention on biological diversity |
| | | The framework convention on climate change |
| | | The principles for the sustainable management of forests |
| | | Rio declaration |
| | | Agenda 21 |
| | | Key commitments: |
| | | Integration of environment and development in decision making |
| | | Recognition of common but differentiated responsibilities |
| | | Application of the precautionary approach to decision making |
| | | Provision for polluters to pay for costs of pollution |
| Brundtland Report | *Environmental degradation*; social and economic development | The report sought for solutions to parallel problems of global environmental degradation and global lack of social and economic development by asking for these challenges to be addressed in an integrated way in the interests of present and future generations |
| MDGs | Poverty; education; gender equality; health; *environmental sustainability*; partnership | The MDGs are time-bound goals and targets to improve human well-being |
| Johannesburg Plan of Implementation | Poverty eradication; changing unsustainable patterns of consumption and production; *protecting and managing the natural resource base* of economic and social development | Designed as a framework for action to implement the commitments originally agreed at the Earth Summit |
| | | Set out specific timetable to address some issues |
| | Globalization; health; small island developing states; Africa; regional initiatives; means of implementation | Strengthened the role of the commission on sustainable development in continuing international oversight monitoring progress on sustainability agreements |
| | Institutional framework for sustainable development | |

disaster-resilient communities in view of the post-2015 development agenda. It describes ecosystem management strategies that link ecosystem protection and disaster risk reduction, elucidates the challenges in advancing the use of Eco-DRR and linking it to policy, and identifies opportunities for increased uptake in the post-2015 development agenda.

## 8.2 Disaster Risk Reduction, Sustainable Development, and Ecosystem Management in the Post-2015 Development Agenda

Global consultation processes have taken place to guide and support deliberations on a post-2015 framework for DRR. In parallel, the post-2015 sustainable development agenda and goals have been discussed. The consideration of ecosystem management in these international frameworks is examined in the following.

### *8.2.1 Post-2015 Framework for DRR*

According to the HFA's mid term review, Priority 4 made the least progress so far (UNISDR 2011). The Rio+20 Outcome document has called for the acceleration of the implementation of the HFA and emphasizes the imperative of reducing risk and building disaster resilience for poverty eradication, addressing the impacts of climate change, and achieving sustainable development. Despite its direct impact on each of the HFA's priority areas, ecosystems are not prominently highlighted as a crosscutting issue. At the 2013 Global Platform consultations, discussions revolved around issues and proposals on:

(i) the importance of community-level involvement,
(ii) targeting and including the most vulnerable populations,
(iii) women as leaders,
(iv) children and youth,
(v) health,
(vi) integrating climate change adaptation, development, and disaster risk reduction,
(vii) the role of science,
(viii) knowledge sharing and education,
(ix) capacity building: financing, risk assessment, preparedness, and early warning,
(x) private sector involvement in disaster risk reduction,
(xi) political will and leadership,
(xii) governance, accountability, transparency, and inclusiveness (UNISDR 2013b).

Environmental degradation (such as deforestation, erosion, and loss of biodiversity) was mentioned only because it will be affected by climate change and will have far-reaching consequences for food and water security. Moreover, environmental risk assessments were proposed to be integrated to risk assessments.

Similarly, the key issues and proposals that came out at the Asia Pacific consultations include (i) building on the HFA for a new framework for DRR; (ii) integrating DRR, climate change, and sustainable development; (iii) local-level action; (iv) turning vulnerability into resilience; (v) multi-stakeholder engagement; (vi) risk governance and accountability; (vii) knowledge-based decision making; and (viii) what kind of new framework (UNISDR 2013a). Again environment was not highlighted.

Despite the low emphasis on environment during consultations, environment is included in Priority 3 (investing in economic, social, cultural, and environmental resilience) of the zero draft of the post-2015 framework for DRR. Ecosystem management is identified as a key development area which requires strengthened sustainable use and management of ecosystems and DRR-integrated environmental and natural resource management approaches (UN 2014).

According to UNISDR (2013c), it is essential that the post-2015 framework for DRR considers enhancing current risk management practices in development planning and investment in order to manage risks inherent to development which manifests through disasters, climate change and variability, financial and economic crises, and other consequences for the economy, society, and the environment. Disaster risk management should aim for development that manages risks, sustainably seizes opportunities, and strengthens resilience to ensure sustainable development. It should not focus on the reduction of disaster loss but on encouraging sustainable development and human welfare and well-being (Lavell and Maskrey n.d.). A new framework for DRR would ideally be composed of (i) the post-2015 framework for disaster risk reduction and its monitoring system and period review process; (ii) the voluntary commitments of stakeholders, as leading examples of assumption of responsibility, vision, and readiness to act; and (iii) the political declaration (UNISDR 2013c).

### *8.2.2 Sustainable Development Goals*

The process to develop a set of Sustainable Development Goals (SDGs) was agreed at the Rio+20 Conference. The SDG consultation process consists of two tracks which will converge in September 2014 into one intergovernmental process: a member state-led (UN General Assembly) intergovernmental process to develop SDGs and the UN secretary-general-led discussions on what should replace the MDGs, supported by global stakeholder consultations. In the Rio+20 Outcome Document, member states agreed that SDGs must:

- Be based on Agenda 21 and the Johannesburg Plan of Implementation.
- Fully respect all the Rio Principles.
- Be consistent with international law.
- Build upon commitments already made.

- Contribute to the full implementation of the outcomes of all major summits in the economic, social, and environmental fields.
- Focus on priority areas for the achievement of sustainable development, being guided by the outcome document.
- Address and incorporate in a balanced way all three dimensions of sustainable development and their interlinkages.
- Be coherent with and integrated into the United Nations development agenda beyond 2015.
- Do not divert focus or effort from the achievement of the Millennium Development Goals.
- Include active involvement of all relevant stakeholders, as appropriate, in the process.

Additionally, it was agreed that the SDGs should be (i) action oriented, (ii) concise, (iii) easy to communicate, (iv) limited in number, (v) aspirational, and (vi) global in nature and universally applicable to all countries. A 30-member Open Working Group (OWG) of the General Assembly, tasked with preparing a proposal on the SDGs, released a zero draft of the proposed Sustainable Development Goals to be attained by 2030. Selected proposed indicators related to DRR and ecosystem management are listed in Table 8.3.

Integrating DRR into the post-2015 development agenda has been a priority issue in the consultations as they are so closely aligned. To synchronize the post-2015 Framework for DRR with the Post-2015 Sustainable Development Framework, Lavell and Maskrey (n.d.) suggest that the new HFA should be inside the SDGs and disaster risk management be made implicit in all the SDGs.

## 8.3 Ecosystem-Based Disaster Risk Reduction

Sudmeier-Rieux and Ash (2009) provide a definition of ecosystem-based disaster risk reduction (Eco-DRR) in the following:

> Ecosystem-based disaster risk reduction refers to decision-making activities that take into consideration current and future human livelihood needs and biophysical requirements of ecosystems, and recognize the role of ecosystems in supporting communities to prepare for, cope with, and recover from disaster situations.

In arguing for Eco-DRR, Sudmeier-Rieux et al. (2006) cite several reasons to integrate ecosystem-based management in DRR and development planning such as:

1. It can decrease vulnerability to natural disasters.
2. Natural disasters have a high cost.
3. It costs less to prevent disasters than it does to fix the damage they cause.
4. At-risk populations depend on ecosystems for their livelihoods.
5. Natural disasters and the responses to them have a negative impact on biodiversity.

**Table 8.3** Proposed SDGs and indicators related to DRR and ecosystem management

| Proposed SDG | Indicators |
|---|---|
| 11. Build inclusive, safe, and sustainable cities and human settlements | By 2030, reduce the environmental impacts of cities and improve the quality of environment in cities |
| | By 2020, increase by x% the number of human settlements adopting and implementing policies and plans towards resilience and adaptation to climate change and natural disasters |
| 13. Promote actions at all levels to address climate change | Build resilience and adaptive capacity to climate induced hazards in all vulnerable countries |
| | By 20xx, integrate climate change adaptation and mitigation strategies into development plans and poverty reduction strategies |
| | Improve education, awareness raising and human and institutional capacity on climate change impact reduction and early warning |
| 14. Attain conservation and sustainable use of marine resources, oceans, and seas | By 2030, reduce by x% marine pollution of all kinds, including from land-based activities |
| | By 2020, sustainably manage, restore and protect marine ecosystems from destruction, including by strengthening their resilience, and support relevant scientific research address and prevent further ocean acidification; ensure the full implementation of existing regional and international regimes for managing oceans and seas by their state parties |
| | By 2020, eliminate illegal, unreported and unregulated (IUU) fishing and destructive fishing practices |
| | By 2020, conserve at least 10 % of coastal and marine areas, including through establishing effectively managed marine protected areas, consistent with international law and based on best available scientific information; implement integrated and participatory coastal management to increase resilience of coastal ecosystems |
| 15. Protect and restore terrestrial ecosystems and halt all biodiversity loss | By 2020 halt the loss of all biodiversity, and protect and prevent the extinction of threatened species |
| | By 2020 ensure conservation and sustainable use of ecosystems, with particular attention to wetlands, including through restoration of at least 15 % of degraded ecosystems |
| | Maintain genetic diversity of both cultivated plants, farmed and domesticated animals and their wild relatives including through effective cooperation of national institutions |
| | By 2030, ensure the implementation of sustainable management of all types of forests and of mountain ecosystems |
| | By 2030 reverse the loss of and enhance forest cover worldwide, increase reforestation by x%, including by providing adequate incentives for developing countries |
| | By 2030, halt and prevent land degradation, reclaim land affected by desertification and drought, and improve land productivity and soil quality |
| | Introduce measures to prevent the introduction and significantly reduce the impact of invasive alien species on land and water ecosystems |
| | By 2020 control or eliminate the priority invasive species; ensure free prior informed consent of indigenous peoples and local communities in decision making and natural resources management, and promote the use of their traditional knowledge |
| | Integrate natural resources and biodiversity values into national and local planning, development processes, and accounts |

Source: http://sustainabledevelopment.un.org/focussdgs.html

Despite these reasons, there are still relatively very few concrete examples of Eco-DRR. This section reviews the link between ecosystems and DRR; examines the tools, approaches, and strategies applied in Eco-DRR; and describes the common elements present in Eco-DRR initiatives.

### *8.3.1 Ecosystems and Disaster Risk Reduction: A Review*

It seems logical that healthy ecosystems can protect communities from the impacts of disasters. Ecosystems increase resilience levels and help deliver development benefits which can help vulnerable people cope with the impacts of more frequent and intense disasters. Ecosystems contribute to increasing resilience through its provisioning and regulating functions (Munang et al. 2009; Saikia et al. 2013). The decline of these protective and regulatory functions can exacerbate and magnify the impacts of hazards. Conversely, ecosystem degradation can be exacerbated by disturbances caused by natural disasters. Thus, any loss or damage in ecosystem reduces its capacity to carry out these functions, reducing the resilience of both human communities and the ecosystem itself (Saikia et al. 2013). Losses from disasters contribute to the decline in ecosystems, social welfare, and economic growth and, thus, undermine local development and national economic growth (UNISDR 2013d).

Literature on the important role of ecosystems in DRR has grown substantially over the years. MA (2005) concludes that 60 % of the ecosystems are not being sustainably used or are in a state of ongoing degradation. It stressed the link between environmental degradation and increased impacts of disasters and the role of ecosystems in reducing risk from disasters such as flooding and forest fires. A study by Danielsen et al. (2005) also shows that the deterioration and clearing of mangroves and other types of coastal vegetation along many coastlines have increased their vulnerability to storm and tsunami damage. Recognizing this, the 2009 and 2011 Global Assessment Reports identify ecosystem decline as one of the underlying drivers of risk (UNISDR 2009, 2011). Land degradation, for instance, increases agricultural drought risk (UNISDR 2013d). Similarly, Beck et al. (2012) point to environmental degradation as a significant risk factor that reduces the capacity of societies especially vulnerable populations to deal with disaster risk.

Healthy ecosystems matter to disaster risk management because (i) human well-being depends on ecosystems that enable people to withstand, cope with, and recover from disasters; (ii) ecosystems, such as wetlands, forests, and coastal systems, can provide cost-effective natural buffers against hazard events and the impacts of climate change; (iii) there are clear links between resource degradation and disaster risk; (iv) healthy and diverse ecosystems are more robust to extreme weather events; and (v) ecosystem degradation reduces the ability of natural systems to sequester carbon, exacerbating climate change impacted disasters (Sudmeier-Rieux and Ash 2009). As presented in Table 8.4, numerous studies have shown the many benefits of healthy and well-managed ecosystems for DRR. Beck et al. (2012)

**Table 8.4** Selected studies demonstrating the role of ecosystems in DRR

| Hazard | Finding of study | References |
|---|---|---|
| Tsunami | Coastal forests and trees (e.g., mangroves) protected lives, resources, and infrastructure during the 2004 Indian Ocean Tsunami | Braatz et al. (2007), Chang et al. (2006), Danielsen et al. (2005), Forbes and Broadhead (2007), and Yanagisawa et al. (2009) |
| | In Thailand, poorly planned tourist developments and fishing communities built close to the shore on flat, low-lying land and in wide, exposed bays with no coral reefs were the worst hit during the 2004 Indian Ocean Tsunami | UNEP-WCMC (2006) |
| | Tsunami damage reached only 50 m inland and waves were only 2–3 m high compared to other areas in Hikkaduwa, Sri Lanka, due to coral reefs in a marine park | World Bank and United Nations (2010) |
| Cyclone/storm/hurricane | Statistical evidence from a sample of 409 villages demonstrated how mangroves reduced death toll during the 1999 super cyclone in Orissa | Das and Vincent (2009) |
| | Wetlands reduce flooding associated with hurricanes in the United States to an average of USD 8,240 per hectare per year, with coastal wetlands estimated to provide USD 23.2 billion a year in storm protective services | Costanza et al. (2008) |
| | A combination of infrastructure and relatively well-preserved natural ecosystems (semi-altered ecosystems) offer a good protection service against the impact of hurricanes in terms of human lives | Perez-Maqueo et al. (2007) |
| Flood | Sri Lanka's Muturajawia marsh, a 3,100 ha coastal peat bog that buffers and regulates flood water discharge into the sea, is estimated to provide protective services at more than USD 5 million or USD 1,750 per hectare | Emerson and Bos (2004) |
| Avalanche | Forests have an estimated economic value in preventing avalanches ranging from less than USD 100 per hectare per year for some of the landscapes in the Swiss Alps to more than USD 170,000 per year for tourist venues and towns | ProAct Network (2008) |

and Renaud et al. (2013) identify benefits such as (i) serving as natural infrastructure to prevent hazards or buffer hazard impacts; (ii) helping reduce the exposure of people and their productive assets to hazards; (iii) sustaining human livelihoods and providing for basic needs, such as food, shelter, and water, before, during, and after hazard events; and (iv) supporting better the post-disaster recovery needs of communities. In addition, these natural protection structures can (i) enhance community ownership of DRR; (ii) adapt to changing conditions, including recovery after a major damage-causing event; (iii) be more readily applied in poor countries as they are more cost-effective; (iv) be maintained with less external assistance; and (v) prevent and reverse environmental degradation. Finally, ecosystems provide many co-benefits such as increased areas for recreation, aesthetic improvements, habitat protection, reduction of human and technological errors or failures associated with structural works, or the spurring of economic growth (Kousky 2010). Recognizing the various benefits provided by ecosystems in risk reduction, IPCC (2012) cites investing in ecosystems as a "low-regrets" measure. In addition, ecosystem-based approaches have been recognized as a key climate change adaptation strategy in the UNFCCC negotiations since the Conference of Parties (COP) in Copenhagen in 2009.

Disasters not only affect people but also ecosystems producing negative consequences to the ecosystem services that they provide. Some of the environmental impacts include (i) direct damage to the natural resources and infrastructure, affecting ecosystem functions; (ii) acute emergencies from the uncontrolled, unplanned, or accidental release of hazardous substances, especially from industries; and (iii) indirect damage as a result of post-disaster relief and recovery operations that fail to take ecosystems and ecosystem services into account (PEDRR 2010). With climate change and the new risks posed by extreme climate events, ecosystems' critical role in reducing the impacts of climate extremes and disasters is lessened. Table 8.5 describes the impacts of climate extremes on ecosystems.

## *8.3.2 Experiences of Eco-DRR*

### 8.3.2.1 Tools and Approaches in Eco-DRR

PEDRR (2010) identifies the following core elements on implementing Eco-DRR: (i) recognize the multiple functions and services provided by ecosystems, including natural hazard protection or mitigation; (ii) link Eco-DRR with sustainable livelihoods and development; (iii) combine investment in ecosystems with other effective DRR strategies, including hard engineering options; (iv) address risks associated with climate change and extreme events and reduce their impact on ecosystem services; (v) enhance governance capacities for Eco-DRR through multi-sector, multidisciplinary platforms; (vi) involve local stakeholders in decision making; and (vii) utilize existing instruments and tools in ecosystems management and enhance their DRR value. As an example of existing instruments and tools which can be enhanced for their DRR value, the convention on biological diversity (CBD) adopts the ecosystem approach as a strategy for the integrated management of land, water, and living resources that promotes conservation and sustainable use in an equitable way. Five points of operational guidance, which can be applied to Eco-DRR, were

**Table 8.5** Consequences of climate extremes on ecosystems

| Extreme climate event | Consequences on ecosystems |
|---|---|
| Drought | Affect forestry and terrestrial and aquatic ecosystems |
| Heatwave | Can directly impact by constraining carbon and nitrogen cycling and reducing water availability, potentially decreasing production or even causing species mortality |
| | Extreme temperature conditions can also shift forest ecosystems from being a net carbon sink to a net carbon source |
| Dzud | Puts heavy pressure on ecosystem services and infrastructure and social services. It lasts all year round and causes dramatic socioeconomic impacts, including significant loss of livestock, unemployment, poverty, and mass migration from rural to urban areas |
| Flood | Impact ecosystems, including species populations |
| Hurricane and storm | Can impact forest ecosystems, particularly in pre-alpine and alpine areas. Saltmarshes, mangroves, and coral reefs can also be vulnerable to such climate extremes |
| Oceanic warming and acidification | Have a negative impact on marine ecosystems particularly for coral reef ecosystems. Anthropogenic oceanic changes may contribute indirectly to damage to coral atolls, by affecting the health of the surrounding reef system |
| Permafrost | Melting of massive ground ice and thawing of ice-rich permafrost can lead to landslides, slope instabilities, subsidence of the ground surface, and the formation of uneven topography known as thermokarst. Such changes have implications for ecosystems, landscape stability, and infrastructure performance |

Source: Seneviratne et al. (2012)

developed following the 12 principles of the ecosystem approach to aid implementation and facilitate wider adoption. Table 8.6 shows lists of CBD's points of operational guidance for ecosystem approach and shows how this guidance has been adopted by the International Union for the Conservation of Nature (IUCN), Wetlands International, and the Department for Environment, Food and Rural Affairs (DEFRA) of the United Kingdom. These steps for implementation of the ecosystem approach place emphasis on holistic approach, structure and function of the ecosystem, spatial scale; risks and vulnerabilities, land and natural resource use, economic issues, stakeholder involvement, and engagement of multiple sectors.

Moreover, Sudmeier-Rieux (2013) lists available tools that can be used in Eco-DRR including (i) environmental assessment tools (e.g., environmental impact assessment (EIA)), strategic environmental assessment (SEA), and rapid environmental assessment (REA), (ii) integrated risk and vulnerability assessments, (iii) spatial planning at regional and local scales, and (iv) integrated ecosystems management (e.g., integrated water resources management, integrated coastal zone management, integrated fire management, protected area management and community-based ecosystem, and disaster risk management. In harmonizing

**Table 8.6** CBD's ecosystem approach as adopted by selected organizations

| CBD's five points of operational guidance | IUCN's five steps (Shepherd 2004) | Wetlands International's five steps (Wetlands International 2013) | DEFRA's six principles (DEFRA 2010) |
|---|---|---|---|
| 1. Focus on the functional relationships and processes within ecosystems | 1. Determining the main stakeholders, defining the ecosystem area, and developing the relationship between them | 1. Assessing risks and vulnerabilities, including the (environmental) root causes of risk | 1. Taking a more holistic approach to policy-making and delivery, with focus on maintaining healthy ecosystems and ecosystem services |
| 2. Enhance benefit sharing | 2. Characterizing the structure and function of the ecosystem, and setting in place mechanisms to manage and monitor it | 2. Identifying risk reduction scenarios and related costs and benefits | 2. Ensuring that the value of ecosystem services is fully reflected in decision making |
| 3. Use adaptive management practices | 3. Identifying the important economic issues that will affect the ecosystem and its inhabitants | 3. Ensuring that risk reduction measures are planned at multiple spatial scales – locally at the community level but also across wider areas (river basins, landscapes) | 3. Ensuring environmental limits are respected in the context of sustainable development, taking into account ecosystem functioning |
| 4. Carry out management actions at the scale appropriate for the issue being addressed, with decentralization to lowest level, as appropriate | 4. Determining the likely impact of the ecosystem on adjacent ecosystems | 4. Designing and implementing ecosystem-inclusive risk reduction measures in partnership with multiple sectors | 4. Taking decisions at the appropriate spatial scale recognizing the cumulative impacts of decisions |
| 5. Ensure intersectoral cooperation | 5. Deciding on long-term goals and flexible ways of reaching them | 5. Addressing the root causes of risk by ensuring sound land use and natural resource use policies, ensuring that ecosystem services are sustained | 5. Applying adaptive management of the natural environment to respond to changing pressures, including climate change |
| | | | 6. Identifying and involving all relevant stakeholders in the decision and plan-making process |

environmental and disaster reduction policies, Sudmeier-Rieux et al. (2006) recommend these actions:

(i) assess the environmental causes of vulnerability;
(ii) assess environmental actions that reduce vulnerability;
(iii) monitor natural processes (e.g., drought and flood) and establish early warning systems;
(iv) consider the effects on ecosystem services (e.g., the impacts of draining wetlands on flood regimes) in decision-making processes;
(v) establish partnerships for and regional approaches to land use and nature conservation;
(vi) establish alternatives to conflicts over the alternative uses of resources;
(vii) provide advice and information to involve people in enhancing ecosystem protection (e.g., community stewardship of mangrove forests);
(viii) consider the economic benefits of the services that ecosystems provide to disaster risk reduction (e.g., the benefits of investing in wetland restoration as a buffer for floods);
(ix) create economic and legal incentives to include ecosystem services in disaster risk reduction (e.g., incentives or disincentives to avoid exploitation of resources from protective sand dunes, mangrove forests, and coral reefs);
(x) enforce environmental regulations, particularly those that may reduce population vulnerability (e.g., zoning laws, protection of key ecosystems, solid waste management); and
(xi) strengthen ecosystem management to include disaster risk reduction (e.g., watershed management, integrated coastal management, protected area management).

In addressing climate-related risks, UNEP (n.d.b) mentions four complementary strategies that are required in implementing an ecosystem approach: (i) political commitment to raise the profile of ecosystems in climate change policy setting at local, national, and international levels; (ii) investment related to ecosystem management and protection, especially as part of a global climate change fund; (iii) incentives to reduce emissions, ease existing pressures on ecosystems, and support changes that increase environmental resilience and resource sustainability; and (iv) comprehensive information that foster closer links between ecosystem management, CCA, and disaster risk reduction communities as well as between science, economics, politics, and policy.

#### 8.3.2.2 Eco-DRR Strategies

There is a wide range of ecosystem management strategies and actions that can be applied for DRR in different ecosystems at local, national, and regional levels. The following describes some of the Eco-DRR strategies that are widely utilized. Table 8.7 provides a summary of these strategies and examples of initiatives undertaken.

**Table 8.7** Examples of Eco-DRR initiatives in different ecosystems and levels

| | Ecosystem-based DRR initiatives | | |
|---|---|---|---|
| Ecosystem | Local | National | Regional |
| Forest | Community-based forest management; community-based DRR/adaptation | DRR plans and programs | REDD+ |
| | | Under the Amazon protected areas program, Brazil has created a mosaic of over 30 million hectares of biodiversity-rich forests reserve of state, provincial, private, and indigenous land, resulting in potential reduction in emissions estimated at 1.8 billion tons of carbon, through avoided deforestation | |
| | | In Muminabad, Tajikistan a Swiss Development Cooperation project adopted an integrated approach to risk through reforestation and integrated watershed management | |
| Watershed/deltas | River basin management | Vietnam's strategic environmental assessment in land use planning projects and hydropower development for the Vu Gia-Thu Bon river basin, including climatic disaster risks | Transboundary/river delta initiatives (Mekong River) |
| | | European countries affected by severe flooding, notably the United Kingdom, the Netherlands, and Germany, have made policy shifts to make space for water by applying more holistic river basin management plans and integrated coastal zone management | |
| Coastal (mangrove, beach, reef, sea grass, salt marsh) | Coastal zone management; mangrove restoration and rehabilitation; coral restoration; protected area management | Philippines' National Greening program; programs after Indian Ocean tsunami | Coral triangle initiative; mangroves for the future |

### Forest Management

Forest management balances demand for forest products with the ecological requirements of forests, while ensuring other key benefits for livelihoods, notably by stabilizing steep slopes and reducing soil erosion. DEWGA (2008) identifies these actions for sustainable forest management: (i) protect and improve the forest environment through increased vegetation; (ii) help alleviate poverty by generating income through increased tree cover and related activities, (iii) increase forest resources, (iv) establish community-driven economic activities based on forest plantation, (v) increase multiple uses for land, and (vi) create popular awareness about sustainable forest management. In addition, forests in potential avalanche release areas can reduce the risk of avalanches because trees break up snow cover, prevent wind-blown snow drifts, and keep snow under shade and therefore colder and firmer and their fallen boles and boughs tend to anchor snow and prevent it from moving (ProAct Network 2008).

### Protected Area Management

Stolton et al. (2008) identify three direct roles that protected areas can play in preventing or mitigating disasters arising out of natural hazards such as (i) maintaining natural ecosystems (e.g., coastal mangroves, coral reefs, floodplains, and forest) that may help buffer against natural hazards, (ii) maintaining traditional cultural ecosystems that have an important role in mitigating extreme weather events (e.g., agroforestry systems, terraced crop growing, and fruit tree forest in arid lands), and (iii) providing an opportunity for active or passive restoration of such systems where they have been degraded or lost.

### Watershed Management

The physical and biological resources of watersheds provide a wide range of ecosystem goods and services to people such as water protection, attenuation of disasters by regulating runoff, protection of coastal resources and fisheries, protection of the environment, and protection of productive lowlands. For these reasons, watershed management is important for agricultural, environmental, and socioeconomic development. Actions for effective management of the watershed include:

(i) when located in floodplains, structures should be built to withstand flood damage, to prevent floodwater contamination, and to avoid disruption to river courses, river banks and vegetation;
(ii) intensive agricultural activity should not to be permitted on slopes greater than a specified percentage reflecting land stability;
(iii) clear-cutting of forests should be limited with forest conservation and sustainable forest management prioritized;
(iv) institutional bodies, such as River Basin Organizations, should be formally established to address land use conflicts and staff trained in conflict-resolution;

(v) public participation of both men and women should be increased in management decisions;
(vi) effective management plans and enforcement of environmental and zoning regulation are critical; and
(vii) regional environmental impact assessments are needed to ensure that cumulative impacts of economic activities are sustainable (DEWGA 2008).

### Coastal Zone Management

Coastal ecosystems such as mangroves, salt marshes, beach vegetation, seagrass beds, and coral reefs are effective buffers against many coastal natural hazards while providing significant social and economic benefits (MA 2005). Barrier islands formed by offshore drift and sedimentation buffer storm surges and waves as well (ProAct Network 2008). These ecosystems are under pressure by coastal development, and thus, coastal zone management actions must consider the continuum of inland areas, coasts, and oceans through actions such as (i) replanting coastal forests and restoration of mangroves, which have been taken up as a part of the environmental recovery process, (ii) restoring and maintaining the health of the coral reefs and seagrass beds, (iii) maintaining and/or developing mangrove belts as buffer zones for coasts and coral reefs, and (iv) protecting wetlands and watersheds to minimize sedimentation (DEWGA 2008).

### Mangrove Restoration and Rehabilitation

Mangroves generally slow the flow of water as the surge moves inland and reduce the waves riding on top of the surge, lowering water levels and reducing damage behind the mangroves. They reduce the magnitude of storm surges and related inundation by absorbing storm energy, reducing flow depths and velocities, and holding sediments in place within root systems (ProAct Network 2008). As with coastal zone management, coastal protection against hazards is recognized as one of the benefits that restored mangroves will provide resulting in the increasing interest in the use of mangroves as coastal defense against hazards such as storm surges (McIvor et al. 2012; Gedan et al. 2011; Shepard et al. 2011; Zhang et al. 2012).

### Coral Restoration

Healthy reef systems provide a buffer zone for the shoreline during extreme surge and wave events thus mitigating erosion and inundation. They are also a source of carbonate sand and gravel for atolls, which are delivered to shore by storms and swell (CDKN 2012). Beck et al. (2012) estimate that there are 200 million people who benefit from risk reduction from coral reefs alone or may have to bear higher costs of disasters if the reefs are degraded. This population lives in low, risk-prone coastal areas (below 10 m elevation) and within 50 km of coral reefs in villages, towns, and cities.

#### 8.3.2.3 Common Elements in Eco-DRR Initiatives

Slocombe (1998) explains that, in general, an ecosystem approach (i) describes parts, systems, environments, and their interactions; (ii) is holistic, comprehensive, and trans disciplinary; (iii) includes people and their activities in the ecosystem; (iv) describes system dynamics; (v) defines the ecosystem naturally; (vi) looks at different levels/scales of system structure, process, and function; (vii) recognizes goals and takes an active management orientation; (viii) incorporates actor-system dynamics and institutional factors in the analysis; (ix) uses an anticipatory and flexible research and planning process; (x) entails an implicit or explicit ethics of quality, well-being, and integrity; and (xi) recognizes systemic limits to action.

In addition, seven core elements associated with implementing Eco-DRR are outlined in PEDRR (2010), namely:

1. Recognize the multiple functions and services provided by ecosystems, including natural hazard protection or mitigation.
2. Link ecosystem-based risk reduction with sustainable livelihoods and development.
3. Combine investments in ecosystems with other effective DRR strategies, including hard engineering options.
4. Address risks associated with climate change and extreme events and reduce their impact on ecosystem services.
5. Enhance governance capacities for ecosystem-based DRR through multi-sector, multi disciplinary platforms.
6. Involve local stakeholders in decision making.
7. Utilize existing instruments and tools in ecosystem management and enhance their DRR value.

Other common elements mentioned in the literature include (i) integration of ecological, sociocultural, economic, and institutional factors; (ii) consideration of ecosystem integrity; (iii) use of environmental planning and management tool, strategy, or system, (iv) practice of adaptive management (Uy and Shaw 2012); (v) cost-effectiveness (Kousky 2010); and (vi) local accessibility (PEDRR 2010).

## 8.4 Challenges and Limits to Implementing Eco-DRR

While ecosystem management is not a new concept, further evidence is needed to build the case and demonstrate how ecosystem management can be maximized for DRR and thus facilitate uptake (PEDRR 2010). The main challenge at present is improving the evidence base for Eco-DRR. To respond to this, the challenges on knowledge and research and institutions and policy, in particular, need to be addressed to increase appreciation of Eco-DRR and guide implementation especially at the local level.

### 8.4.1 Knowledge and Research

Essentially, there is a need for additional research due to lack of understanding of the potential of natural buffers (ProAct Network 2008; Kousky 2010). Challenges exist in developing understanding of an ecosystem; identifying appropriate spatial and temporal scales for analysis, planning, and management; and governance and institutional jurisdiction which determine the issues and opportunities to address and the different information and management tools required. Among the gaps in knowledge and research on Eco-DRR include (i) ecology on multiple scales, (ii) monitoring and evaluation, (iii) "benchmarks" of ecosystem condition, (iv) human dimensions of natural resource use, (v) ecological restoration technology development, (vi) quantifying uncertainty and assessing risk, and (vii) adaptive management process. In addition, insufficient recognition of the economic and social benefits of ecosystem services under current risk situations, let alone under potential changes in climate extremes and disaster risks, lack of interdisciplinary science and implementation capacity for making informed decisions associated with complex and dynamic systems, inability to estimate economic values of different ecosystem services, and lack of capacity to undertake careful cost and benefit assessments of alternative strategies to inform choices at the local level are challenges to increasing investments in ecosystem-based solutions (CDKN 2012).

### 8.4.2 Institutions and Policy

Kousky (2010) identifies uninterested decision makers and political opposition as potential challenges to increasing adoption of the use of natural capital to reduce risks. Linking Eco-DRR to policy and institutional mandates is necessary to facilitate implementation. Also, having a champion advocating for Eco-DRR would ensure that it is prioritized (UNEP 2009). In this light, institutional barriers need to be addressed such as (i) fragmentation and specialization in administration and research [e.g., data and monitoring on ecosystem status and risk are often dispersed across agencies at various scales and are not always accessible at the sub national or municipal level, where land use planning decisions are made (CDKN 2012)], (ii) competition within and between agencies and governments, (iii) overlapping efforts, (iv) narrow focus, (v) lack of standardization, (vi) politically defined management units, (vii) short-term and self-interested politics, and (viii) economic determinism.

Lastly, it needs to be understood that there are many factors that may limit the ecosystem's ability to provide protection against hazards. It is important to note that the nature of the relationship between ecosystems and disaster risk reduction depends strongly on the characteristic of hazard and the type and state of ecosystem – that ecosystem functions are very complex and disaster risk is influenced by many factors (IPCC 2012). It would be necessary, therefore, to consider hybrid solutions such as combining hard engineering with soft ecosystem approaches. As Feagin et al. (2010) note, the use of ecosystems as bioshields is not a panacea for decreasing vulnerability and must be combined with other measures.

## 8.5 Way Forward and Opportunities in the Post-2015 Development Agenda

The discussion and elaboration of three international frameworks and instruments (i.e., post-2015 framework on DRR, SDGs, and post-Kyoto global climate agreement) provide a unique opportunity to integrate ecosystem approaches to disaster risk reduction into a harmonized post-2015 paradigm. Eco-DRR brings together three distinct communities, environment, DRR, and climate change adaptation, along with their knowledge, expertise, experience, and resources subsequently contributing to multiple development priorities (Renaud et al. 2013). Many groups are starting to recognize the merits of Eco-DRR as a way to link ecosystem management, disaster risk reduction, and sustainable development to achieving environmental, social, and economic goals. To this end, the post-2015 development agenda offers important opportunities for Eco-DRR as discussed below.

**Increasing Focus on Environmental and Ecosystem Service Degradation and Climate Change as Underlying Drivers of Risk in DRR Activities**. Given the risks posed by climate change and increasing disaster losses globally, Eco-DRR provides an integrated solution for reducing disaster risk through ecosystem management and climate change adaptation. UNEP (n.d.a) summarizes the opportunities for a renewed focus on environment by:

(i) engaging environmental managers fully in national disaster risk management mechanisms;
(ii) including risk reduction criteria in environmental regulatory frameworks;
(iii) assessing environmental change as a parameter of risk;
(iv) utilizing local knowledge in community-based disaster risk management;
(v) engaging the scientific community to promote environmental research and innovation;
(vi) protecting and valuing ecosystem services;
(vii) considering environmental technologies and designs for structural defenses;
(viii) integrating environmental and disaster risk considerations in spatial planning;
(ix) preparing for environmental emergencies; and
(x) strengthening capacities for environmental recovery.

**Increasing Understanding of Eco-DRR Through Documentation, Dissemination, and Capacity Development**. The lack of awareness and capacity on Eco-DRR often presents a barrier to its uptake. Improving understanding of Eco-DRR requires proper documentation, dissemination, and capacity development. Evidence of Eco-DRR needs to be documented particularly for monitoring and evaluation. To aid the effective implementation, replication, and scaling up of Eco-DRR, information that are well understood and tools that are user friendly need to be developed and disseminated for practitioners and decision makers to learn the value added of ecosystem approaches.

**Integration of Eco-DRR into Development Planning**. The post-2015 framework for disaster risk reduction is in a strong position to introduce the necessary changes to enhance current risk management practices in development planning and investment (UNISDR 2013c). Risks to natural capital compromise future wealth (UNISDR 2013d). Achieving well-being and sustained prosperity will require development pathways that respect ecological limits and restore ecosystem health while optimizing the contribution of the environment to economic progress (IRF 2013). One way of doing this is to integrate DRR into existing development instruments and mechanisms and protect ecosystems through employing participatory valuation and management of ecosystem services and mainstreaming of ecosystem approaches in DRR (UNISDR 2013d).

More efforts would be required in building a case for Eco-DRR, improving its evidence base, and linking it to policy. The post-2015 development agenda is expected to put back environment into the limelight along with its multidisciplinary and intersectoral linkages. It would be important for stakeholders especially decision makers to be able to appreciate Eco-DRR for its many benefits as well as its importance in sustainable development.

## References

Beck MW, Shepard CC, Birkmann J, Rhyner J, Welle T, Witting M, Wolfertz J, Martens J, Maurer K, Mucke P, Radtke K (2012) World risk report 2012. Alliance Development Works, Berlin

Braatz S, Fortuna S, Broadhead J, Leslie R (eds) (2007) Proceedings of the regional technical workshop: coastal protection in the aftermath of the Indian Ocean tsunami: what role for forests and trees? Food and Agriculture Organization, Khao Lak

Chang SE, Adams BJ, Alder J, Berke PR, Chuenpagdee R, Ghosh S, Wabnitz C (2006) Coastal ecosystems and tsunami protection after the December 2004 Indian Ocean tsunami. Earthq Spectra 22(3):863–887

Costanza R, Pérez-Maqueo OM, Martínez ML, Sutton P, Anderson SJ, Mulder K (2008) The value of coastal wetlands for hurricane protection. Ambio 37:241–248

Danielsen F, Sorensen MK, Olwig MF, Selvam V, Parish F, Burgess ND, Hiraishi T, Karunagaran VM, Rasmussen MS, Hansen LB, Quarto A, Suryadiputra N (2005) The Asian tsunami: a protective role for coastal vegetation. Science 310:643

Das S, Vincent JR (2009) Mangroves protected villages and reduced death toll during Indian super cyclone. Proc Natl Acad Sci 109(18):7357–7360

Department for Environment, Food and Rural Affairs, United Kingdom (2010) What nature can do for you: a practical introduction to making the most of natural services. Assets and resources in policy and decision-making. DEFRA, London

Climate and Development Knowledge Network (CDKN) (2012) Managing climate extremes and disasters for ecosystems: lessons from the IPCC SREX report. Available at www.cdkn.org/srex

DEWGA (2008) Stocktaking paper: linking disaster risk reduction, environmental management and development practices and practitioners in Asia Pacific Region: a review of opportunities for integration, working paper. Available at www.dewga.net/Data/Publication/Stocktaking%20Paper_Version%206%20080825.pdf

Feagin RA, Mukherjee N, Shanker K, Baird AH, Cinner J, Kerr AM, Koedam N, Sridhar A, Arthur R, Jayatissa LP, Seen DL, Menon M, Rodriguez S, Shamsuddoha M, Dahdouh-Guebas F (2010) Shelter from the storm? Use and misuse of coastal vegetation bioshields for managing natural disasters. Conserv Lett 3:1–11

Forbes K, Broadhead J (2007) The role of coastal forests in the mitigation of tsunami impacts. Food and Agriculture Organization Regional Office for Asia and the Pacific, Bangkok

Gedan KB, Kirwan ML, Wolanski E, Barbier EB, Silliman BR (2011) The present and future role of coastal wetland vegetation in protecting shorelines: answering recent challenges to the paradigm. Clim Change 106:7–29

Independent Research Forum (2013) Post-2015: framing a new approach to sustainable development. IRF2015 Secretariat, Washington, DC

IPCC (2012) Managing the risks of extreme events and disasters to advance climate change adaptation. In: Field CB, Barros V, Stocker TF, Qin D, Dokken DJ, Ebi KL, Mastrandrea MD, Mach KJ, Plattner GK, Allen SK, Tignor M, Midgley PM (eds) A special report of Working Groups I and II of the Intergovernmental Panel on Climate Change. Cambridge University Press, Cambridge/New York

Kousky C (2010) Using natural capital to reduce disaster risk. J Nat Res Policy Res 2(4):343–356

Lavell A, Maskret A (n.d.) The future of disaster risk management: an-ongoing discussion

McIvor AL, Spencer T, Möller I, Spalding M (2012) Storm surge reduction by mangroves. Natural Coastal Protection Series: report 2. Cambridge Coastal Research Unit working paper 41. Published by The Nature Conservancy and Wetlands International, 35 pp. ISSN 2050-7941. Available at: http://www.naturalcoastalprotection.org/documents/storm-surge-reduction-by-mangroves

Millennium Ecosystem Assessment (2005) Ecosystems and human well-being: synthesis. Island Press, Washington, DC

Munang R, Rivington M, Liu J, Thiaw I (2009) Ecosystem management: part of the climate change solution. UNEP, Nairobi

Partnership for Environment and Disaster Risk Reduction (PEDRR) (2010). Demonstrating the role of ecosystem-based management for disaster risk reduction. Partnership for environment and disaster risk reduction

Perez-Maqueo O, Intralawan A, Martinez ML (2007) Coastal disasters from the perspective of ecological economics. Ecol Econ 63:273–284

ProAct Network (2008) The role of environmental management and eco-engineering in disaster risk reduction and climate change adaptation. ProAct Network, Tannay

Renaud FG, Sudmeier K, Estrella M (eds) (2013) The role of ecosystems in disaster risk reduction. UNU Press, Tokyo

Saikia A, Davis I, Ariyabandu MM (2013) Ecosystem based disaster risk reduction. Ecosyst Based Disaster Manag Plan 93:16

Seneviratne SI, Nicholls N, Easterling D, Goodess CM, Kanae S, Kossin J, Luo Y, Marengo J, McInnes K, Rahimi M, Reichstein M, Sorteberg A, Vera C, Zhang X (2012) Changes in climate extremes and their impacts on the natural physical environment. In: Field CB, Barros V, Stocker TF, Qin D, Dokken DJ, Ebi KL, Mastrandrea MD, Mach KJ, Plattner G-K, Allen SK, Tignor M, Midgley PM (eds) Managing the risks of extreme events and disasters to advance climate change adaptation. A special report of Working Groups I and II of the Intergovernmental Panel on Climate Change (IPCC). Cambridge University Press, Cambridge, pp 109–230

Shepard C, Crain C, Beck MW (2011) The protective role of coastal marshes: a systematic review and metaanalysis. PLoS One 6(11):e27374

Shepherd G (2004) The ecosystem approach: five steps to implementation. IUCN, Gland

Slocombe DS (1998) Lessons from experience with ecosystem-based management. Landsc Urban Plan 40:31–39. Elsevier

Stolton S, Dudley N, Randall J (2008) Natural security: protected areas and hazard mitigation. WWF, Gland

Sudmeier-Rieux K (2013) Ecosystem approach to disaster risk reduction: basic concepts and recommendations to governments, with a special focus on Europe. Council of Europe, Strasbourg

Sudmeier-Rieux K, Ash N (2009) Environmental guidance note for disaster risk reduction: healthy ecosystems for human security, Revisedth edn. IUCN, Gland

Sudmeier-Rieux K, Masundire H, Rizvi A, Rietbergen S (eds) (2006) Ecosystems, livelihoods and disasters, an integrated approach to disaster risk management. IUCN, Gland

United Nations (2012) The future we want. Outcome of the Rio+20 United Nations conference on sustainable development. Rio de Janeiro, 20–22 Jun 2012. Available at http://www.un.org/en/sustainablefuture/

United Nations (2014) Post-2015 framework for disaster risk reduction. Zero draft submitted by the co-chairs of the preparatory committee. UN, Geneva

United Nations Development Programme (UNDP) (2007) Human development report 2007/2008. Fighting climate change: human solidarity in a divided world. New York

United Nations Environment Programme (UNEP) (n.d.a) Environment and vulnerability: emerging perspectives. UNEP, Nairobi

United Nations Environment Programme (UNEP) (n.d.b) Ecosystem management: part of the climate change solution. UNEP, Nairobi

United Nations Environment Programme (UNEP) (2009) The role of ecosystems management in climate change adaptation and disaster risk reduction. Copenhagen discussion series

United Nations Environment Programme-World Conservation and Monitoring Center (UNEP-WCMC) (2006) In the front line: shoreline protection and other ecosystem services from mangroves and coral reefs. UNEP-WCMC, Cambridge

United Nations International Strategy for Disaster Reduction (UNISDR) (2005) Hyogo framework for action 2005–2015. UNISDR, Geneva

United Nations International Strategy for Disaster Reduction (UNISDR) (2009) Global assessment report on disaster risk reduction 2009. UNISDR, Geneva

United Nations International Strategy for Disaster Reduction (UNISDR) (2011a) Global assessment report on disaster risk reduction 2011: revealing risk, redefining development. UNISDR, Geneva

United Nations International Strategy for Disaster Reduction (UNISDR) (2011b) Hyogo framework for action 2011–2015 mid-term review. UNISDR, Geneva

United Nations International Strategy for Disaster Reduction (UNISDR) (2013a) Asia pacific synthesis report: consultations on the post-2015 framework for disaster risk reduction (HFA2). UNISDR, Bangkok

United Nations International Strategy for Disaster Reduction (UNISDR) (2013b) Post-2015 framework for disaster risk reduction (HFA2): report from 2013 global platform consultations. UNISDR, Geneva

United Nations International Strategy for Disaster Reduction (UNISDR) (2013c) Proposed elements for consideration in the post-2015 framework for disaster risk reduction by the UN special representative of the secretary-general for disaster risk reduction. UNISDR, Geneva

United Nations International Strategy for Disaster Reduction (UNISDR) (2013d) Global assessment report on disaster risk reduction 2013: from shared risk to shared value-the business case for disaster risk reduction. UNISDR, Geneva

Uy N, Shaw R (2012) The role of ecosystems in climate change adaptation and disaster risk reduction. In: Uy N, Shaw R (eds) Ecosystem based adaptation. Emerald Group Publishing Limited, Bingley, pp 41–59

Wetlands International (2013) Working with nature: towards integrated approaches to disaster risk reduction. Wetlands International, Wageningen

World Bank and the United Nations (2010) Natural hazards, unnatural disasters: the economics of effective prevention. The World Bank, Washington, DC

Yanagisawa H, Koshimura S, Goto K, Miyagi T, Imamura F, Ruangrassamee A, Tanavud C (2009) The reduction effects of mangrove forest on a tsunami based on field surveys at Pakarang Cape, Thailand and numerical analysis. Estuar Coast Shelf Sci 81:27–37

Zhang K, Liu H, Li Y, Xu H, Shen J, Rhome J, Smith TJ III (2012) The role of mangroves in attenuating storm surges. Estuar Coast Shelf Sci 102–103:11e23

# Chapter 9
# Access, Equity and Hazards: Highlighting a Socially Just and Ecologically Resilient Perspective on Water Resources

**M. Usman Mirza and Daanish Mustafa**

**Abstract** Historically there was a general trend towards infrastructural and physical investment in supply side water related initiatives for the provision of clean drinking water and livelihood needs such as irrigation and agriculture. The critical missing link was the absence of a social/human aspect to water resources and its relation to the human society. Access to water resources revolved predominantly around the health and livelihood needs of the society. Multiple values that a society could derive from its access to water were ignored. This limited focus on access to water coupled with a growing problem of water scarcity gave birth to a new phenomenon of considering water as an 'economic good'. This commoditization of water meant water was provided based on the ability to pay and efficiency of use, thus further alienated the social value of water. Furthermore, the link between water and society can also be viewed from a hazards perspective. With the increasing awareness of climate change and water related hazards, view of water based upon assumption of average normal conditions is no longer tenable. Building resilience and adaptation capacity to address water hazards must involve a fundamental shift towards a planning paradigm that works inwards from extremes rather than outward from means. With this background, the objective of the chapter is to review water research literature through the tri-focal lens of Access, Equity and Hazards and attempt to identify the gaps – when the water resources field is viewed through this tri-focal lens. To set the stage, the chapter will first briefly discuss the rationale for the choice of our tri-focal analytical lens before delving into the international academic and policy literature to address the aforementioned objectives.

**Keywords** Social justice • Ecological resilience • Water • Equity • Risk

---

M.U. Mirza (✉)
Leadership for Environment and Development (LEAD) Pakistan, Islamabad, Pakistan
e-mail: umirza@lead.org.pk

D. Mustafa
Department of Geography, King's College, London, UK



J.I. Uitto, R. Shaw (eds.), *Sustainable Development and Disaster Risk Reduction*, Disaster Risk Reduction, DOI 10.1007/978-4-431-55078-5_9

## 9.1 Introduction

Water resources research was mostly concerned with technical supply side, managerial and policy problems through the 1960s–1980s (e.g., see White 1968; Michel 1967; Wescoat et al. 2000). The 1980s brought with it the political wave of neo-liberalism. Together with the growing problem of water scarcity from the mid-1980s onwards and the ideological setting of neo-liberalism – a new phenomenon of considering water as an 'economic good' emerged in the 1990s (Bakker 2005). This new shift of viewing water as a commodity resulted in the commercialization of water management and governance. The commoditization of water further alienated the social value of water and instead access to water was deemed to be linked to the ability to pay and efficiency of use. Through all this transition, the actual improvement in water quality and progress towards social justice in distribution, particularly for the most needy was nonexistent. Starting with the 1990s there was an emerging concern with water under the post-cold war neo-liberal global regime to questions such as access to water for drinking and sanitation, water for environmental quality, conflict over water and irrigation water distribution (Swyngedouw 1997, 1999; Wescoat 1987; Lightfoot 1996; Bates et al. 1993; Mageed and White 1995; Homer-Dixon 1994). This chapter is an attempt at critiquing the dominant neo-liberal view of water as simply a commodity subject to the market mechanism. Furthermore, the argument presented in this chapter seeks to highlight the potential of more socially just and ecologically resilient use of water through recognition of a broader set of values of water and different paradigmatic approaches to understanding its socio-geographical distribution.

With the beginning of twenty-first century, water resources have attracted research attention with a renewed urgency. The projected effects of climate change have changed the future scenarios for all natural resources and water is no different. With the increasing awareness of climate change impacts and the associated risks involved we can no longer afford to have future planning based upon extrapolated historic average trends. The increasing awareness of ongoing global climate change and associated hazards and, realization of the finite nature of fresh drinking water resources has propelled social scientific water research into areas relating to water hazards, access to water and water distribution issues.

This literature review of major social scientific research in the water resources field is organized around the three themes of access, equity and hazards, to capture the strengths and gaps in the water literature since the turn of the century. The review attempts to point out the gaps, when the water resources field is viewed through the tri-focal lens. We shall briefly discuss the rationale for the choice of themes that define the tri-focal analytical lens and then delve into the international academic and some major policy literature that has emerged since the turn of the twenty-first century.

## 9.2 Defining the Tri-focal Analytical Lens

Access is often considered substitutable with equity. We however, interpret the concept in its expansive sense of access for satisfying multiple values and uses of water. It is a broader concept going beyond the distributive agenda of equity, towards accounting for multiple values and uses of water – normally not considered in water policy. Water has an intrinsic value with respect to its indispensability for all life, presence in nature, its sanctity and spiritual significance in most religions and its link with human cultural heritage. Access to water, hence, warrants a broader debate in a social context (Llamas 2003). Questions in this context typically revolve around the concepts of water as a common good, water's connection to human dignity, water as a basic need of life, water having cultural significance, and water as part of an ecosystem. A reductionist approach to limit the utility of water to only health and livelihood does not appreciate the multiple uses for which access to water has always been valued in human society. A root cause for this reductionist approach is the increasing consumptive demands on water resources, leading to the popular perception of water scarcity. Water scarcity agenda drives a wedge between water efficiency argument and the intrinsic value of water for the society. This intrinsic value of water reflects the link of water use to the religious, community and cultural needs a society derives from water (Pradhan and Meinzen-Dick 2003).

Equity is a distributive and a procedural concept. It is concerned about how water might be distributed between users and through which procedures. It has a normative component in the sense that it alludes to notions of distributive and procedural justice and transparency. Water is a source of life and its distribution and management should abide by the values of equity and social justice. Equity in the case of water implies the right to a minimum amount of water for basic human needs plus the assurance of procedural fairness and transparency in the provision of this basic human right (Mustafa 2013). Speaking about water management there is also a distributive angle to the question of equity. According to Bates et al. (1993), equitable water resource management should distribute water to users commensurate with their stake in it.

This inclusion of social justice as a fundamental component of equity leads us to the realm of fairness and ethics. Murray-Rust et al. (2000) conceptualized equity as fairness where a fair distribution for the society may not necessarily be strictly equal. For the purpose of this chapter, we define the notion of equity in water resources as incorporating both the concepts of fairness as in procedural and distributive justice, as well as treating water as a basic human right. Lack of equity and social justice in water distribution can be termed as the human induced aspect of water scarcity. Water scarcity is, in part, socially constructed i.e. inequitable distribution diminishing the right to water use. One of the main concerns in operationalizing the principles of equity in policy level water planning is the reliance on conventional economic analysis and treating water as an economic commodity as opposed to a considering water in a broader social context (Syme and Nancarrow 2008). Whilst the access concern outlined above stresses the range of societal values, the equity aspect focuses on the distributive and procedural aspect of how those values are realized within a society.

Contemporary water resource planning frameworks put a premium upon normal conditions continuing into the future. This is, in part, also based on the humans socially constructed trait of discounting extreme events and averaging out risks in future planning. Water related hazards and extreme events are not as integral to the standard planning process as they should be. In this context a hazards approach calls to refocus the attention to water related extremes, such as droughts and floods, and also promotes a paradigm shift to replace the contemporary planning frameworks with frameworks treating extremes a part of the normal continuum of human environment relations. A major cause of water related hazards is climate change. Climate change has increasingly become the centre of attention for research and policy around the globe. The increasing understanding of climate related changes and the related more pronounced extreme events has led us to seriously consideration of climate impacts on human populations. One of the key areas of concern worldwide is the impact of climate change on water resources. Climate change and the associated unpredictability of temperature and precipitation will have physical effects on water quantity and quality (IPCC, 2013). Considering the volatile nature of climate change events, the aforementioned effect will also be non-linear and unpredictable (Frederick and Major 1997). There is a need to re-centre water hazards related policy and research to address the vulnerabilities likely to be accentuated in a climate change future. Despite the urgent need to address climate change vulnerabilities related to water, very little work has been done to address these vulnerabilities especially in low income countries/areas (Muller 2007). A paradigmatic shift towards hazards in water resources research will point towards newer pathways for addressing water related challenges and hazards. In this chapter we will see the largely unrealized potential of such a hazards approach to water resource planning in mainstream water research.

## 9.3 Water Social Power and Equity Problematique

The recognition of water as a basic human right has its roots in the 1970s when it emerged on the international agenda (Mirosa and Harris 2012). For example in the 1977 UN Water Conference in Mar del Plata, Argentina, there was a strong emphasis on equitable distribution of water for all. In the 1980s, a general trend began towards investment in water related infrastructure for the provision of clean drinking water. The inadequacy of a dominant infrastructural and physical approach to water provision was soon realized. The critical missing link was the absence of social/human aspect to water supply (Mehta and Mirosa 2004). The 1980s also brought with it the political wave of neo-liberalism. Together with the growing problem of water scarcity in the mid 1980s onwards and the ideological setting of neo-liberalism – a new phenomenon of considering water as an 'economic good' emerged in the 1990s (Swyngedouw 1997; Loftus and McDonald 2001; McDonald and Pape 2002; Bakker 2010, 2003a, b; McDonald and Ruiters 2005). This new shift of realizing water as commodity resulted in the increasing privatization of water management, and decentralization of governance from the state (Kaika 2003; Brannstrom et al. 2004; Budds 2004). Through all this transition, the actual improvement in water quality and progress towards social justice in distribution, particularly for the most needy, has been marginal (Mirosa and Harris 2012). It is against this backdrop – the equity problematique is gaining increased traction with respect to its practical implications and conceptual importance.

Water as a basic human right is the starting point in any discourse on the equitable distribution of water resources. Hence, viewing equity from a human rights perspective – every human has a right to get clean drinking water to sustain healthy living regardless of the ability to pay (Langford 2005). Moreover, water as a human right considers the nation state as the prime entity responsible for the realization of this basic right. Notwithstanding the propriety of considering equity in terms of a basic human right to water, it does pose some difficult questions. Coupled with the human right to water are always complexities in terms of policy questions and political consequences. Most of these complexities stem from the, so called, problem of water scarcity. According to a UNDP report (2006) water scarcity is a consequence of institutional mismanagement and should not be linked, and in effect rather conveniently, to the unequal physical distribution of water resources. Essentially water scarcity is mediated through institutional distortions, therefore the solution will also be found at the institutional level. Equity is a key concept that can help reorient institutions towards addressing socially constructed water scarcity. Hence there is a need to focus on international and local policy level debates to develop mechanisms delivering the human right to water (Parnell 2007).

The 1990s onslaught of neo-liberalism and the accompanied commoditization of water created a new tension between viewing water as an economic good and water as a human right. The shift to privatization, water markets and efficiency centred management paradigms were at the centre of the water related research discourse at the time (Serageldin 1995; Gleick 2000; Rogers et al. 2002). The purpose here is to highlight the effects of privatization of water on the notions of equity and social justice for the most needy. Privatization changes the perspective on water from being a basic human right to a human need mediated by the forces of market demand and supply (Hughes 2010). The result of distributing water based on the ability to pay has limited this indispensable resource to the more affluent urban areas and industries. Mustafa and Reeder (2009) document with an elaborate case of Belize City, Belize, the failed experiment of privatization and its inability to materialize the expected benefits in terms of better quantity and quality of water for the general population. The problem to an extent lies in the separation of nature from society under a liberal framework (Smith 1990). The commercialization of natural resources – such as water – through the introduction of market principles has prioritized and separated the economic efficiency of water from its social equity imperative (Roberts 2008).

Shifting our focus to local issue of water distribution, the discussion revolves around the concept of Hydrosolidarity. The concept of Hydrosolidarity induces, what otherwise has been neglected i.e. notions of equity, fairness, human welfare and conflict reduction in water distribution decisions (Gerlak et al. 2011). Hydrosolidarity aims to understand the dynamic of human control of water flows with the ecological consideration plus incorporating the ethical considerations of equity for the powerless. Operationalizing the concepts of Hydrosolidarity requires adapting integrated management of resources and ensuring equity of information and participation through involving a greater and diverse set of stakeholders (Gerlak et al. 2011). Despite the conceptual attraction of Hydrosolidarity, the practical application will require supportive water laws, policy framework, embedded incentives, and strong institutions (Jagerskog 2002; Wouters 2000). Water issues need to be assessed, reviewed and resolved in an overall societal and developmental context (Biswas 2008). Appropriate incentives coupled with a robust legal institutional infrastructure will be needed to actualize the positive results of Hydrosolidarity.

Taking the discussion to a more practical basis, one of the most widely cited and implemented approach to incorporate equity into water resource management is Integrated Water Resource Management (IWRM). The philosophy of IWRM constrains the definition of equity and equitable water distribution to the ultimate objective of sustainability. IWRM does not incorporate equity in a social justice perspective. According to Hefny (2009), IWRM should be underpinned by an ethical framework recognizing cultural rights from a social equity angle. Taking the discussion forward, a similar concept made popular in the 1990s, linking water to local level development and incorporating the concept of equitable distribution is 'sustainable development'. Sustainable development as an approach takes into account three major factors: social, environmental and economic (Mukheibir 2010). Demands of an equitable distribution should not only benefit (even partially) from water resource management frameworks like Hydrosolidarity or IWRM but also an approach like sustainable development – rigorously defined and politicized – which aims to link all three aspects of society, economy and the environment.

Discussions around an equitable distribution at the local level, while calling for greater participation of stakeholders, empowerment and better governance, is subservient to the differentials in social power at the local level. Michel Foucault defines power as compelling force wielded through knowledge, discourse and behaviours (Foucault 1980). Power through knowledge can define what is normal, acceptable or deviant. Power has a central role to play in equity and social justice with respect to water resource distribution. Power plays can be realized through a range of sources such as social capital, political clout, economic means or geographic position. These sources define the rules of the game and categorize what is normal, fair and just from perspective of the powerful. According to Mustafa (2002) power relations play a vital role in determining the distribution and management of water resources. Here power refers to the ability to act, dictate and influence an outcome into ones favour. Looking from the power-equity nexus the social power differentials will have very real geographical outcomes in the perspective of equitable water resource distribution.

Exploring the literature on water, power and equity, we cannot complete the discussion without looking at gender as a key cross cutting area. There is a need to realize the basic human right to water at the local level to make a significant difference in the lives of women. Challenges impeding an equitable distribution of water for women involve local level factors such as cultural, religious, institutional and social norms. Water rights, an important ingredient of effective allocation and distribution of water resources, were a key pillar of water distribution alongside institutions and infrastructure (Zwarteveen 1997). A gendered approach to water is needed to achieve an equitable allocation of water for women.

Moving the discussion towards an international and regional perceptive, power plays are some of the most salient undercurrents determining the basic equity questions of who gets access to water resources, and to their degree of control over the resources. These kinds of asymmetric power relations and the resulting exploitations have been captured in the term hydro-hegemony. Hydro-hegemony attempts to explain how groups of powerful entities (hegemons) tend to maintain their control over the distribution of shared water resource and their ability to define the 'rules of the game' (Zeitoun and Warner 2006). Also in determining water sharing mechanisms across national boundaries, it is important for policy makers to consider the asymmetric power relations. A common theme in asymmetric power relations is the ability of the more powerful riparian to secure and sustain a favourable (inequitable) share of transboundary waters. Hence the overall unfortunate outcome of these hydro-hegemonic relations is the absence of principled and equitable sharing of water resources (Zeitoun and Allan 2008). For practitioners and policy makers to address transboundary equity issues, it is absolutely critical to understand and consider social power and its influence on sharing water resources.

This section has attempted to highlight key topics in international literature addressing the equity void in water related discourse. Equitable access to water as a basic human right is closely linked to the human right to life, health and food. Despite the immense significance of water for life; the research on the interplays of water with notions of social justice and equity is contested and emergent at best. A lot has been written about the valence and importance of the idea but a methodical attempt to embed it into an operational framework is missing. The logical consequence is the promotion of inequalities through the current forms of institutional arrangements governing the use and distribution of waters. Overall, equity concerns are being neglected at our own peril – till they become a large scale problem. This might be primarily because currently equity issues are only felt by the power less and vulnerable sections of the society. The social power realized through water is important in analyzing water sharing relations and how operational concepts such as equity are addressed to secure the rights of the most vulnerable.

## 9.4 From Livelihoods and Health to Access to Satisfy Multiple Values

Water is not a commercial product to be traded and used devoid of social values. Indeed the commercial aspect of water is itself a deeply social process, imbricate with complex patterns of historic structures and political economy. Humans and by extension the society, does not only need water for health and livelihood purposes, but they also need water to satisfy multiple uses i.e. from a social, cultural, esthetic and spiritual perceptive (Pradhan and Meinzen-Dick 2003). A narrow approach to water management and governance without accounting for the multiple values, a community puts on water, will lead to outcomes with high social costs. To worsen the problem, there is a general lack of research attention on the linkages between the multiple values a society derives from water and access to water resources (Finn and Jackson 2011). This also highlights a void in the understanding of policy makers and by extension key water policy documents to satisfy the full spectrum of values realized from the access to water.

One of the primary uses of water is for health purposes. This includes clean water for drinking use and water for sanitation and hygiene uses. As part of a broader strategy to meet basic needs of the poor, for many years there has been a focus by the governments to achieve better health benefits through the provision of clean drinking water (Nicol 2000). Failure to access safe and hygienic source of water is a basic form of deprivation, and it threatens human life and undermines human dignity.

According to the World Water Development Report – giving the poor access to better managed water services can make a substantial contribution to poverty reduction (WWAP 2003). This assertion captures well the link between equity in water and complex multidimensional concept of poverty. Extending the argument we can identify water as one of the key determinants of a person's livelihood. The word livelihood here refers to the person's asset profile, risk and challenges they face and the institutional environment they survive in (Hope and Gowing 2003). The role water plays in the livelihood of the poor is mediated by the person's access to water for agriculture, food, livestock and other modes of income generation.

The above discussion gives a flavour of the type of topics addressed when we look at access. Access can be defined as the ability to benefit from a thing – water in this instance (Ribot and Peluso 2003). Despite the importance of health and livelihood benefits derived from water, the current literature is lacking in addressing a range of values the society puts on water. More broadly, access implies a concern with ability of different actors to access water for multiple uses, ranging from the obvious livelihood and health uses to aesthetic, cultural, community, ecological and spiritual values – a society expects to realize from water. This can be explained through the interaction of Foucault's concept of power with access to water. Power as knowledge, discussed in the previous section, has prioritized the health and livelihood uses of water as the ones most pertinent to wield power. This power attracting potential of livelihood and health uses of water has neglected the communal, spiritual, aesthetic and cultural uses of water.

Water has cultural and spiritual significance for the indigenous populations who build their societies around water. Disputes about water are invariably value laden and thus cannot be mediated through recourse to scientific discourse or supply side infrastructural solutions alone (Hoekstra 1998). It is also essential not to split the relationship between land and water as done by most of the Western classifications of the natural world (Sheehan 2001). In reality water plays a very significant role in the lives, religious beliefs and cultural practices of many societies (Pradhan and Meinzen-Dick 2003). Despite the age old development of these cultural and spiritual values, only few researchers have shown interest in this aspect of water access (Strang 2004). This cultural and spiritual relationship between water and society has been understated in water related research and almost neglected in water management and governance (Pradhan and Meinzen-Dick 2003).

Water is part of an ecosystem and needs to be protected, valued and used with due care. Concerns with water do not end with the quality of water itself but also with the health of the environment it serves. Water has the ability to sustain life and hence all life depends on it. We as a society need access to water and to secure it special attention is required to protect the surrounding ecology. Taking this step further – protecting the ecology is one of the key values a society needs to satisfy for sustainable access to water – particularly for the poor.

This section has attempted to highlight the importance of a range of values associated with the access to water other than the usual health and livelihood uses. Human society and water resource related issues are linked at multiple levels. Decisions regarding water resource distribution and giving access to water for any society should consider the cultural, spiritual and ecological value of water before any consideration of commercial or economic interests (Moench et al. 2003). Water development, devoid of social values, globally results in many unrecognized side effects like destruction of local aquatic habitats, displacement of communities, severing of hydro-social cultural ties of the dependent indigenous community etc. In reality the extent to which societal values and interest are included in formal water laws and distribution mechanisms remains extremely limited. Water management and governance agencies should broaden their capacities and consider the multiple uses of water, as discussed above, while providing access to water for its users.

## 9.5 Destabilizing Normality: From Physical Hazards to the Hazards Approach

There is a weak link, at best between water related hazards and water planning. Water management decisions consider by default the normality condition continuing into the future. This status quo has been maintained for too long and is etched into the thinking of water planners. Furthermore, the elite and powerful either live protected from environmental hazards (floods and droughts) or have the resources to guard themselves from the negative consequences of these hazards (Mustafa and Reeder 2009). Here the concept of power knowledge can, in part, explain the lack of focus on hazards in water resource planning. Hazards and extreme events are conveniently ignored to the detriment of the powerless.

Water related hazards are composed of a population's physical exposure to risk and also its social and economic vulnerability (Wisner et al. 2004). Both physical and socio-economic vulnerabilities are important to address in developing a hazards approach (Adger 2006). The physical vulnerabilities relate with the physical location of the human population at risk and the availability and use of resources (Burton et al. 1993). In managing physical vulnerabilities, the technical and institutional factors mediate resilience to hazards. Looking from a purely physical aspect to managing hazards, the risk level of a population will only depend on its geographic characteristics. The actual composition of the society and economic conditions will not have any effect on the population's vulnerability. To build a more holistic approach, we need to incorporate the socio-economic vulnerabilities into developing a hazards approach. The socio-economic vulnerabilities relate to the social and political aspect of a human population as risk (Cutter 1996). A population consisting of the poor and marginalized will demonstrate very different resilience to hazards compared to an economically robust society (Hewitt 1983; Watts 1983). Likewise women in the society are considered to be more vulnerable to hazards due to their restricted movement and domestic roles (Fordham 2003). Incorporating both physical and socio-economic aspects of vulnerability, a hazards approach will need to address deficient knowledge, lack of social learning, substandard technology, lack of political power, underdeveloped social capital, frail infrastructure, cultural barriers, income disparities, gender inequalities and marginalized section of the society (Cutter et al. 2003; Cutter 2001; Tierney et al. 2001; Putnam 2000).

Reactive responses to water related hazards are not uncommon. To move away from a 'reactionary response' approach, efforts are directed to understand the link between human societies and critical resources such as water. The aim is to reduce vulnerabilities, incorporate adaptation and enhance the resilience of this interlinked hydro-social system. Water related hazards such as floods and droughts pose significant risks. The increased awareness of disasters and the high risks involved has prompted a general trend towards management by social-learning and adaptation in different fields, as opposed to controlling complex and unpredictable ecosystem (Pahl-Wostl 1995, 2005; Levin 1998; Hartvigsen et al. 1998; Berkes et al. 2002). Adaptation has usually been considered as a defensive mechanism in a technological sense without any social and political associations. According to Pelling and Manuel-Navarrete (2011), adaptation should be seen as a progressive phenomenon with political, social and cultural considerations for the society. Incorporating adaptation in policy should foster socio-political and economic development; improve power relations; built ecological sensitivity; cater for the interest of future generations and protect the vulnerable and marginalized members of the community. Such an approach to adaptation is absent in the mainstream water related research literature and not practiced in water resource planning and management. There is a need to understand the complex nature of water related hazards and extreme events in order to adapt and plan better in our water resource management efforts. There is a need to think outside the box by considering alternate futures through exploratory modelling and building robust adaptation strategies based on foresight and flexibility (Lempert et al. 2003). The normality paradigm in water resource planning is no longer functional in the context of water related hazards and hence a hazards approach to water planning and governance is required to better deal with hydrologic extremes and idiosyncrasies.

Climate change is one of the key determinants of water related hazards like floods and droughts. According to Arnell and Gosling (2013), climate change impacts on water scarcity has proven to be very sensitive to the changing climatic scenarios. However, the effect of climate change on the water (hydrological) cycle has received relatively less attention in the literature (Stocker and Raible 2005). This focus needs to change to realize the critical link between climate change hazards and water availability. Climate change and associated global warming will have significant effects on the hydrological cycle globally (IPCC, 2013). Sudden flooding, prolonged droughts, coastal inundation and similar unexpected environmental events indicate the nonlinear, unpredictable and extreme nature of climate change impacts (IPCC, 2007). In the face of this unpredictability, we do not have the luxury to predict and plan for future water resource availability simply by extrapolating averages. Hence, water management and planning need to view climate change as a new reality and take into account the associated unpredictability.

The dominant approach on adaptation has focused on the social actor(s) as the primary agent responding to hazards and working towards reducing vulnerability; while the resilience approach on the other hand takes a broader systems view in building the capacity to resist change and retain the original functional form (Nelson et al. 2007). Adaptation, more holistically, refers to the capacity of a social actor or system to show resilience to a water hazard event and then adapt to prepare for future such events (Adger 2006; Folke 2006). It includes the notions of decreasing vulnerability, enhancing resilience/robustness and then transforming in response to an adverse climatic event (Gallopin 2006; Smit and Wandel 2006; Gober 2013). A hazards approach to water resource planning and management thus should incorporate both adaptation and resilience in addressing these aforementioned water hazards and related uncertainties. A key ingredient in developing a holistic hazards approach is to develop Institutional mechanisms as a complement to infrastructural developments (Muller 2007).

If we look at water resources planning from a hazards approach another important area in research literature deals with the role of social learning in adaptation. Impacts from hazards are also a function of historically determined social systems and their ability to learn and reduce vulnerabilities. This ability to learn and reduce vulnerabilities necessitates demands on social learning to be a vital component of hazards approach to water resource planning. Social learning can be defined as learning within social groups or a society through peer to peer social interaction (Haas 2004; Siebenhuner 2008). The scale of learning is societal and hence social learning usually deals with transforming societal values, assumptions and worldviews as opposed to individual values and beliefs. The water ecology and the related climate change hazards are too complex to predict. Keeping in view the complex nature of climate change, social learning is expected to play a key role in changing and revising overall societal understanding of water related hazards and facilitate in better water planning and management (Berkes 2009; Muro and Jeffrey 2008; Pahl-Wostl and Hare 2004). Social learning changes the focus from simple prediction to a planned approach based on adaptation (Pahl-Wostl 2007). Reaching out to stakeholders at the community level is important to coordinate any water related planning. In the process of social learning, the stakeholders are empowered to step forward and take responsibility of appropriate actions in building resilience (Paavola and Adger 2006).

This section has attempted to highlight the shortcoming of a historical approach to water planning and management based on forecasting averages. The assumption of a linear trend in water resource planning does not hold true anymore. The advent of climate change has changed the world in unexpected ways. Now the importance of factoring in the onslaught of climate change is critical in adapting our responses to the upcoming water related hazard(s). There is enough evidence for us to start thinking about the inherent uncertainties and unexpected nature of climatic events. Water is a vital resource on all counts and with population pressures and an increasing problem of water scarcity, it is vital to manage efficiently and effectively the water resources available to us. In addition to managing our water resources we also need to build the capacity of an area or community to cope up with the destructive force of water. Hence to aim for better water resource management and protecting ourselves from water related vulnerability we need a comprehensive hazards approach to water (Pahl-Wostl 2007). The hazards approach will aim to build resilience and include adaptation strategies by incorporating both physical and socio-economic vulnerabilities with the ultimate aim of improved water resource planning /management and better coping with water related climate change disasters. With a comprehensive hazards approach we can better respond to the existing water related vulnerabilities and better plan for future water demands in a world with increasing populations and climate change.

## 9.6 Conclusion and Identifying Gap

The chapter has attempted to highlight key trends in water resource literature when looking from our analytical lens of access, equity and hazards. We have also used Foucault's concept of Power/Knowledge to explore the three thematic foci of Access, Equity and Hazards. The confluence of power through power knowledge is a slow process – but once the goal post, as to what is normal, is defined it is very difficult to challenge the status quo.

Understanding the influence of 'power' on water resource management in the context of access, equity and hazards; and following the argument made in the chapter – we have identified three significant gaps in water related literature. First, despite a significant amount of research literature on notions of equity and social justice in water resources, the concept has not been internalized or operationalized in policy. Criteria of equity have to be embedded in water resource management and governance to realize on ground outcomes. Second, there is an over emphasis on health and livelihood as the only uses derived from the access to water. Multiple values a society derives from water receives sparse attention and research foci at best. Third, hazards are dealt only as episodic events and not addressed as part and parcel of the planning process. This translates into an absence of a holistic hazards approach to water resource planning; i.e., the need to develop a resilience thinking approach and integrating proactive adaptation strategies in our response to water related climate change hazards.

First, the chapter argues for a move towards main streaming of equity issues in water resource management and a consensus on the status of water as being a basic human right. Despite the progress, we see scant efforts/a gap in terms of research on topics relating to equity in water resources and a focus on actualizing the idea of an equitable distribution. The notions of equity and social justice are of immense importance for the needy and the marginalized. Be it an individual struggling within an unresponsive political system or a nation contesting against the more powerful riparian in a bid to secure water rights; in both cases water is a vital lifeline and hence needs to be managed/governed on the principles of equity and justice.

Second, the chapter brings to our attention the gap pertaining to access to water for multiple uses and the over emphasis of livelihoods and health as the two main uses of water. Looking beyond the obvious uses of water for health and livelihood purposes there is not much emphasis on the multiple of uses, a society derives from access to water. This lack of emphasis on taking a broader view of the hydro-socio relationship has resulted in water resource planning and development devoid of any social considerations. Hence, it is important to consider the multiple values a society recognizes with water and incorporate them in future water development and planning.

Third, the chapter calls for a hazards approach to water resource management. Conservation of water, sustainable management and protecting ourselves from water related hazards are needed in a world impacted by climate change. The hazards approach defined in this chapter incorporates resilience thinking and adaptation strategies to reduce vulnerabilities of climate change associated with water. We need to protect ourselves from water related hazards and also at the same time understand the complexities induced from climate change impacts on our future water resources. To address both these needs, research efforts are required to incorporate a hazards approach to water resource management, thus reducing the climate change vulnerabilities related to water and improving planning for our future water needs.

## References

Adger WN (2006) Vulnerability. Glob Environ Chang 16:268–281
Arnell NW, Gosling SN (2013) The impacts of climate change on river flow regimes at the global scale. J Hydrol 486:351–364
Bakker K (2003a) An uncooperative commodity: privatizing water in England and Wales. Oxford University Press, Oxford
Bakker K (2003b) Archipelagos and networks: urbanization and water privatization in the South. Geogr J 169:328–341
Bakker K (2005) Neoliberalizing nature? Market environmentalism in water supply in England and Wales. Ann Assoc Am Geogr 95(3):542–565
Bakker K (2010) Privatizing water: governance failure and the world's urban water crisis. Cornell University Press, Ithaca
Bates SF, Getches DH, MacDonnell LJ, Wilkinson CF (1993) Searching out the headwaters: change and rediscovery in western water policy. Island Press, Washington, DC
Berkes F (2009) Evolution of co-management: role of knowledge generation, bridging organizations and social learning. J Environ Manag 90:1692–1702
Berkes FL, Colding J, Folke C (2002) Navigating social-ecological systems: building resilience for complexity and change. Cambridge University Press, Cambridge
Biswas AK (2008) Integrated water resources management: is it working? Int J Water Resour Dev 24(1):5–22
Brannstrom C, Clarke J, Newport M (2004) Civil society participation in the decentralisation of Brazil's water resources: assessing participation in three states. Singap J Trop Geogr 25:304–321
Budds J (2004) Power, nature and neoliberalism: the political ecology of water in Chile. Singap J Trop Geogr 25:322–342
Burton I, Kates RW, White GF (1993) The environment as hazard. Guilford, New York
Cutter SL (1996) Vulnerability to environmental hazards. Prog Hum Geogr 20:529–539
Cutter SL (2001) American hazardscapes: the regionalization of hazards and disasters. Joseph Henry Press, Washington, DC
Cutter SL, Boruff BJ, Shirley WL (2003) Social vulnerability to environmental hazards. Soc Sci Q 84:242–261
Finn M, Jackson S (2011) Protecting indigenous values in water management: a challenge to conventional environmental flow assessments. Ecosystems 14:1232–1248
Folke C (2006) Resilience: the emergence of a perspective for social-ecological systems analyses. Glob Environ Chang 16:253–267
Fordham M (2003) Gender, disaster and development: the necessity for integration. In: Pelling M (ed) Natural disasters and development in a globalizing world. Routledge, London, pp 57–74
Foucault M (1980) Power/knowledge: selected interviews and other writings, 1972–1977. Pantheon Books, New York
Frederick KD, Major DC (1997) Climate change and water resources. Clim Chang 37:7–23
Gallopin G (2006) Linkages between vulnerability, resilience and adaptive capacity. Glob Environ Chang 16:293–303
Gerlak AK, Varady RG, Petit O, Haverland AC (2011) Hydrosolidarity and beyond: can ethics and equity find a place in today's water resource management? Water Int 36:251–265
Gleick P (2000) The changing water paradigm: a look at twenty first century water resources development. Water Int 25:127–138
Gober P (2013) Getting outside the water box: the need for new approaches to water planning and policy. Water Resour Manag 27:955–957
Haas PM (2004) When does power listen to truth? A constructivist approach to the policy process. J Eur Public Policy 11:569–592
Hartvigsen G, Kinzig A, Peterson G (1998) Use and analysis of complex adaptive systems in ecosystem science. Ecosystems 1:427–430

Hefny M (2009) Water management ethics in the framework of environmental and general ethics: the case of Islamic water ethics. In: Llamas R, Martinez-Cortina L, Mukherji A (eds) Water ethics. Taylor & Francis, London, pp 25–45
Hewitt K (1983) Interpretations of calamity from the viewpoint of human ecology. Allen and Unwin, Boston
Hoekstra A (1998) Appreciation of water: four perspectives. Water Policy 1:605–622
Homer-Dixon TF (1994) Environmental scarcities and violent conflict: evidence from cases. Int Secur 19(1):5–40
Hope RA, Gowing JW (2003) Managing water to reduce poverty: water and livelihood linkages in a rural South African context. Alternative Water Forum. University of Bradford, Centre for International Development, Bradford
Hughes RA (2010) Pro-justice ethics, water scarcity, human rights. J Law Relig 25:521–540
IPCC (2007) Climate change 2007: the physical science basis. Contribution of Working Group I to the fourth assessment report of the Intergovernmental Panel on Climate Change [Solomon, S., D. Qin, M. Manning, Z. Chen, M. Marquis, K.B. Averyt, M. Tignor and H.L. Miller (eds.)]. Cambridge University Press, Cambridge/New York, 996 pp
IPCC (2013) Climate change 2013: the physical science basis. Contribution of Working Group I to the fifth assessment report of the Intergovernmental Panel on Climate Change [Stocker, T.F., D. Qin, G.-K. Plattner, M. Tignor, S.K. Allen, J. Boschung, A. Nauels, Y. Xia, V. Bex and P.M. Midgley (eds.)]. Cambridge University Press, Cambridge/New York, 1535 pp
Jagerskog A (2002) Contributions of regime theory in understanding interstate water cooperation lessons learned in the Jordan River Basin. In: Turton A, Henwood R (eds) Hydropolitics in the developing world: a South African perspective. African Water Issues Research Unit, Pretoria, pp 73–78
Kaika M (2003) Constructing scarcity and sensationalizing water politics. Antipode 35:919–954
Langford M (2005) The United Nations concept of water as a human right: a new paradigm for old problems? Int J Water Resour Dev 21:273–282
Lempert RJ, Popper SW, Bankes SS (2003) Shaping the next one hundred years: new methods for quantitative, long-term policy analysis. RAND Corporation, Santa Monica
Levin SA (1998) Ecosystems and the biosphere as complex adaptive systems. Ecosystems 1:431–436
Lightfoot DR (1996) Moroccan Khettara: traditional irrigation and progressive desiccation. Geoforum 27(2):261–273
Llamas R (2003) Ethical considerations in water management systems. Water Nepal 10:13–27
Loftus A, McDonald DA (2001) Of liquid dreams: a political ecology of water privatization in Buenos Aires. Environ Urban 13:179–199
Mageed YA, White GF (1995) Critical analysis of existing institutional arrangements. Int J Water Resour Dev 11(2):103–112
McDonald DA, Pape J (2002) Cost recovery and the crisis of service delivery in South Africa. Zed Books, London
McDonald DA, Ruiters G (2005) The age of commodity: water privatization in Southern Africa. Earthscan, London
Mehta L, Mirosa O (2004) Financing water for all: behind the border policy convergence in water management. Institute of Development Studies, Brighton
Michel AA (1967) The Indus rivers: a study of the effects of partition. Yale University Press, New Haven
Mirosa O, Harris LM (2012) Human right to water: contemporary challenges and contours of a global debate. Antipode 44:932–949
Moench M, Dixit A, Caspari E (2003) Water, human rights and governance: issues, debates and perspectives. Water Nepal 10:1–9
Mukheibir P (2010) Water access, water scarcity, and climate change. Environ Manag 45:1027–1039

Muller M (2007) Adapting to climate change water management for urban resilience. Environ Urban 19:99–113
Muro M, Jeffrey P (2008) A critical review of the theory and application of social learning in participatory natural resource management processes. J Environ Plan Manag 51:325–344
Murray-Rust H, Lashari B, Memon Y (2000) Water distribution equity in Sind province. International Water Management Institute, Colombo
Mustafa D (2002) To each according to his power? Participation, access, and vulnerability in irrigation and flood management in Pakistan. Environ Plan D: Soc Space 20:737–752
Mustafa D (2013) Water resource management in a vulnerable world. I.B. Tauris, New York
Mustafa D, Reeder P (2009) 'People is all that is left to privatize': water supply privatization, globalization and social justice in Belize City, Belize. Int J Urban Reg Res 789–808
Nelson DR, Adger WN, Brown K (2007) Adaptation to environmental change: contributions of a resilience framework. Annu Rev Environ Resour 32:395–419
Nicol A (2000) Adopting a sustainable livelihoods approach to water projects: implications for policy and practice. Overseas Development Institute, London
Paavola J, Adger NW (2006) Fair adaptation to climate change. Ecol Econ 56:594–609
Pahl-Wostl C (1995) The dynamic nature of ecosystems: chaos and order entwined. Wiley, Chichester
Pahl-Wostl C (2005) Towards sustainability in the water sector: the importance of human actors and processes of social learning. Aquat Sci 64:394–411
Pahl-Wostl C (2007) Transitions towards adaptive management of water facing climate and global change. Water Resour Manag 21:49–62
Pahl-Wostl C, Hare M (2004) Processes of social learning in integrated resources management. J Commun Appl Soc Psychol 14:193–206
Parnell S (2007) Urban governance in the South: the politics of rights and development. In: Cox K, Louw M, Robinson J (eds) A handbook of political geography. Sage, London, pp 1857–1876
Pelling M, Manuel-Navarrete D (2011) From resilience to transformation: the adaptive cycle in two Mexican urban centers. Ecol Soc 16(2):11
Pradhan R, Meinzen-Dick R (2003) Which rights are right? Water rights, culture, and underlying values. Water Nepal 9/10:37–61
Putnam RD (2000) Bowling alone: collapse and revival of the American community. Simon & Schuster, New York
Ribot JC, Peluso NL (2003) A theory of access. Rural Sociol 68:153–181
Roberts A (2008) Privatizing social reproduction: the primitive accumulation of water in an era of neoliberalism. Antipode 40:535–560
Rogers P, de Silva R, Bhatia R (2002) Water is an economic good: how to use prices to promote equity, efficiency, and sustainability. Water Policy 4:1–17
Serageldin I (1995) Water resources management: a new policy for a sustainable future. Water Int 20:15–21
Sheehan J (2001) Indigenous property rights and river management. Water Sci Technol 43:235–242
Siebenhuner B (2008) Learning in international organizations in global environmental governance. Glob Environ Polit 8:92–116
Smith N (1990) Uneven development: nature, capital and the production of space, 2nd edn. Basil Blackwell, Oxford/Cambridge
Smit B, Wandel J (2006) Adaptation, adaptive capacity, and vulnerability. Glob Environ Chang 16:282–292
Stocker TF, Raible CC (2005) Climate change: water cycle shifts gear. Nature 434:830–833
Strang V (2004) The meaning of water. Berg, Oxford
Swyngedouw E (1997) Power, nature, and the city. The conquest of water and the political ecology of urbanization in Guayaquil, Ecuador. Environ Plan 29:311–332
Swyngedouw E (1999) Modernity and hybridity: nature, regeneracionismo, and the production of the Spanish waterscape, 1890–1930. Ann Assoc Am Geogr 89(3):443–465

Syme GJ, Nancarrow BE (2008) Justice and the allocation of benefits from water. Soc Altern 27:21–25

Tierney KJ, Lindell MK, Perry RW (2001) Facing the unexpected: disaster preparedness and response in the United States. Joseph Henry Press, Washington, DC

Watkins K (2006) Human development report 2006 beyond scarcity: power, poverty and the global water crisis. UNDP, New York

Watts M (1983) On the poverty of theory: natural hazards research in context. In: Hewitt K (ed) Interpretations of calamity for the viewpoint of human ecology. Allen and Unwin, Boston, pp 231–262

Wescoat JL (1987) The 'Practical Range of Choice' in water resources geography. Prog Hum Geogr 11:41–59

Wescoat J, Halvorson S, Mustafa D (2000) Water management in the Indus basin of Pakistan: a half-century perspective. Int J Water Resour Dev 16(3):391–406

White GF (1968) With the committee, water and choice in the Colorado Basin: an example of alternatives in water management. National Research Council Committee on Water, National Academy of Sciences, Washington, DC

Wisner B, Blaikie P, Cannon T, Davis I (2004) At risk. Routledge, London

Wouters P (2000) The relevance and role of water law in the sustainable development of freshwater: from "hydrosovereignty" to "hydrosolidarity". Water Int 25(2):202–207

WWAP (2003) Water for people, water for life. United Nations World Water Assessment Programme. UNESCO, Paris

Zeitoun M, Allan JA (2008) Applying hegemony and power theory to transboundary water analysis. Water Policy 10:3–12

Zeitoun M, Warner J (2006) Hydro-hegemony – a framework for analysis of trans-boundary water conflicts. Water Policy 8:435–460

Zwarteveen MZ (1997) Water: from basic need to commodity: a discussion on gender and water rights in the context of irrigation. Gend Prop Rights 25:1335–1349

# Chapter 10
# Sustainable Development and Coastal Disasters: Linking Policies to Practices

**Rajarshi DasGupta and Rajib Shaw**

**Abstract** Coastal areas are the driver of economic development, yet, over the last decade, more than 60 % of disaster losses have been confined to the coastal area. While on the one hand, the world has witnessed some deadly geo-hydro hazards in the past decade, rapid onset of climate change is also likely to intensify a range of hydrometeorological disasters impairing the world's coast. Undoubtedly, this has severe developmental consequences. Furthermore, continual coastward migration, land development, urbanization, and loss of vital ecosystem services trigger a new set of worries in order to sustainably develop and manage coastal areas. Against this backdrop, this chapter highlights the key linkages between sustainable development and disaster risk reduction in coastal areas and addresses the emerging challenges in coastal zone management. The chapter provides an integrated framework and rationale for sustainable development of coastal areas with effective incorporation of disaster risk reduction. In conclusion, the chapter identifies some of the existing hindrances and provides an advocacy of ecosystem-based risk reduction in coastal areas.

**Keywords** Sustainable development • Coastal zone management • Coastal disasters • Disaster risk reduction

## 10.1 Introduction

For the last three decades, the idea and notion of sustainable development were mostly concentrated to environmental conservation and practices that are linked to effective resource utilization. As a result, global communities expected the

R. DasGupta (✉)
International Environment and Disaster Management Research Group,
Graduate School of Global Environmental Studies, Kyoto University,
Yoshida Honmachi, Sakyo-ku, Kyoto 606-8501, Japan
e-mail: rajarshidg1@yahoo.co.in

R. Shaw
Graduate School of Global Environmental Studies, Kyoto University, Sakyo-ku, Kyoto, Japan
e-mail: shaw.rajib.5u@kyoto-u.ac.jp



J.I. Uitto, R. Shaw (eds.), *Sustainable Development and Disaster Risk Reduction*,
Disaster Risk Reduction, DOI 10.1007/978-4-431-55078-5_10

industrial and production sectors to perform at a higher efficiency and to conserve natural resources as much as reasonably possible. However, as argued by Pearce et al. (1989), the idea of sustainable development sounds so impressive that everyone has his own expectations. One good reason for this has been the broadness of the terminology which does not restrict to link almost all crucial human elements to such concept (Giddings et al. 2002). In fact, the process of addressing environmental and socioeconomic concerns under the same umbrella was advocated by the Brundtland Commission itself. Consequently, over the last three decades, the broader context of sustainable development heavily differed among places, governments, and communities. While in developed countries, industrial sectors received the majority of limelight, and serious effort has also been made to interlink a host of socio-economic issues, including poverty reduction, gender balance, and quality of life. Recently, the sustainable development concept received further attention when the Rio+20 Conference agreed to develop the sustainable development goals (SDGs) which would replace the existing Millennium Development Goals in post-2015. Needless to mention, the proposed goals would encompass a variety of environmental, social, and economic issues including poverty reduction, education, health, and sustainable production vis-à-vis consumption pattern, although the SDGs are still being debated in various forums; however, at the time of writing this chapter, the proposed 17 SDGs exclude "disaster risk reduction" as a major component or as a stand-alone indicator in its post-2015 development agenda.

Disasters, whether natural or man made, can be catastrophic in the ongoing development process. They have the potential to rapidly jeopardize a wide range of social and economic aspirations staggering developmental priorities. The world has witnessed some of the historic mega-disasters with havoc loss of lives and properties in the last decade. From the Indian Ocean Tsunami in 2004 to the East Japan Earthquake and Tsunami in 2011 to Typhoon Haiyan in 2013, disasters strongly evolved as an important component that requires immediate and adequate attention. In all the cases, local governments suffered huge moral and economic damage that had strong implications to the country's GDP as well as other developmental aspirations. Although the adoption of the Hyogo Framework for Action in 2005 has emphasized the need for integrating DRR as national priority, the outcome, so far, is far behind the expectations. Therefore, while the global communities remain vastly occupied in drafting the new sustainable developmental goals and agendas, it is perhaps highly imperative to include the disaster risk reduction component into the mainstream developmental agenda. In line with such agreements, this chapter provides an advocacy by establishing the strong and direct linkages between disaster, environment, and developmental goals against the backdrop of coastal areas. The chapter further attempts to explore the shortcomings of past coastal development strategies, examines their environmental and social sustainability, and recommends future strategies for sustainable development of the coastal area.

## 10.2 The Significance of the World's Coast

The Earth is essentially a coastal planet with 1,634,701 km long coastline that marks the boundary between the land and the sea (Burke et al. 2001; Martínez et al. 2007). Human civilization evolved from the coast, and over the years, it attracted huge population due to its enormous biological diversity and ease of mobility. It is well known today that the world is extremely crowded along its coasts despite the fact that there are contradicting opinions about the exact population size in coastal areas. Moreover, the definition of coastal zone varies extensively based on the local geomorphology and geographic features. Yet, on a rough estimation, approximately 41 % of the world population lives within the 100 km from the coast (Martínez et al. 2007). This is apparently for the benefits that the coastal areas provide, ranging from its huge ecological and economical productivity to ease of navigation and industrial development. For example, a study conducted by Fleisher and Chen (1997) in China revealed that the non-coastal provinces are twice less productive than the coastal provinces despite high investment rates relative to provincial GDP. This is often associated with high rate of foreign investment, skilled manpower, and job opportunities in coastal areas. Consequently, many countries observed huge coastward migration over the past. Some of the world's most crowded cities are therefore located near the coast. As a matter of fact, 12 of the world's 16 cities with populations more than ten million are located within 100 km from the coast (Nicholls et al. 2007). World's coastal population is further estimated to rise up to 3.1 billion by the end of 2025, an increase of approximately 34 % from its earlier assessment in 2000 (Duxbury and Dickinson 2007). However, despite the rich coastal services, there are also huge disadvantages for living in the coastal areas. Because of a variety of natural processes such as coastal erosion, tropical storms, flooding, saltwater intrusion, subsidence, and tsunamis, coastal areas across the world are critically exposed to a range of natural hazards. For example, hydro-meteorological disasters such as flooding, storm surges, and wet mass movement constitute the largest share accounting for almost 90 % of total catastrophic events encountered during the last decade (Helmer 2006). Importantly, much of the impact of such disasters has been concentrated along the coast. The two most costly disasters in recent years, i.e., Hurricane Katrina (2005) and East Japan Earthquake and Tsunami (2011), have also a coastal connection. Therefore, coastal dwellers remain critically exposed to multiple hazards compared to the inland dwellers (Nicholls and Small 2002). Furthermore, climate change is expected to compound the problems by massive alteration of the existing geological settings. Especially, small islands, be it in the tropics or at higher latitudes, have the characteristics which make them exceptionally vulnerable to the effects of climate change (Mimura et al. 2007). Climate change is also believed to impact the trends and patterns of natural disasters and largely in a negative way. For example, the mean maximum speeds of tropical cyclones are predicted to increase by 2–11 % by the next century; rainfalls are likely to intensify by 20 % within 100 km of the tropical cyclone center (Knutson et al. 2010). Additionally, scientists estimated an accelerated rise ($3.3 \pm 0.4$ mm/year) in global sea level since 1993 compared to their earlier estimation (1.7 ± 0.3 mm/year). Despite certain amount of suspicions, it is believed that global sea level will continue to rise and may increase by 30–180 cm by the end of 2100 (Nicholls and Cazenave 2010). Undoubtedly, such predictions would considerably vary across regions; however, coastal areas, particularly small island communities, would find it extremely challenging to adapt to such adverse changes. In addition, discontinuation of coastal ecosystem services due to land conversion, water withdrawal, industrial pollution, and overharvesting will result in a complex socio-environmental problem that is bound to magnify disaster impacts. Consequently, in the next few years, coastal managers and policy planners may find it extremely difficult to address a wide set of developmental challenges covering both natural and human dimensions.

## 10.3 Human Impact on the Coastal Areas

Human settlements in the coastal areas historically depended on the coastal ecosystem services for a wide range of products. Apart from that, coastal ecosystems have huge role to play in climate change mitigation. For example, the quaintly carbon captured by the coastal ecosystems such as mangroves, salt marshes, and sea grass bed is estimated to 120–130 million tonnes/annum (http://www.cbd.int/gbo3/?pub=6667andsection=6709). The amount of captured carbon is equivalent to the greenhouse gas emitted by Japan. However, over the years, unsustainable human practices led to tremendous adverse impacts on coastal ecosystems and their functioning. For example, all the major coastal ecosystems have suffered significant losses over the past decades (see Table 10.1). Even today, coastal areas across the world are under severe stress as a result of over fishing, environmental pollution, unplanned land development, urbanization, and expansion of agricultural facilities.

Over the past, rapid development at the cost of environmental degradation has led the coastal areas to lose its natural resilience and resulted in increased vulnerability of the coastal communities. With the onset of climate change and an increase in the frequency of other hydrometeorological disasters, the extent of their vulnerability also increased enormously. Most importantly, when it comes to coastal disasters, the natural ability of the coastal ecosystems to mitigate disaster impacts reduced drastically. The Indian Ocean Tsunami in 2004 served as a typical example of the preventive role of coastal ecosystems. For example, several research works carried out just after the Tsunami strongly established the role of mangrove forests in wave attenuation and safeguarding of lives and properties. Kathiresan and Rajendran (2005) and EJF (2006) reported that the Simeulue Island which is merely 41 km away from the epicenter of the earthquake was saved partly due to substantive mangrove cover. Similar instances were also recorded in India, and researcher established an extrinsic linkage between loss of human lives and abundance of mangrove forests. In following years, a new concept of "ecosystem-based disaster risk reduction" emerged strongly as several commitments were obtained from the global

**Table 10.1** Degradation of coastal ecosystem

| Type of ecosystems | Ecosystem services | Loss in recent past | Projected loss |
|---|---|---|---|
| Mangrove forests | Wood, non-wood forest products, nursery area for fishing, vital energy barrier for protection of coastline and over 70 identified ecosystem services | About one fifth of all mangroves are thought to have been lost since 1980; South and Southeast Asia lost over 1.9 million ha during the same period (Spalding et al. 2010) | Mangroves are being lost nearly 1–2 % per year and may completely annihilate in next 100 years or so with the present rate of extinction (Duke et al. 2007) |
| Sea grass beds | Support for commercial fisheries, stabilization of sediments, removal of carbon dioxide from the atmosphere | 29 % of sea grass habitats have disappeared since the nineteenth century with a sharp acceleration in recent years (http://www.cbd.int/gbo3/?pub=6667andsection=6709) | The rate of loss is 110 sq km/year, almost similar to the mangroves |
| Salt marshes | Natural buffer and storm barriers, bird and wildlife habitats, removal of carbon dioxide from the atmosphere | Salt marshes have lost 25 % of the areas of their original occupancy (Macreadie et al. 2013) | The current rate of loss is estimated at 1–2 % per year |
| Coral reefs | Contribute to tourism-based livelihood, protection from tsunami and tides, support 1–3 million species | The world has lost 19 % of the original area of coral reefs mainly due to fishing and natural disasters. The rise of carbon dioxide is at present leading to ocean acidification and may lead to significant decrease of coral reefs at near future (Wilkinson 2008) | Approximately 15 % are seriously threatened with predicted loss within the next 10–20 years; and 20 % are under threat of loss in 20–40 years (Wilkinson 2008). With the climate change consideration, 80 % of the coral reefs may be lost within decades (Nellemann et al. 2008) |
| Shellfish reefs | Maintaining water quality, habitats for fish, crab, and birds | 85 % of oyster reefs have been lost already (Beck et al. 2011) | The shell fish reefs are functionally extinct (Beck et al. 2011) |

communities as well as the national governments in order to preserve and maintain healthy coastal ecosystems. However, despite some initiatives, the continuity of such programs was largely interrupted by economic developmental priority. This clearly served as an example of policy paralysis in order to balance the developmental needs and ecosystem conservation. Therefore, the coastal areas remain at a crucial juncture as the trade-offs between environment and development are yet to be firmly established. The continual increase of coastal population and their demands largely challenges the carrying capacity of the coastal ecosystems, and consequently degradation and loss of ecosystem services are becoming inevitable. This is further accelerated by the human greed for short-term economic gain compared to long-term resilience. This was evident from conversion of mangroves to aquaculture ponds, developmental of eco-fragile coastal lands, over fishing, and unsustainable tourism development, etc.

## 10.4 Environment, Development, and Disaster Linkages in Coastal Areas

Many researchers have, so far, highlighted that environment, disaster, and development are closely associated. In fact, such association is more prominent and dynamic in coastal areas. The reason for such close association is the existence of vast and diverse range of ecosystem services and also the intrinsic nature of coastal areas. Coastal areas are naturally prone to change, driven by the constant interactive force between the land and the sea. Therefore, coastal environment is intrinsically more robust compared to other inland ecosystems. It has the capacity to bear the hostile nature such as saline water, strong currents, etc. A variety of ecosystems such mangrove forests, coral reefs, sea grass, etc. essentially maintain and protect the ecological balance of the coastal areas. Apart from their direct impact on coastal risk reduction, they serve as an impeccable source for enormous and precious coastal resources. For example, communities living in coastal rural areas characteristically depend on fisheries and agriculture, which have direct dependence of the ecosystem services of coastal areas. Hence, loss of ecosystem services severely impairs community's livelihood and well-being. However, as mentioned earlier, unplanned coastal development deteriorates the smooth functioning of the coastal ecosystems. Needless to mention, in majority of the world, this coastal ecosystem serves as a major resource of community's well-being. This is particularly true for coastal areas of developing countries. Pomeroy et al. (2006) mentioned that the social and economic status of coastal communities in the developing countries is very much fragile due to severe interdependence between depleting environmental services. As argued by them, during the Indian Ocean Tsunami in 2004, resource-dependent coastal communities such as fishermen in four major affected countries, i.e., India, Indonesia, Sri Lanka, and Thailand, suffered the bulk of economic losses. Even though some of the affected communities were able to change their livelihood, majority of them suffered irreversibly due to discontinuation of coastal ecosystem services.

As discussed earlier, coastal areas due to their proximity to the sea and other economic benefits have been subjected to widespread development despite the known threats of disasters. This has resulted in development of ports and harbors, coastal cities, waterfronts, and business centers. Subsequently, coastal waters, which host the highest amount marine biodiversity, was subjected to severe coastal pollution. Environmental degradation of coastal waters has resulted in dead waters in the major industrialized coasts, especially in Europe and Japan. It has been observed that there exit strong linkages between areas with high densities of industrial activities and zones of seasonal oxygen-depleted coastal waters (UNEP 2008). On the contrary, the non-developed coastal areas are suffering from overexploitation of fisheries and conversion of coastal lands for agriculture and aquaculture. For example, massive demand for artificially cultivated shrimps has led to explosive growth of aquaculture activities in developing countries, especially in South and Southeast Asia. Unfortunately, the scale of aquaculture farming has increased from a rural and individual basis to organized business over the past decade. Consequently, conversion of coastal land to aquaculture continues to destroy mangroves, coral reefs, and a host of other coastal ecosystems.

Ecosystem degradation and unplanned coastal development essentially serve as two major components that trigger severe adverse impact of coastal disasters. Post-disaster analysis from across the world revealed that even if the disaster could not be avoided but the impact could have been significantly reduced, had we conserved the coastal ecosystems and maintained the environmental and logical basis of the coastal development. However, little regard has been put so far on coastal disaster and ecosystem linkages. Consequently, the world has witnessed dramatic losses from coastal disasters in the last decades. The specter of loss, unfortunately, ranges from highly developed to least developed coasts – irrespective of the engineered barriers that could only provide a false impression of safety and preparedness. This was partly evident from the East Japan Earthquake and Tsunami in 2011 when communities had a false perception of security due to the existence of heavy engineering structures as sea dikes.

## 10.5 Evolution of Integrated Coastal Zone Management Concepts

The last three decades has witnessed several emerging policies to manage highly challenging coastal areas irrespective of country boundaries. The most discussed of the lot was the integrated coastal zone management (ICZM) concept which evolved during the Earth Summit in 1992. Subsequently, the Agenda 21 dedicated a full chapter (*Chapter 17: Protection of the oceans, all kinds of seas, including enclosed and semi-enclosed seas, and coastal areas and the protection, rational use and development of their living resources*) on coastal issues. The chapter advocated the utility of integrated management and sustainable development of coastal and marine areas, including the exclusive economic zone (EEZ). Since then, significant advocacy was made for countries to adopt and adhere the ICZM concept. Importantly, the major thrust of ICZM has been the conservation and sustainable utilization of coastal and marine resources. It essentially envisaged a holistic sectoral approach integrating human, environment, and developmental components. The advocacy of the Agenda 21 gave a broader pathway for sustainable development of the coastal areas through the incorporation of integrated coastal zone management framework. Moreover, it mentioned that coastal manager should integrate broader stakeholders including the local resource-dependent communities such as fishermen and other beneficiaries from coastal resources in designing the management strategies. In a way, ICZM was also probably the first decentralized coastal zone management tool that considered coastal dwellers as an integrated component of the larger "socio-ecological" system. However, in general, the concept does not directly focus on the DRR component in coastal areas. Although part of it focused on the contingency planning for human-induced and natural disasters, the document itself did not mention about the action required for disaster risk reduction in coastal areas. As a matter of fact, many countries adopted the ICZM concept to manage their coast in a more efficient manner in the subsequent years with or without some modifications. However, only a few of them incorporated disaster risk reduction as an element of ICZM. Yet, some of the country recommendations in their respective coastal zone management plan have strong implications in disaster risk reduction. For example, greenbelt, mangrove, coral reef, and regulatory zones all are mostly considered as environmental issues, but all these elements have direct link to disaster risk reduction.

## 10.6 Framework for Integrating Disaster Risk Reduction and Sustainable Development of the Coasts

Since the incorporation of ICZM concept in the Rio Summit, the world has witnessed some of the historical mega-disasters along the coast. The Indian Ocean Tsunami in 2004, Hurricane Katrina in 2005, The East Japan Earthquake and Tsunami in 2011, and Typhoon Haiyan (strongest tropical cyclone ever recorded) in 2013 are only a few of the long list. These disasters not only impacted coastal areas dissolving the developed and developing country's barrier but also left the governments perplexed to design appropriate action plan for future disasters. In most of the cases, the communities living at the coast were severely affected not because that they were living in the coast but because of their lack of knowledge and preparation. The ICZM concepts led to modification of several coastal regulations in the respective countries; however, it mostly did not include the disaster risk reduction as a crucial component of coastal zone management. As the world is witnessing several high-intensity coastal disasters and perhaps will continue to be impacted in near future, it is highly imperative that disaster risk reduction (DRR) be incorporated in the coastal zone management. These can only be done with risk-sensitive developmental and environmental planning under the broader framework of sustainable development of coastal areas.

Essentially, there are three major components that need to be highlighted during the formulation of coastal zone management plan, i.e., environmental conservation, disaster risk reduction, and risk-sensitive development (Fig. 10.1). There are several interlinked subcomponents that actually bridge the three main dimensions. The first most important component is the one that already received adequate considerations in conservation of ecosystems and rejuvenating environmental benefits. As mentioned earlier, coastal areas all across the world are the recipient of heavy amount of pollutants from land-based sources, especially from the upstream megacities. In order to maintain adequate environmental quality, there should be provisions to control environmental pollution of coastal areas. This should be followed by carrying capacity assessment of coastal ecosystem. In many parts of the world, communities characteristically live on coastal ecosystems; however, there is no scientific assessment on the sustainable limits for such exploitation. Once such assessment is

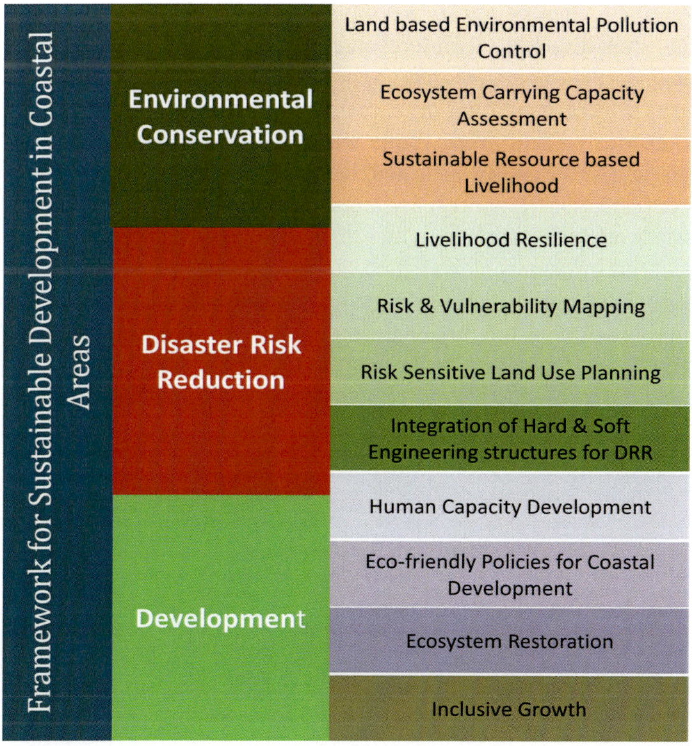


**Fig. 10.1** Framework for sustainable development in coastal areas

conducted, then the coastal resources can be exploited with the self- regeneration limits. These would further promote sustainable resource-based livelihood in coastal areas. The next important component is livelihood resilience of the coastal communities. As discussed earlier, coastal disasters are particularly detrimental to resource-based livelihood, precisely in developing countries. Therefore, development of sustainable livelihood such as fishing and other resource exploitation is extremely crucial for a disaster-resilient community. Initiatives for disaster risk reduction need to be properly integrated with the environmental conservation and developmental goals. The DRR initiative for coastal areas starts with defining hazard-prone areas through mapping and demarcating risk zones. Based on such categories, the most important component is to develop a risk-sensitive land use plan for the coastal areas. The concept of risk-sensitive land use plan broadly includes the locations for greenbelt and residential and industrial infrastructure in order to reduce the impact on the natural and human environment. This further includes the protective and corrective measures that should be incorporated for effective disaster risk reduction. In the case of developed countries, a variety of engineered structures are presently being used for disaster risk reduction such as dikes and seawall. These served historically to reduce disaster impacts, particularly from wet mass movement. However, this has also resulted in poor water qualities in coastal areas resulting in ecologically non-productive waters. On the other hand, ecosystem-based disaster risk reduction such as planting mangrove forests and developing coastal buffers has been proved effective in recent disasters. Therefore, careful integration is required between hard and soft engineering (bioengineering) approaches so that the ecological conditions of the coastal waters should not deteriorate, and further the impact of disasters is also simultaneously minimized. While it can be argued that countries like the Netherlands have successfully mitigated the seaward hazards by their extensive dike systems, ecosystem-based approaches are especially applicable for the developing or the least developed countries, which has, otherwise, limited capacity to erect huge engineering structures along the coast.

It is projected that in the next few decades, coastal areas all across the world will receive huge investment for further development. While development will make communities more economically resilient and significant infrastructure such as early warning, faster evacuation routes, and better connectivity, it is also imperative that significant amount is also diverted for ecosystem restoration and human capital development. Most importantly, development of the coastal areas should include all relevant stakeholders and should follow an inclusive growth model.

## 10.7 Way Forward

The concept of "disaster risk reduction" has the highest relevance when it comes to coastal areas. As discussed in this chapter, the linkages between disaster, environment, and development are well established and widely acknowledged. Consequently, there have been several policy advocacies for risk-sensitive development of coastal areas. For example, the Indonesian government developed coastal greenbelt policy in Banda Aceh following the Indian Ocean Tsunami in 2004. Similarly, policies have also been adopted in many parts of the world; however, in many cases such initiatives are not legally binding. Therefore, implementations of such projects are vastly challenged and often led only by short-term realization and long-term ignorance. In most cases, local level initiatives fail when it comes to long-term sustainability due to lack of funding and absence of formal implementation mechanism. Therefore, sustainable development of the coast includes not only holistic planning but also a strong implementing framework that ensures significant funding on ecosystems, environment, and disaster risk reduction. Moreover, sustainable development of coastal areas requires capacity development of coastal managers and policy planners who are not always aware about the disaster risk reduction components. Therefore, human capacity development in conjugation with natural and environmental capacity development remains extremely crucial.

**Acknowledgment** The first author greatly acknowledges the Ministry of Education, Sports, and Culture of Japan (MEXT) scholarship provided by the Japanese government for conducting research in the Graduate School of Global Environmental Studies of Kyoto University. The authors also acknowledge the support of GOCE-ARS (Sustainability/survivability science for a resilience society to extreme weather conditions) and Studies on the Connectivity of Hilltop, Human and Ocean (CoHHO) program of Kyoto University.

## References

Beck MW, Brumbaugh RD, Airoldi L, Carranza A, Coen LD, Crawford C, Defeo O, Edgar GJ, Hancock B, Kay MC, Lenihan HS, Luckenbach MW, Toropova CL, Zhang G, Guo X (2011) Oyster reefs at risk and recommendations for conservation, restoration, and management. Bioscience 61(2):107–116

Burke L, Kura Y, Kasem K, Revenga C, Spalding M, McAllister D (2001) Coastal ecosystems. World Resources Institute, Washington, DC

Duke NC, Meynecke JO, Dittmann S, Ellison AM, Anger K, Berger U, Cannicci S, Diele K, Ewel KC, Field CD, Koedam N, Lee SY, Marchand C, Marchand C, Nordhaus I (2007) A world without mangroves? Science 317(5834):41–42

Duxbury J, Dickinson S (2007) Principles for sustainable governance of the coastal zone: in the context of coastal disasters. Ecol Econ 63(2):319–330

EJF (2006) Mangroves: nature's defence against tsunamis: a report on the impact of mangrove loss and shrimp farm development on coastal defence. Environmental Justice Foundation, London

Fleisher BM, Chen J (1997) The coast–noncoast income gap, productivity, and regional economic policy in China. J Comp Econ 25(2):220–236

Giddings B, Hopwood B, O'brien G (2002) Environment, economy and society: fitting them together into sustainable development. Sustain Dev 10(4):187–196

Helmer M (2006) Natural disasters and climate change. Disasters 30:1–4

Kathiresan K, Rajendran N (2005) Coastal mangrove forests mitigated tsunami. Estuar Coast Shelf Sci 65(3):601–606

Knutson TR, Mcbride JL, Chan J, Emanuel K, Holland G, Landsea C, Held I, Kossin JP, Srivastava AK, Sugi M (2010) Tropical cyclones and climate change. Nat Geosci 3:157–163

Macreadie PI, Hughes AR, Kimbro DL (2013) Loss of 'blue carbon' from coastal salt marshes following habitat disturbance. PLoS One 8(7), e69244. doi:10.1371/journal.pone.0069244

Martínez ML, Intralawan A, Vázquez G, Pérez-Maqueo O, Sutton P, Landgrave R (2007) The coasts of our world: ecological, economic and social importance. Ecol Econ 63(2):254–272

Mimura N, Nurse L, McLean RF, Agard J, Briguglio L, Lefale P, Payet R, Sem G (2007) Small islands. Climate Change 2007: impacts, adaptation and vulnerability. Contribution of Working Group II to the fourth assessment report of the Intergovernmental Panel on Climate Change, Parry ML, Canziani OF, Palutikof JP, van der Linden PJ, Hanson CE (eds), Cambridge University Press, Cambridge, pp 687–716

Nellemann C, Hain S, Alder J (eds) (2008) Dead water: merging of climate change with pollution, over-harvest, and infestations in the world's fishing grounds. UNEP/Earthprint, Arendal

Nicholls RJ, Cazenave A (2010) Sea-level rise and its impact on coastal zones. Science 328:1517–1520

Nicholls RJ, Wong PP, Burkett VR, Codignotto JO, Hay JE, McLean RF, Ragoonaden S, Woodroffe CD (2007) Coastal systems and low-lying areas. Climate Change 2007: impacts, adaptation and vulnerability. Contribution of Working Group II to the fourth assessment report of the Intergovernmental Panel on Climate Change, Parry ML, Canziani OF, Palutikof JP, van der Linden PJ, Hanson CE (eds), Cambridge University Press, Cambridge, pp 315–356

Nicholls RJ, Small C (2002) Improved estimates of coastal population and exposure to hazards released. EOS Trans Am Geophys Union 83(28):301–305

Pearce D, Markandya A, Barbier EB (1989) Blueprint for a green economy. Earthscan, London

Pomeroy RS, Ratner BD, Hall SJ, Pimoljinda J, Vivekanandan V (2006) Coping with disaster: rehabilitating coastal livelihoods and communities. Mar Policy 30(6):786–793

Spalding M, Spalding M, Kainuma M, Collins L (2010) World atlas of mangroves. Earthscan, London

UNEP (2008) An overview of the state of the world's fresh and marine waters, 2nd edn. UNEP, Nairobi

Wilkinson C (ed) (2008) Status of coral reefs of the world: 2008. Global Coral Reef Monitoring Network, Townsville, p 298

# Chapter 11
# Climate Change and Integrated Approach to Water Resource Management in the Murray-Darling Basin

**Erika Onagi**

**Abstract** Provision of water resources is one of the most major elements to secure sustainable development for agriculture, industry, energy, and society. Climate change has raised concerns about the threat to water resources and increased international awareness of the importance of cross-border water resource management to confront such borderless problems. Integrated approach to water resource management for climate change is still a new field of study to develop an effective management framework even in the developed countries. This chapter presents one case study from the Murray-Darling Basin of Australia, which has recently engaged to apply the Basin Plan. The Basin Plan is the first case to introduce the integrated approach to water resource management in the Basin region. One of the major purposes of the Basin Plan is to restore long-term sustainable water quality and environmental development. The main objectives of this chapter are (1) to analyze mechanism of the integrated approach to the water resource management in the region of the Murray-Darling Basin under the federal political system and (2) to examine how this federal political system affects the process of negotiation within the Basin Plan. This chapter also raises several questions in order to provide some lessons from the case study and suggest applicable implications to other situation of the transboundary river management.

**Keywords** Transboundary rivers • Negotiation • Decision-making • Top-down approach

E. Onagi (✉)
Department of International Studies, Graduate School of Frontier Sciences, The University of Tokyo, 5-1-5 Kashiwanoha, Kashiwa-shi, Chiba 277-8563, Japan
e-mail: elly_23fs@mac.com



J.I. Uitto, R. Shaw (eds.), *Sustainable Development and Disaster Risk Reduction*, Disaster Risk Reduction, DOI 10.1007/978-4-431-55078-5_11

## 11.1 Introduction

Climate change is extensively recognized as one of the largest obstacles to attain sustainable development which the global society faces in the twenty-first century. A wide range of scientific research into climate change has carried out and provided an understanding of key elements and scenarios of possible projected outcomes. Researchers raise important questions that climate change considerably affects the Earth's weather and living life but also causes degradation of global climate cycle.

Water resource has been considered a hot topic of climate change issues and is significantly subjected to influence the problems raised by climate change. Responding to the problems, many different projects and programs have been undertaken not only at the regional and national level but also at the cross-border level of the governance. Transboundary rivers including international and domestic river basins are now faced with a difficulty of advancing integrated water resource management with different roles of stakeholders. The current situation does not achieve a satisfactory level of management even in the transboundary rivers in a developed country. This chapter highlights "transboundary rivers" as the key element and argues with a problematic situation of water resource management in Australia.

This chapter proposes one case study from the Murray-Darling Basin in Australia as a major topic of the analysis. Recently, Australia is at the center of the water reform to evaluate water allocation and restore adequate supply and demand balance of water for sustainable river development by implementing the Basin Plan. The nation has encountered a difficult situation on whether or not to adopt this Plan. Firstly, the chapter reviews the aspect of climate change and examines historical context of water reforms. Secondly, the chapter analyzes the case of the Basin Plan and addresses several research questions. Did the Commonwealth government promote the right direction for implementing the Basin Plan? Are there any preferable alternatives to the process of decision-making? What if the Murray-Darling Basin Authority (MDBA) proposed less quantity of water than 3000–4000 GL at the first attempt of the Plan? How about 2000 GL? What is the main reason of low transparency and accountability between the basin states? The chapter lastly provides some useful lessons and guidance to the future integrated water resource management.

## 11.2 Background

There has been a wide range of researchers studying and tackling the climate change issues over the past few decades. There is unquestionable evidence that increased emission of greenhouse gas due to human activity causes global warming as well as climate change. According to the Intergovernmental Panel on Climate Change (IPCC), the scientific-based research showed that climate change is occurring across the world, rising air and ocean temperature, melting of snow and ice, and increasing average sea level. The IPCC analyzed that global surface temperature has been increased at 0.74 °C (0.56–0.92 °C) since the last 100 years (1906–2005). By comparing to the trend (1901–2000) of 0.6 °C (0.4–0.8 °C) reported in the Third Assessment Report (TAR) (Fig. 11.1), the increase of global temperature has become greater and significant. In addition, the IPCC Special Report provided possible future scenarios of global climate change. According to the survey (IPCC 2007):

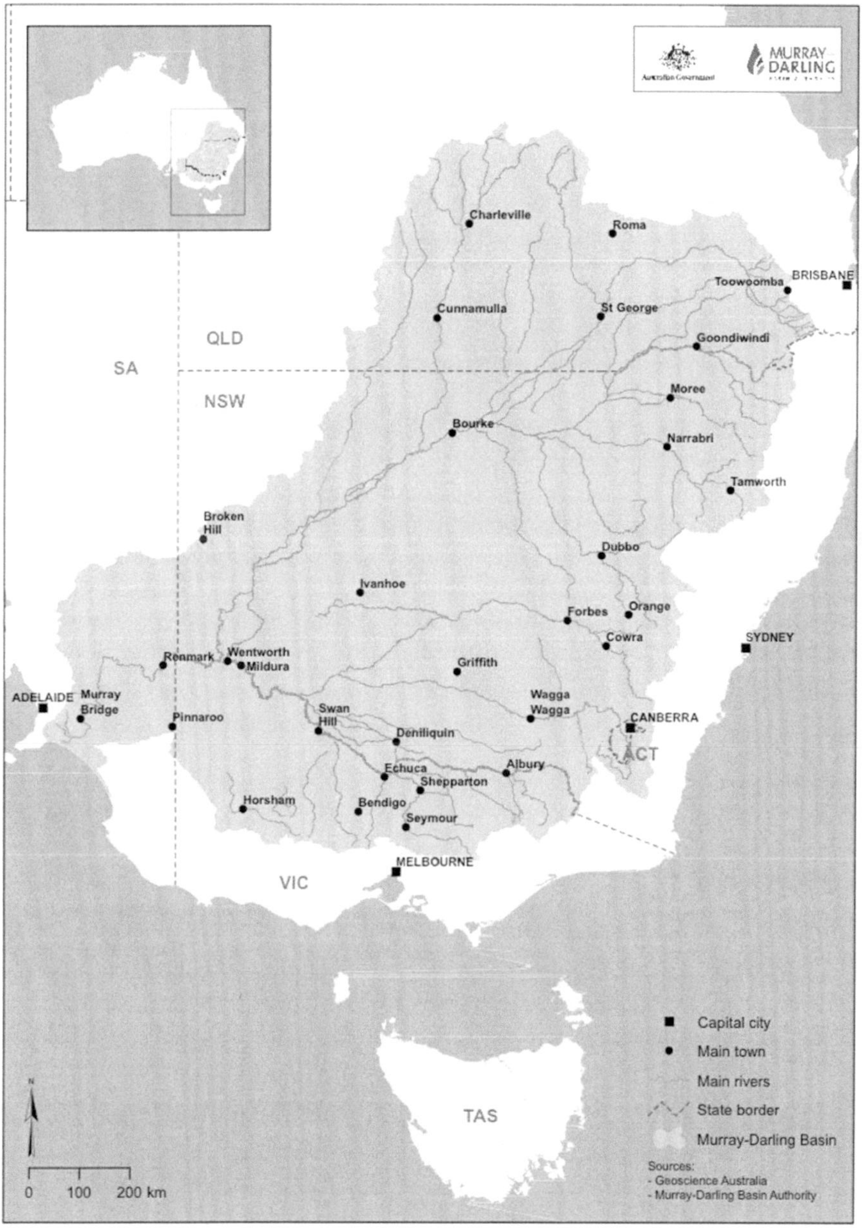


**Fig. 11.1** Murray-Darling Basin boundary (Source: http://www.mdba.gov.au/sites/default/files/images/8_Murray-Darling_Basin_Boundary.jpg)

> Continued GHG emissions at or above current rates would cause further warming and induce many changes in the global climate system during the 21st century that would very likely be larger than those observed during the 20th century.

Global warming could change the whole climate system such as atmosphere, land surface, water, environment, fauna and flora, and human activities. Weather-related disasters would also increase.

## 11.3 International Movement for Water Management

Climate change is one of the factors that cause water scarcity. According to the United Nations Department of Economic and Social Affairs (UNDESA), under the existing scenario of climate change, about 50 % of the population will live in the area of "high water stress" by 2030, and the demand for water will increase all over the world (UNDESA 2013).

Climate change is undoubtedly recognized as one of the primary issues on which the world needs to work together in order to avoid and reduce the future threats to water resources. The idea of the integrated approach for water resource management has been well known since the United Nations Conference on Environment Development (UNCED), also called the "Earth Summit." The World Water Forum is widely recognized as the largest international event in the field of water resource management. Recently, the 6th World Water Forum was held in Marseille in 2012. One hundred forty-five representative countries and more than 35,000 participants gathered to raise issues and find solutions (World Water Forum 2013). "Respond to climate and global changes in an urbanising world" is one of the priorities of the Forum (World Water Forum 2013).

## 11.4 The Basin Plan

In October 2010, the Murray-Darling Basin Authority published the *Guide to the Proposed Basin Plan* in order to "assist all interested parties in understanding the basis of the proposed Basin Plan before the formal, legislated consultation process begins" (MDBA 2010). The release also provided opportunity for the public to present their opinions and interests and then to give appropriate feedback to the MDBA. After the release of the Basin Plan, the news were widely broadcasted (Wahlquist 2011). It has become a controversial issue that a large number of people living in Australia have a strong view against this Plan.

## 11.5 Objectives

The main objective of this chapter is twofold: (1) analyze mechanism of the integrated approach of the water resource management in the region of the Murray-Darling Basin under the federal system and (2) examine the negotiation process to implicate preferable framework of the Basin Plan.

## 11.6 Methodology

Literature review is the main research methodology of this chapter. Interviews with officers of the Murray-Darling Basin Authority (MDBA) and professors from the Australian National University were also arranged in September 2012 in Canberra, Australia.

## 11.7 Climate Change in Murray-Daring Basin of Australia

This section illustrates critical aspects of climate change in Australia and analyzes historical movement of the water reforms in the region of the Murray-Darling Basin.

Climate in Australia represents unique factors. In other words, climate varies from region to region in the country. The land contains many different climate zones: The northern part of the land is wet tropics, with dry region in inland part and alpine climates in the southeast (Bureau of Meteorology 2012a). The average annual amount of rainfall for 2012 was 476 mm contrasting with 2011 annual amount of 699 mm (Bureau of Meteorology 2012b). Based on the IPCC's report (IPCC 2007), climate change will affect the future Australian society in a variety of ways:

- Degradation of biodiversity is projected to occur by 2020.
- In southern and eastern Australia, water security problems are projected by 2030.
- Southern and eastern Australia is projected to have a decrease in agricultural production by 2030.

Climate change will have an impact on demand for agricultural water and change global agricultural distribution (FAO 2012). In Australia, agriculture is one of the most important industries.

The Murray-Darling Basin is a catchment for the Murray, Darling, and Murrumbidgee rivers. One of the biggest river basins in southeastern Australia, it covers over 1,059,000 km$^2$ and represents 14 % of the total area of the land (ABS 2013). The Murray-Darling Basin is a transboundary river system which includes four states, Queensland, South Australia, New South Wales, and Victoria, and ACT (the Australian Capital Territory). A major part of the Basin is located in New South Wales (56 %) and Queensland (24 %). The Basin is significant for agricultural production. Within the Murray-Darling Basin, 84 % of the land use is related to agriculture (MDBA 2013a).

Approximately two million people are living in the Murray-Darling Basin which is 10 % of the national population. A large number of people lives in New South Wales (39 %) and Victoria (29 %). About 10 % of the population works as a farmer or in agricultural-related work, compared to 3 % at the national level.

## 11.8 Major Actors and Water Reforms

Australia has a long history of water reforms since the foundation of the nation. Throughout the history, large amounts of water have been used for agriculture. Since Australia introduced federal political system, state governments uniquely possess a solid independent legal power. Consequently, power balance between Commonwealth and states has become complex and controversial even in the field of water resource management. In other words, relationships between Commonwealth and each basin state government are critically complicated with consequent problems.

The complicated and ambiguous features of governance in the Murray-Darling Basin are a critical point of this discussion. The following gives an overview of major actors involved in the management of the Murray-Darling Basin.

Founded in December 2008, the Murray-Darling Basin Authority aims to manage water resource in the Basin in order to reflect the national interest. It is the first "single agency" that is legally responsible for providing integrated management to this region. The Water Act 2007 requires the MDBA to prepare the Basin Plan as their main objective. About 300 staff who are specialized in various fields of the study work at the office. Before the establishment of the MDBA, Murray-Darling Basin Commission (MDBC) was in charge of the management. The Ministerial Council and the Basin Officials Committee also take their roles in providing advice and making decision to the MDBA (2013b).

Although the basin states have been faced with continual struggles for more than 100 years, the main actor of water resource management tends to be the hand of state governments. Water resource management has always been one of the main objectives in this country. However, people's interest toward water was quite low until the late 1980s, and the idea of the integrated approach of water resource management was barely considered under the federal and state governments (Kondo 2006). Aggregated damage from drought in the 2000s and excessive use of water and water rights by the basin states have since become a serious problem. As a result, the improvement of traditional frameworks for water resource management was required. Table 11.1 shows recent movement of the water reforms.

**Table 11.1** Recent movement of water reforms in Australia

| Year, month | Name of water reforms |
|---|---|
| 2004, June | Intergovernmental Agreement on a National Water Initiative |
| 2004, July | Australian Government Water Fund |
| 2007, Jan | A National Plan for Water Security |
| 2007, April | National Climate Change Adaptation Framework |
| 2007, November | Water Act 2007 |
| 2008, August | Establishment of the Murray-Darling Basin Authority (MDBA) |
| 2010, October | Proposed Basin Plan |

## 11.9 Findings and Discussions

The Basin Plan has been recently accepted as a law by the Federal Water Minister Tony Burke in November 2012 after long stymied discussion. It has just entered a new stage and been asked to look closely to see consequences and result of the implementation.

This chapter closely traces the recent movements of the Murray-Darling Basin and analyzes how the integrated water reform is managed in Australia. In addition to literature survey, several interviews with the officers of the MDBA and professors of the Australian National University (ANU) were carried out in Canberra in September 2012. The purpose of the interviews was to find out the ongoing process of the Basin Plan.

The first part of the chapter questions are as follows: Did the Commonwealth government promote the right direction for implementing the Basin Plan? Are there any preferable alternative processes of decision-making?

### *11.9.1 Problems of Traditional Governance Framework*

The Basin Plan made under the Commonwealth Water Act 2007 is the first case to introduce the integrated approach to water resource management in the Basin region (Connell 2011a). A choice as to whether to implement the Basin Plan will change a direction of the future of the nation. As a matter of fact, Australia is now at the turning point of the history of water resource management. Looking into the past, a traditional approach of decision-making method required "unanimous agreement" among all governments. Consequently, this method was often unable to function and caused difficulties to have a consensus of all governments. In addition, it resulted in delays in discussion and implementation (Connell 2011a). By recognizing a need for improvement in the traditional framework of the governance, the Basin Plan is developed under the Water Act 2007.

Australia has now reached the "third wave" of major water reform since the 1990s (Alexandra 2012). The earlier two waves were the Council of Australian

Governments (CoAG) water reforms known as the National Water Initiative (NWI) in 1994 and 2004. The CoAG program required all governments' agreement to improve environmental sustainability in the Murray-Darling Basin and implemented water trading across the basin state borders to boost water markets. In contrast, the Water Act 2007 is the most recent attempt (Connell 2011a). As the third wave, it put emphasis on the improvement of environment and sustainable use of water resources. It is obvious that the characteristic of water reforms has shifted from market-based framework to concerning environmentally friendly and sustainable development of the Murray-Darling Basin. In other words, the third wave of the water reform has tried to rebalance water resources, reflecting changes on Australia's national interest.

Additionally, across the Murray-Darling Basin, Australian government has challenged a wide range of other water reforms. Examples of the reform are listed below:

- Salinity and Drainage Strategy in 1989
- The Natural Resources Management Strategy in 1990
- The 1994 Council of Australian Governments (CoAG) Water Reform Framework
- The Cap in 1995
- The Integrated Catchment Management (ICM) Policy Statement in 2000
- The Living Murray First Step during 2003–2004
- The National Water Initiative (NWI) in 2004

The latest attempt of the water reform is called the Water Act 2007. Key elements of the Act are (Australian Government 2013):

- Establishment of the MDBA with the power of enforcement
- Preparation of the Basin Plan
- Establishment of the Commonwealth Environmental Water Holder
- Implementation of the Australian Competition and Consumer Commission (ACCC) with enforcement of "water charge and water market rules"
- Monitoring of water-related information by the Bureau of Meteorology

### 11.9.2 Conflict in Water Resource Management

From a historical perspective, during the time of pre-federation, management of the River Murray was under the problematic situation between the colonies of New South Wales, Victoria, and South Australia. Conflict was caused by the boundary between states, and the use of water for irrigation was a controversial issue. The River Murray Waters Agreement was enacted in 1915 by the Australian governments, New South Wales, Victoria, and South Australia (MDBC 2013). In response to the Agreement, the Murray-Darling Basin was managed until the early 1980s under the control of River Murray Commission (MDBC). The River Murray Waters Agreement was in operation over 90 years. In spite of the changes provided by the Agreement and expanding the power of the MDBC, there were increased difficulties in the management of the Basin. Water resource management by "individual agencies within the separate states" was confronted with a new type of problems such as environmental degradation and increased salinization in the early 1980s (MDBC 2013). It was gradually noted as a serious problem that the "identical legislation" adopted between the Commonwealth, New South Wales, Victoria, South Australia, Queensland, and the Australian Capital Territory no longer provided effective development for the Basin (Connell 2011a).

Under the judicial arrangement in the Water Act 2007, the Commonwealth government enforces Commonwealth constitutional powers. In this background, state governments failed to engage effective water policy and control over-allocations of irrigation water as they agreed to follow the rules of the NWI 1994 and 2004 (Byron 2011). In the end, the Commonwealth government set a top-down approach to water resource management. However, problems still remain that this single legislation system within a top-down approach is also complex and complicated. And the question is how the Commonwealth top-down approach helps to solve the recent situation of the Basin.

### *11.9.3 Integrated Approach and Climate Change*

Arrangement of the Basin Plan creates a new framework of the integrated approach to manage water resource and to tackle with climate change in the Murray-Darling Basin. First of all, the meaning of the term "integrated approach" remains obscure and might cause misunderstanding of its implications. Hence, by clearly representing the meaning of the integrated approach in the case of the Basin Plan, it expected the integrated water resource management within the bound of Commonwealth government's top-down strategy to overcome the difficulties and restore sustainable water in the Murray-Darling Basin. The role and responsibility of the Commonwealth government are important because coordination for facing with broader issues such as threat of environmental degradation and water security is necessary.

### *11.9.4 Federal System in Water Resource Management*

The introduction of the Basin Plan eventually changed the power balance of the Commonwealth government and state governments in terms of water resource management. Ever since the foundation of the nation, Australia is under the federal political system: State governments are not subordinated bodies but have independent rights. The question is how the change in balance of power would affect the Basin Plan and what is the benefit from the change. On the first sight, the top-down approach somehow seems to be a disadvantage for the state government and local communities. Yet, it is also an advantage for them to have opportunities to provide local knowledge to improve "environmental, cultural, and socioeconomic values" and to encounter climate change impacts (Hatton et al. 2011). In other words, each state government has had a close relationship with their local communities such as farmers and irrigators. In addition, taking advantage of conducting a large number of programs and projects throughout the basin state history, it is clearly proved that there is an advantage for the state governments in terms of water resource management. As a proof of this, from the interviews with the officers of the MDBA, they were aware of their lack of knowledge and ability to conduct technical skills. Regarding the water resource management, transboundary rivers including the Murray-Darling are often face problems such as lack of the process of decision-making, low transparency and accountability, high transaction cost, and conflict between different stakeholders (Connell 2011b). Once the integrated approach is adopted, there is a need to have a "good balance" of relations between states and the Commonwealth government. Yet, the questions is, what is the main reason of low transparency and accountability?

### 11.9.5 Information Sharing

Information sharing is another key factor of the discussion and one of the complicated problems. As discussed before, under the federal system, the power of state governments tends to be sovereign and independent. Each state has their local relationships with their local farmers and irrigators. As a result, there has been a localized network within the basin states as if it is the hereditary system of information sharing. On the one hand, local network is an advantage of the state governments. On the other hand, it also creates difficulties to manage equal and cooperative information sharing with other basin states and the Commonwealth government. To tackle with the borderless problem such as climate change, there is a need for comprehensive and integrated system for information sharing.

### 11.9.6 Process of Preparing the Basin Plan

Since the MDBA released the *Guide to the Proposed Basin Plan* in October 2010, it has become a controversial issue and was broadcasted widely across the states. Under the Water Act 2007, it clearly states that consultation process is required with the basin states, Basin Officials Committee, and Basin Community Committee in order to arrange the Plan. According to the interviews with the MDBA's officers, MDBA held several meetings with farmers before releasing the *Guide to the Proposed Basin Plan*. However, until the release of the proposed Plan, it seems that the public hardly enabled to catch the information of the Plan due to the fact that MDBA seemed to remain silent until the release (Wahlquist 2011). From the aspect of the journalist, the proposed plan was suddenly appeared to the public. Journalist Margaret Simons stated that (Wahlquist 2011):

> The plan, two years in the making, is the result of the first exercise ever in asking the vital question: what is sustainable use of this nation's major river system?... And we knew, or should have known, that the answer would be 'something very different to what we are doing now.'

The Commonwealth government and the MDBA failed to engage with the public and communities in the early stage of preparation for the Plan. It is important to provide clear and accurate information to the public and encourage participation of communities and people in the process of preparation. When the proposed Basin Plan was firstly released, the media and opponents mainly criticized and pointed out the matter of reduction of water use in the Murray-Daring Basin. However, one of the main purposes of the Basin Plan is to restore healthy river environment and to adopt sustainable water resource management to tackle with future threat of water. The first step would be the hardest but most important process for the Basin Plan. If the keyword "environment" was in the front page of the *Guide to the Proposed Basin Plan*, it would not be impossible to have smoother situation of the process.

### *11.9.7 Process of Negotiation*

As soon as the *Guide to the Proposed Basin Plan* was issued, the Commonwealth and state governments put effort into managing negotiations. The MDBA provided consultation meetings with basin communities including Basin Community Committee, national peak bodies, some scientists and technical experts, indigenous representatives, and local government representatives. In response to this consequence, MDBA revised the proposed Basin Plan (MDBA 2013c). Table 11.2 outlines timeline related to the Basin Plan.

**Table 11.2** Timelines related to the Murray-Darling Basin Plan

| | |
|---|---|
| August 2007 | Establishment of the MDBA was introduced by the Howard government and taking control over water rights (used to be controlled by the states) |
| March 2008 | Enforcement of the Water Act |
| October 2010 | The release of the guide to the proposed Basin Plan with arguing cuts of 4,000 GL water allocation from the Basin, raising a great number of protests by rural communities |
| October 2011 | The release of the revised draft plan with proposing cuts of 2,750 GL and starting 20-week public consultation |
| May 2012 | The release of the third version of the Basin Plan taking into consideration the comment and suggestions from the public |
| November 2012 | Passing the law, after receiving all individual views |

Sara Phillips, 'Murray-Darling Basin reflect the failure of the government', *ABC Environment*, 31 May 2012, viewed 23 January 2013, http://www.abc.net.au/environment/articles/2012/05/31/3514567.htm

In November 2012, the Basin Plan was passed into a law after longstanding difficulties and controversies with the basin states. Until the very late moment of time, basin states and related stakeholders, including farmers and irrigators, opposed the Plan. The question is how and why basin states turned their opinions and accepted the propose.

One of the significant purposes of the Basin Plan under the Water Act 2007 is to guarantee environmentally sustainable use of water in the Murray-Darling Basin. This draft plan required cutting 3000–4000 billion liters (GL) of water allocation. Based on the "hydrological indicator site method," the percentage of water reduction was around 27–37 %. In response to the voices of stakeholders including opponents, the MDBA revised the Plan and new version required 2750 GL cut of water. Under the negotiations, the MDBA compromised the amount of water cut, since one of the opponents, the Wentworth Group of Concerned Scientists, proposed 2800 GL. The question is, what if the MDBA proposed less quantity of water than 3000–4000 GL at the first draft Plan? How about 2000 GL?

It seems that the quantity of water reduced from rivers is not the main point of this controversial discussion. As long as the Commonwealth government decides to limit the use of water, no matter how hard they attempt to deal with the issue, the situation would remain the same as before or end up with deadlocked negotiation unless they find out an alternative way to solve the situation.

### *11.9.8 Sustainable Environment and Use of Water*

There is an alternative. Since new version of the Basin Plan emphasizes a threat of climate change and requires recovery of sustainable water resource, the environment has been taken into a serious consideration throughout the assessment to recover healthy rivers. The following table shows major points of the changes applied to the final Basin Plan (MDBA 2013d):

- The Sustainable Diversion Limit (SDL) adjustment mechanism
- Apportionment
- Climate change
- Groundwater
- Water trading

In the end, the Basin Plan proposed 2750 GL as the baseline target for the water reduction in order to recover the Murray-Darling environment. The first point of "SDL adjustment mechanism" means that the establishment of 2750 GL would be changeable. It would be possible to reduce the target of 2750 GL by a contribution of effective use of water resource for the environment. On the other hand, it would also be possible to increase the limit to 3230 GL.

Until the last moment, there was no agreement between the basin states due to the fact that how to share and be responsible for the 971 GL of downstream components shared among the basin states. Right after signing the agreement, the Federal Water Minister Tony Burke stated that (Vidot 2012):

> The environment, when it turned up to the negotiating table, turned out to be more ruthless and less compromising than any of the states; the environment turned up at the negotiating table and said, 'if you're going to manage the rivers this way then none of you can have the water'...

It is considered to be a strategic approach of the Commonwealth government and the MDBA to reinforce the importance of the "environment" in the Murray-Darling Basin to stimulate people's incentive to have a consensus and conclude the negotiation.

## 11.10 Conclusion and Implications

It is generally assumed that informed decision-making on water issues can enhance local development practices and is linked to larger water policy issues of the country. Sustainable development of a country may be realized through efforts of this sort in various sectors and regions within the country.

Some conclusions and implications can be drawn from the discussion of the Basin Plan in the Murray-Darling Basin. It is suggested that the right procedure to attain sustainable development may not be attained automatically even in a developed country.

It was witnessed that the Basin Plan adopted a strategy of the top-down approach to manage the Murray-Darling Basin. Although state governments are not subordinated bodies but have independent rights, the Commonwealth government leads all involved stakeholders into achieving the nation's goal under the Water Act 2007. Yet, the Commonwealth government and the MDBA still have a trouble with conducting local communities. Implications observed in this context include:

- Take an advantage and cooperate with state governments in order to improve regionally specific information and values to encounter climate change impacts
- Create comprehensive water resource network to share the information and full a gap between different stakeholders

The MDBA also needs to improve the first attempt at releasing the *Guide to the Proposed Basin Plan*. It should:

- Engage with the public and communities in the early stage of preparation
- Provide clear and accurate information to the public and encourage participation of communities and people in the process of preparation
- Highlight priorities of the Plan, in this case, environment and sustainable water resource management

Under the process of the negotiation, the MDBA and the Commonwealth government experienced a hard time to reach agreement. The concerns of opponents were not the amount of the water to reduce but something else. The negotiation could have been completed, by discussing what is the best reasonable option and stimulating incentive to have a consensus.

The following lessons are also learned from the case study in this chapter: Under the leadership of the Commonwealth government, it is necessary to take an advantage and cooperate with state governments in order to improve regionally specific information and values to encounter climate change impacts under the comprehensive networks among all basin states and stakeholders. It is also very important to engage with the public and communities in the early stage of preparation and provide clear and accurate information to the public and encourage participation of communities and people in the process of preparation. Highlighting priorities of the Plan, in this case, environment and sustainable water resource management, is also another key aspect. To avoid deadlocked negotiation, discussing what is the best reasonable option is a crucial point to stimulate incentive to have a consensus and end the negotiation.

**Acknowledgment** The survey carried out for this article was partly funded by Grants in Aid for Scientific Research (KAKENHI 24310027) by the Government of Japan.

## References

ABS (2013) Water and the Murray-Darling Basin – a statistical profile, 2000-01 to 2005-06. http://www.abs.gov.au/ausstats/abs@.nsf/mf/4610.0.55.007. Accessed 22 Jan 2013

Alexandra J (2012) Australia's landscapes in a changing climate – caution, hope, inspiration, and transformation. Crop Pasture Sci 63:227

Australian Government (2013) Australian Government, Department of Sustainability, Environment, Water, Population and Communities, 'Water legislation'. http://www.environment.gov.au/water/australia/water-act/index.html#water-act. Accessed 23 Jan 2013

Bureau of Meteorology (2012a) Rainfall deficiencies. http://www.bom.gov.au/climate/drought/. Accessed 22 Jan 2013

Bureau of Meteorology (2012b) Twelve-monthly rainfall totals for Australia. http://www.bom.gov.au/jsp/awap/rain/index.jsp?colour=colour&time=latest&step=0&map=totals&period=12month&area=nat. Accessed 22 Jan 2013

Byron N (2011) What can the Murray-Darling Basin plan achieve? Will it be enough? In: Connell D, Quentin Grafton R (eds) Basin future water reform in the Murray-Darling Basin. ANU E Press, Canberra, p 389

Connell D (2011a) The role of the commonwealth environmental water holder. In: Connell D, Quentin Grafton R (eds) Basin future water reform in the Murray-Darling Basin. ANU E Press, Canberra

Connell D (2011b) Water reform and the federal system in the Murray-Darling Basin. Water Resour Manag 25:3998

FAO (2012) FAO water report, coping with water scarcity, an action framework for agriculture and food security, p 15

IPCC (2007) Climate change 2007: synthesis report, summary for policymakers. http://www.ipcc.ch/pdf/assessment-report/ar4/syr/ar4_syr_spm.pdf. Accessed 21 Jan 2013

Kondo M (2006) Water reform in Australia: its general overview. Research Center for Sustainability and Environment, Shiga University, Research Paper, 2006, vol 3, p 52

Macdonald DH, Bark R, Garrick D, Banerjee O, Connor J, Morrison M (2011) Multiple benefits through the life cycle of the Basin Plan. In: Connell D, Quentin Grafton R (eds) Basin future water reform in the Murray-Darling Basin. ANU E Press, Canberra, p 269

MDBA (2010) Guide to the proposed Basin Plan technical background. http://download.mdba.gov.au/Guide-to-proposed-BP-vol2-0-12.pdf. Accessed 22 Jan 2013

MDBA (2013a) Quick facts. http://www.mdba.gov.au/explore-the-basin/about-the-basin. Accessed 8 Jan 2013

MDBA (2013b) Governance of the Murray-Darling Basin Authority. http://www.mdba.gov.au/about/governance/ministerial_council. Accessed 22 Jan 2013

MDBA (2013c) Process for the proposed Basin Plan. http://www.mdba.gov.au/have-your-say/whats-next/process-for-PBP. Accessed 23 Jan 2013

MDBA (2013d) Change to the Basin Plan. http://www.mdba.gov.au/basin-plan/changes-to-the-basin-plan. Accessed 23 Jan 2013

MDBC (2013) A brief history of the Murray-Darling Basin agreement, http://www2.mdbc.gov.au/about/history_mdbc.html. Accessed viewed 23 Jan 2013

UNDESA (2013) Water scarcity, International Decade for Action 'Water for Live 2005–2015'. https://www.un.org/waterforlifedecade/scarcity.shtml. Accessed 21 Jan 2013

Vidot A (2012) Murray-Darling Basin Plan signed into law at last. ABC Rural, 22 November 2012. http://www.abc.net.au/rural/news/content/201211/s3638832.htm. Accessed 23 Jan 2013

Wahlquist A (2011) The media and the guide to the Basin Plan'. In: Connell D, Grafton RQ (eds) Basin futures water reform in the Murray Darling Basin. ANU E Press, Canberra, pp 115–134

World Water Forum (2013) Contribute to 'time for Solutions' and share your solutions with the world. http://www.worldwaterforum6.org/fileadmin/user_upload/pdf/Thematic_targets_EN.pdf. Accessed 21 Jan 2013

# Chapter 12
# Vulnerability and Sustainable Development: Issues and Challenges from the Philippines' Agricultural and Water Sectors

**Juan M. Pulhin and Maricel A. Tapia**

**Abstract** The threats of climate change on agriculture and water sectors in the Philippines have added a new dimension and challenge in pursuing sustainable development. Agricultural losses to extreme weather events have seen record-high damages in the recent past amounting to hundreds of thousand of dollars. Impacts on the water sector, on the other hand, have led to water scarcity, agricultural production losses, and decrease in energy supply during El Niño and damage to infrastructure and properties due to flooding brought by intense rainfall. While these adverse effects are more localized and have minimal implications on the national economy and overall economic growth, they present major environmental and social repercussions that thwart the achievement of sustainability goals and inclusive growth. The bundle of adaptation responses to address these impacts should employ integrated approaches to capture the synergies of agriculture and water sectors, and consider social, economic and environment aspects to effectively maximize the desired outcomes. Adaptation should also be scaled up at higher level to create an enabling environment to build the resilience for these sectors to climate change.

**Keywords** Agriculture • Water sector • Climate change • Extreme events • Vulnerability • Adaptation • Economic growth • Sustainable development

## 12.1 Introduction

Agriculture and water sectors are important pillars of the Philippine economy. The former contributed 11 % of the country's Gross Domestic Product (GDP) in 2013 and employed 12.09 million of the 40.43 million labor force in 2012 (BAS 2013). The latter, meanwhile, provides a critical support to the agriculture sector, the major

J.M. Pulhin (✉) • M.A. Tapia
Department of Social Forestry and Forest Governance, College of Forestry and Natural Resources, University of the Philippines Los Baños, College, Laguna 4031, Philippines
e-mail: jmpulhin@up.edu.ph



J.I. Uitto, R. Shaw (eds.), *Sustainable Development and Disaster Risk Reduction*, Disaster Risk Reduction, DOI 10.1007/978-4-431-55078-5_12

user of freshwater, with the reliable access and dependable supply of water resources directly affecting agricultural viability and sustainability (PCARRD-DOST 2009). Both sectors, however, are highly vulnerable to climate change and climate variability as they are directly exposed to these, and their behavior is largely dependent on climatic variables, such as rainfall and temperature (Pulhin et al. 2010; Lansigan et al. 2000; Jose and Cruz 1999; Fazal and Wahab 2013; Israel and Briones 2012), together with other factors.

While much still remains to be done to achieve the desired performances in agriculture and water resources in the Philippines that would aid in realizing economic growth (such as increased productivity, sufficiency and efficiency), the increased demands, however, further put pressure on these critical natural resources. Unsustainable practices add up to the stress experienced by the sectors, which is further worsened by the impacts of climate change and climate variability (ADBI Philippine Report 2013). It is important, therefore, that the pathway and process by which we pursue development, including that of agriculture and water resources, be sustainable so that it would not increase our vulnerability and efforts to improve human welfare would be protected (Anderson 1995).

Agriculture and water are core areas in the discourse on sustainable development (Aquino et al. 2014), as the continued services gained from these sectors rely on the interaction between the human-ecological systems (Kates 2011). Both depend on the ecological services provided by the environment and natural resource systems to realize the benefits derived from them, such as food and clean water. Further, with the significant number of population in the Philippines and around the world dependent on the two sectors for their livelihood, they present important avenues to achieve sustainable development goals and inclusive growth (Aquino et al. 2014).

The threats associated with climate change and climate variability, however, have added a new dimension and challenge in pursuing the goal of sustainable development. Climate change and the associated climate variability serve as additional stressors that threaten livelihoods, destroy the resource base, and impede the country's efforts for economic development. Country's resources otherwise invested for development are instead rechanneled to disaster relief operations and rehabilitations disrupting the efforts towards pursuing the goal of sustainable development. Controlling and limiting the damage from these disasters requires assessment of vulnerability in order to take appropriate actions for reducing the negative impacts, especially before the potential for damages becomes actual (Anderson 1995). It is likewise important to determine the level of vulnerability and when impacts have gone beyond the threshold of survival of individual, community or society, or beyond the carrying capacity of ecosystems and landscapes.

Using one of the most vulnerable countries of the world, the Philippines, as a case, this chapter thus explores how the current and future potential impacts of climate change threaten the contribution of agriculture and water sectors to the country's economic development and the overall welfare of the people. Issues and challenges facing the sectors will be analyzed and potential solutions explored to reduce the adverse impacts of climate change and help achieve the country's quest towards sustainable development.

## 12.2 Impacts of Climate Change on Agricultural Sector

The Philippines has a total area of 9.671 M ha agricultural land based on 2002 Censuses of Agriculture and Fisheries. Of this figure, 4.936 M ha or 52 % is classified as arable land, 4.225 M ha or 44 % as permanent crops, 0.129 M ha or 1 % as permanent meadows/pastures, 0.074 M ha or 1 % as forestland, and 0.307 M ha or 3 % as other lands (BAS 2013). The country's five major crops are: rice, corn, coconut, sugarcane and banana (Aquino et al. 2013). Geographically, nearly 60 % of national rice production comes from the island of Luzon, about 70 % of sugarcane from Visayas, and nearly 60 % of corn and coconut and 80 % of banana from Mindanao (BAS 2013).

In terms of irrigation, the total service area of the national, communal, private and other government agencies (OGA)-assisted irrigations systems was 1.68 M ha in 2013, which is only 56 % of the total irrigable area. Agricultural production statistics also showed that out of the 4.75 M ha area harvested for rice in 2013, 3.24 M ha (68 %) were irrigated and 1.51 M ha (32 %) were rainfed (BAS 2013). The above figures alone are reflective of the potential impacts of climate change and variability to the country's agricultural sector.

FAO (2009) reports that the impact of climate change on global food production is small at least until 2050. However, developing countries like the Philippines are projected to experience decline estimated between 9 and 21 % in overall potential agricultural productivity as a result of global warming. On the other hand, Cline (2007 in Pulhin et al. 2010) modeled the impacts of climate change in 2080 for the Philippines, which indicated an 11.9–23.4 % decrease in agricultural production, with and without carbon fertilization effect. These results uncover a serious threat to the sector, particularly to the country's food security.

Rising temperatures, changes in precipitation patterns and occurrence of extreme weather events and sea level rise will have various impacts on the agricultural sector (Climate Change and the Philippine Agriculture n.d.; Cameron 2014). Long-term changes in the pattern of rainfall and temperature are expected to shift production seasons, pest and disease patterns, and modify the set of feasible crops affecting production, prices, incomes, and ultimately, livelihoods and lives (Rudinas et al. 2013). In addition, other ecosystem functions and services that would be affected by climate change, such as nutrient cycling, waste decomposition, seed dispersal and habitat loss, would all have implications to agricultural production (Cameron 2014).

Lansigan et al. (2000) reported that natural hazards, such as typhoons, floods and droughts, caused 82.4 % of the total Philippine rice losses from 1970 to 1990. Overall, it is typhoon that accounted for the most damage in agriculture, particularly from 1990 to 2006, with total losses equivalent to US$ 199 M (PhP 8.739B).[1] Drought came at second place causing 18 % of the total damage amounting to US$ 51 M (PhP 2.23B), while flooding and flashfloods contributed 5 % of the damage valued at US$ 13 M (PhP 564 M) (Benson 2009).

[1] Following an exchange rate of US$ 1 = PhP 44.

**Table 12.1** Total agricultural area affected by and value of loss to typhoons, floods and drought

| Year | Area affected by extreme events (ha) | Damage to agriculture due to extreme events (M PhP) | Damage equivalent to M US$ | No. of occurrences of typhoons at national level | No. of regions affected by floods | No. of regions affected by drought |
|---|---|---|---|---|---|---|
| Department of Agriculture (DA) assessment of damages to agriculture from natural hazards | | | | | | |
| 2000 | 683,440 | 2,644.15 | 60.09 | | | |
| 2001 | 458,324 | 2,199.96 | 50.00 | 17 | | |
| 2002 | 200,490 | 1,149.51 | 26.13 | 13 | | |
| 2003 | 593,656 | 3,856.89 | 87.66 | 25 | | |
| 2004 | 666,017 | 7,576.04 | 172.18 | 25 | | |
| 2005 | 412,025 | 4,447.16 | 101.07 | 17 | | |
| 2006 | 1,461,608 | 11,012.17 | 250.28 | 20 | | |
| 2007 | 412,362 | 5,310.66 | 120.70 | 13 | 8 | 6 |
| 2008 | 599,355 | 13,683.30 | 310.98 | 21 | 12 | 0 |
| 2009 | 892,841 | 29,518.51 | 670.88 | 22 | 9 | 0 |
| 2010 | 977,208 | 25,484.35 | 579.19 | 11 | 4 | 12 |
| Damages to agriculture brought by typhoons (consolidated from various sources) | | | | | | |
| 2011 | | 17,366.72 | 394.70 | 19 | | |
| 2012 | | 29,120.00 | 661.82 | 17 | | |
| 2013 | | 33,000.00 | 750.00 | 25 | | |

Source: 2000–2010 values from Department of Agriculture as cited by Israel and Briones 2012; 2011 values from National Disaster Risk Reduction Management Council and Emergency Events Database (EM-DAT) as cited by Senate Economic Planning Office 2013. 2012 values are from a news report by Valencia (2012) in Philippine Star. 2013 values are reported by the ABS-CBN News Channel (ANC) in Yahoo Philippines News

Table 12.1 shows the total agricultural area affected by and value of losses to the different natural hazards that hit the country from 2000 to 2013. The list, however, was not complete for regions affected by floods and drought, and 2011–2013 figures were the values of damages recorded from typhoons only. Based on the 2001–2010 assessment of the Department of Agriculture (DA), the year with the largest agricultural area affected by extreme events was 2006, with a total of 1.46 M ha. It can be noted that it is within this year that Typhoons Durian (Reming) and Xangsane (Milenyo) hit the country. However, in terms of value of damages, it only ranked fourth, or equivalent to US$ 250 M (PhP 11B). Meanwhile, the year 2009 had the highest value of losses amounting to US$ 670 M (PhP 29.5B). A total of 22 typhoons visited the country in the said year, and 9 regions were affected by floods. This was the same year when Tropical Storm Ketsana (Ondoy) and Typhoons Parma (Pepeng) and Mirinae (Santi) visited the country, which enhanced the southwestern monsoon and brought torrential rains. The total area affected by the natural hazards in 2009 was 892,841 ha, which sits at third place among the largest agricultural area affected within the 2000–2010 period.

The past 3 years (2011–2013) have demonstrated overall record-high damages to agriculture from typhoons alone, which is higher than the combined damages from different natural hazards. The period has also recorded typhoon events each of which had incurred more than US$ 24 M (PhP 1B) total damages in agriculture. These are: Typhoon Nesat (Pedring) and severe Tropical Storm Washi (Sendong) in 2011, Typhoon Bopha (Pablo) in 2012, and Super Typhoon Haiyan (Yolanda) in 2013. It seemed as though that these extreme events have become the new "normal" that the country has to contend with in the future.

Destruction to agriculture brought by tropical cyclones is caused by the strong winds and heavy rainfall (and the resulting floods), which negatively impact the crops. The Department of Agriculture (2010 in Lansigan 2014) analyzed the estimated yield loss of various crops at different growth stage due to flooding or strong winds. For rice, yield loss due to flooding was largest at the panicle initiation and flowering stages, with up to 100 % and 70 % estimated losses, respectively, under 7 days of floodwater submergence. In the case of corn, the reproductive stage is also the most vulnerable, with 55 % and 60 % estimated loss if subjected to 101–150 kph wind velocity for $<12$ and $\geq 12$ h, respectively, and 80 % and 80–100 % estimated loss at $>150$ kph wind velocity for $<12$ and $\geq 12$ h, respectively. Aside from production losses, damage to facilities and/or infrastructures, damaged farm supply routes and markets, and worse, death or injury to farm workers may be incurred from typhoons (and resulting floods). Nevertheless, typhoons also bring positive impacts to agriculture in terms of increased water supply and improved soil nutrients (Israel and Briones 2012).

The El Niño Southern Oscillation (ENSO) is an oceanic event associated with fluctuation of inter-tropical surface pattern and circulation in the Indian and Pacific Oceans. This inter-annual climate phenomenon is a significant determinant of climate variation in the Pacific Rim region (Roberts et al. 2009). In the Philippines, ENSO or El Niño events are associated with: (a) late onset of the rainy season, (b) early termination of the rainy season, (c) weak monsoon events characterized by isolated heavy rainfall events of short-duration, and (d) weak tropical cyclone activity characterized by less intense cyclones and less number of tropical cyclones occurring within Philippine territory (Lansigan et al. 2000). This event, oftentimes associated with drought, has considerable damaging impacts to the agricultural sector.

In the recent decade, the year 2010 was the driest, with 12 regions affected by drought (Table 12.1). The number of typhoons that visited the country also dropped to only 11, with 4 regions affected by floods. Nevertheless, the damage to agricultural areas (largely due to decreased rainfall) reached 977,208 ha, which translates to US$ 580 M (PhP 25.5B). These figures come second in both categories for the decade under study, emphasizing the severity of the impacts of drought to agriculture despite its occasional occurrence.

Rice is the agricultural crop most affected by drought, as it uses more water than other major crops above. In fact, rice production plunged during strong El Niño events, particularly in 1972–1973, 1982–1983, and 1997–1998 (Comiso et al. 2014). A closer look by Roberts et al. (2009) found that the drop in production was due to decrease in area harvested rather than yield, as farmers abandon rice farming due to low rainfall. The rainfed farm systems also bore much of the losses, compared to the irrigated systems. For other crops, like corn, the impact on its production is not as dramatic as rice. The impact on coconut production, on the other hand, takes time to effect due to the lag between flowering and filling of nuts, during which the crop may be subjected to water stress, and the actual harvest time (Comiso et al. 2014). Some coconut farmers also noticed shrinkage in the size of the fruit during drought (Tapia et al. 2014).

It can be gleaned from the figures in Table 12.1 that higher frequency of extreme events does not necessarily translate to greater damages to agriculture, as exhibited in the years 2003 and 2004 with both 25 typhoons. Oftentimes, it is the intensity and the exposure of the agricultural area to the destructive elements of the natural hazards, such as excessive rains, floods, and extensive and long-term droughts, which dictate the magnitude of the impacts.

Meanwhile, spatial analysis of the three predicted consequences of climate change (i.e., landslides, drought and flooding) showed that approximately 67 % (20 M ha) of the country's total area would be severely affected (Table 12.2). To determine its impacts to the agricultural sector, the coincidence maps was overlaid with the Strategic Agricultural and Fishery Development Zone (SAFDZ, i.e., areas identified for production, agro-processing, and marketing activities to help modernize, with support of government, the agriculture and fisheries sector in an environmentally and socio-culturally sound manner). Results showed that 10.2 M ha or 34 % of the country would be affected, encompassing 85 % of the SAFDZ areas. The co-occurrence of the three events with SAFDZ is estimated to be 162,000 ha (Rudinas et al. 2013).

**Table 12.2** Impacts of climate change to Philippine agriculture

| Code | Description | Hectares | Percent of country |
|---|---|---|---|
| 1 | Drought + flooding + Landslide + SAFDZ | 162,098.58 | 0.54 |
| 2 | Drought + landslide + SAFDZ | 397,715.25 | 1.33 |
| 3 | Flooding + landslide + SAFDZ | 151,605.21 | 0.51 |
| 4 | Drought + flooding + SAFDZ | 2,597,893.53 | 8.66 |
| 5 | Drought + SAFDZ | 3,358,360.89 | 11.19 |
| 6 | Flooding + SAFDZ | 2,720,264.80 | 9.07 |
| 7 | Landslide + SAFDZ | 729,550.58 | 2.43 |
| 8 | Drought + flooding + landslide | 101,732.60 | 0.34 |
| 9 | Drought + landslide | 703,825.30 | 2.35 |
| 10 | Flooding + landslide | 155,947.01 | 0.52 |
| 11 | Drought + flooding | 1,129,297.76 | 3.76 |
| 12 | Dry land only | 4,549,601.28 | 15.17 |
| 13 | Flooding only | 1,560,165.01 | 5.20 |
| 14 | Landslide only | 1,723,463.33 | 5.74 |
| 15 | SAFDZ only (not affected) | 4,248,134.32 | 14.16 |
| | **Total** | **24,289,655.44** | **80.97** |

Godilano 2009, 2010 in Rudinas et al. (2013)

While the above present and future impacts of climate change present a bleak scenario for the Philippine's agriculture sector, Isabel and Briones (2012) alluded that the impacts of typhoons, floods or droughts in specific geographical areas in the Philippines would not significantly affect national production and prices. Therefore, food security at the national level may not be a concern following climatic disasters. Nevertheless, as the impacts of typhoon, particularly on rice production, cause significant losses at the provincial level, it is at this level that diminished rice availability, and therefore food insecurity, would be felt, as well as at the household level.

## 12.3 Impacts of Climate Change on Water Sector

Water resources in the Philippines include rainfall, surface water (rivers, lakes and reservoirs) and groundwater. The mean annual rainfall ranges from 1,000 to 4,000 mm, while the annual average is 2,400 mm. About 50 % of rainfall is collected as water runoff in the 421 principal river basins, 59 lakes and a number of small streams. The dependable surface water supply from rivers, lakes and reservoirs is estimated at 125.8 billion cubic meters. The groundwater reservoir of the country covers an aggregate area of 5 M ha, with a potential to supply 20.2 billion cubic meters, and constantly recharged by rains and seepage. Given the high volume of the total water resources potential of the country, the Philippines should have adequate water supply. However, geographical and seasonal variations make water availability both time- and site-specific. Likewise, weather and climate-related variability and extreme events can lead to water shortages, particularly in highly populated areas (PEM 2003; Greenpeace 2007 in Pulhin et al. 2010; Climate Change Commission n.d.).

Climate change and climate variability are expected to shift the hydrologic process in watersheds that would affect the spatial and temporal distribution of water resources (Comiso et al. 2014). Climate scenarios that project wetter wet season and drier dry season (see for instance the study of Tapia et al. 2014) would have profound effects on streamflow, dam operation and water allocation, domestic water supply, irrigation, hydro-power generation, depth and recharge of aquifers, water quality (e.g., salt-water intrusion), and even on water infrastructures and management systems. These would have detrimental consequences on environmental integrity, food and human security, and the economy (Climate Change Commission n.d.; Comiso et al. 2014).

Recent extreme weather events like typhoons and droughts have shown the vulnerability of the country's water sector to these natural hazards and climate change. For instance, Tropical Storm Ketsana (Ondoy) exposed the deficiencies in water infrastructures and management systems in the country, catching the sector off-guard to extreme climate variability. Pumping facilities to ease floodwaters in Metro Manila were found capable only of handling up to 100 mm of rainfall per hour, leaving the greater part of the metro and adjacent municipalities submerged in floodwaters. The said extreme event also destroyed more than US$ 18.7 M (PhP 820 M) worth of irrigation facilities, including dikes and canals that serviced 53,000 ha of farmland in Central Luzon. Further, it disabled the water supply in the city, affecting more than 100,000 households without piped-in water (Climate Change Commission n.d.). The National Irrigation Administration (NIA) was also forced to open the gates of some water reservoirs, such as La Mesa Dam, Ipo Dam, Ambuklao Dam and Binga Dam, as their water levels had already reached critical status, causing floods to more than 500 barangays or communities in Region III (NDCC 2009).

Meanwhile, drought impacts the major reservoirs in the country as it leads to significant dips in water inflows. This causes shortages in domestic water and irrigation supply (Jose 2002). The 1997–1998 El Niño in the country led to a major reduction in the water level of Angat Dam (from 37 to 22 $m^3$ per second), which supplies more than 90 % of domestic water in Metro Manila. The Metropolitan Waterworks and Sewerage System resorted to water rationing, and supply was only made available for 4 h a day. The Bureau of Soils and Water Management also spent US$ 0.83 M (PhP 36.7 M) in cloud seeding during this El Niño event. Further, indiscriminate use of water wells was rampant, which resulted in groundwater depletion and salt-water intrusion (Juanillo 2011).

Aside from agricultural production losses and water shortage, drought events also have huge implications on the energy condition in the country. The 1989–1990 drought incurred a hydropower generation loss of US$ 7.9 M (PhP 348 M). In 1991–1992, the major multi-purpose dams in Luzon (Angat, Magat and Pantabangan) experienced power generation losses of about 31 %, resulting in greater use of thermal plants to compensate for the deficit. Angat dam also had a hydropower generation deficit of 333 Gwh from second quarter of 1997 to third quarter of 1998, constituting a drop in power generation from 26.4 to 58.9 % (Hilario et al. 2009; Tejada et al. n.d.; Juanillo 2011).

Several studies have simulated the effect of climate change on hydrologic responses in specific watersheds and water reservoirs. Both studies of Jose and Cruz (1999) in Angat Reservoir and Lake Lanao and Combalicer and Im (2012) in Mt. Makiling Forest Reserve (MMFR) have underscored that more than temperature, it is precipitation that influences the variation of discharge, and therefore the availability of surface water, in the watershed. A reduction in rainfall, however, was observed to have more impacts in terms of runoff rather than increased rainfall. In general, the two studies foresee a reduction in runoff in the future for all the studied watersheds. Meanwhile, the projected higher temperatures would lead to higher evaporation rate, which would also result in loss of water. It is also worthy to note, based on the study of Combalicer and Im (2012) of MMFR, that while significant variations in hydrologic systems would be observed regardless of land cover due to the changing climate, it is the land use with less vegetative cover where severe outcomes are expected.

Drought, including the El Niño-driven ones, would be the climate extreme that would have an acute impact on the water sector, especially given the high likelihood that its occurrence would increase in the future (Solomon et al. 2007). Focusing on the Pampanga River Basin, Jaranilla-Sanchez et al. (2011) looked into three different types of drought at the basin scale: (1) meteorological droughts characterized by precipitation deficit; (2) hydrological drought characterized by inadequate streamflows (as determined by discharge and groundwater level deficit); and (3) agricultural drought characterized by soil-moisture deficit insufficient to meet crop requirements. Based on hydrological parameters, such as rainfall, discharge, surface and root-zone soil moisture, and groundwater level, it was found that the above different drought types are time- and area-specific and may happen, as well, in combination. There is also time lag of 1–7 months observed in the basin between the parameters, such that, for instance, a precipitation deficit does not immediately result in groundwater deficit. Knowledge of these hydrological behaviors at the basin scale is therefore useful in preparing for potential impacts.

## 12.4 Climate Change Vulnerability, Economic Growth and Sustainable Development: Issues and Challenges

From the economics viewpoint, natural disasters obstruct the smooth operation of the economic system, adversely affecting assets, production factors, output, employment or consumption (Hallegate and Przyluski 2010 in Israel and Briones 2012). The 1994 ADB report (in Benson 1997) recounted that typhoon alone incurred overall damages of about US$ 55 M (PhP 2.4B) every year which is equivalent to 0.6 % of the country's Gross National Product (GNP). Meanwhile, reports on the damages of natural hazards to agriculture in the recent decade have reached an average of almost US$ 286 M (PhP 12.6B) per annum from 1990 to 2006 (Benson 2009). These figures purport that natural hazards diminish any economic growth achieved and perpetrate poverty in the country, particularly in the already poverty-stricken rural areas engaged in agricultural activities.

However, Philippine economists shared that the impacts of these typhoons to the national economy are minimal (Ang 2014; Larano 2013; Moss 2014). Citing the recent typhoons Haiyan (Yolanda) in 2013 and Rammasun (Glenda) in 2014, the latter had greater (though still not very significant) impact on the economic growth surge of the Philippines as it passed through the country's centers of economic activities (National Capital Region, Region IV and Region V), which collectively contributes 65 % of the total economic production. The major impacts of these typhoons as well were largely on agriculture, of which the share to the country's total output has dropped over the years and the impacts are felt on selected agricultural areas only depending on the typhoon path.

Overall, the Philippine economy is resilient to typhoons, which is more vulnerable to external factors such as the 1998 Asian financial crisis and the 2009 global financial crisis. The remittances from the overseas Filipino workers contribute to the country's economic resiliency, serving as safeguard from major economic blows. However, this should not promote complacency as historical records have shown an increase in the frequency and intensity of typhoons in the Philippines (http://www.typhoon2000.ph; Comiso et al. 2014). The cost of damage from typhoons has been rising as well, based on damage and Gross Domestic Product (GDP) ratio, which reached more than 1 % already in several occasions and 2.7 % in 2009 (Ang 2014; Comiso et al. 2014; Barbon 2013). This should therefore urge the policymakers and decision-makers to take the necessary steps to invest in the climate proofing of the agricultural sector.

Meanwhile, economic impacts of drought are usually expressed in terms of damages to agriculture, with the US$ 50.7 M (PhP 2.23B) of the US$ 50.8 M (PhP 2.237B) losses reported by the Philippine National Disaster Coordinating Council (in Benson 2009) attributed to this sector. Nevertheless, the impacts on water sector, particularly in its hydropower generation, are also huge, with drought (together with the ill-maintenance of the critical watersheds) among the factors causing the inadequate power capacity in the country. Hydroelectric power supports 20 % of the country's energy supply with a total capacity of about 3,500 MW, and comprising 64 % of the country's total renewable energy capacity (Renewable Facts 2011). In Mindanao alone, 55 % of power is generated from hydroelectric power plants, whose dams are approaching critical levels (Rubrico 2010). These have huge implications on the Philippine economy, specifically with the imminent energy crisis in 2015.

While macroeconomic figures on the effects of natural hazards and climate change on agriculture and water sectors are marginal compared to the total economic outputs, this in no way downplays the magnitude of the impacts described above. Several studies (Tapia et al. 2014; Pulhin et al. 2006, 2014) have underscored the reduction in the services or benefits gained from agriculture and water sectors, such as total or partial crop losses, which affects livelihood, food security, health and the general welfare of the communities. The above effects, thus, demonstrate the repercussions on social growth of the impacts of climate change and extreme weather events on agriculture and water.

The documented impacts include children experiencing malnutrition due to unavailability of food. This therefore affects the children's performance in school, and in worse cases, cause them to stop schooling due to inability of parents to financially support their education. Some members of the family, particularly women, look for other sources of income in other provinces and cities (usually as household helpers) to pay for debts incurred in agriculture production but cannot be paid due to losses. Other family members, on the other hand, resort to unsustainable livelihood practices, such as charcoal-making and illegal logging, which further destroy the already critical watersheds. In some cases, it dampens the confidence of the people to government agencies or institutions, who are supposed to provide them with technical assistance to build resilience to climate change and extreme weather events (Tapia et al. 2014; Pulhin et al. 2006, 2014). These realities point out what countless of literatures has already echoed that indeed it is at the local level that the impacts of climate change and the accompanying variability and extremes would be felt. It also presents a major setback in achieving sustainable development goals for these agricultural communities.

The Philippines has experienced strong economic growth rates in the past despite the strong blows from natural calamities, such as Super Typhoon Haiyan. Of late, the country has exhibited exceptional performance as fastest growing economy in the ASEAN region. However, it lagged in translating this development into inclusive growth. Two-thirds of the Filipino population resides in rural areas, which are predominantly agricultural economies (Evangelista 2014). Critical watersheds that supply the domestic, agriculture, hydropower and industrial water requirements of the country (such as Magat, Angat and Pantabangan) are also situated in the same rural areas. It is high time that the plight of these people and the environment, which is their source of livelihood, be paid serious attention.

Economic growth and a resilient economy are not enough measures of development (Rudinas et al. 2013). These achievements also may not last long especially if the capability of land and water ecosystems to provide services that fuel the economy is gradually reduced by natural disasters and climate change (Israel and Briones 2012). We need growth that focus as well on social and environmental dimensions to ensure that the progress achieved is equitable and sustainable.

Aside from considering green and inclusive growth strategies, we also need systemic or holistic approaches to development, including climate change adaptation and disaster risk reduction, that looked into the different scales of the environment, economy and political sector. Piecemeal approaches to solving agricultural and water problems have time and again proven to be ineffective, discriminatory and prostrate some groups (particularly the farmers and natural resources-dependent communities) to further poverty and vulnerability. For instance, rice importation may be an answer to production deficits and food insecurity caused by agricultural losses from natural hazards. However, this puts our local farmers in jeopardy, as they cannot compete with cheaper and sometimes subsidized imported rice. The case of rehabilitating and protecting critical watersheds for continuous supply of water presents another story, with the communities residing within these areas bound by restrictive livelihood opportunities and are expected to participate in rehabilitation efforts. Yet, they do not benefit from the piped-in water supply, irrigation or electricity from these dams, as these are serviced to nearby cities or lowland farms.

Other issues and challenges faced by the agricultural and water sectors to achieve sustainable development in the face of climate change are: raising public awareness on climate change; strengthening the capacity of local government units to champion effective adaptation strategies at the local level; improving the climate change projection; enhancing capacity of researchers to conduct integrated assessment; enhancing science-policy-action interface; building on experience of indigenous communities for effective adaptation strategies; and developing and using adaptation metrics for planning and monitoring purposes (Pulhin et al. 2010).

## 12.5 Potential Adaptation Strategies of Agriculture and Water Sectors

Given the vulnerability of the agriculture and water sectors to climate change, it is imperative therefore that their increased resilience to climate change risks be an important development priority of the Philippines (Israel and Briones 2012). However, there is no one-size-fits-all adaptation strategy that would address the concerns for each sector across the country. Every agricultural region or watershed is different and requires context-specific and localized responses (Cameron 2014; Israel and Briones 2012). While the literature is replete with examples of adaptive responses for agriculture and water resources, the applications of these rest on solid knowledge of their vulnerability to the above natural hazards. It should be clear as well that any planned intervention, particularly at farm level, could also accomplish much in terms of the desired outcomes (Cameron 2014).

For agriculture, there are a number of responses that deals with adapting to the changing biophysical environment through thorough knowledge of crop growth, improved crop rotation system, and efficient weather forecasts. These are also usually accompanied by understanding the watershed's hydrological behavior (as emphasized by Jaranilla-Sanchez et al. 2011) and availability of infrastructure support, such as provision for irrigation, rainwater harvesting systems, postharvest and storage facilities. Climate variability and extremes are among the chief adversaries of crops that also affect soil condition and water availability. Equipped with the know-how on the above subjects, a proper cropping calendar can be generated that would determine the appropriate window for planting (Cameron 2014; Lansigan 2010; Rudinas et al. 2013). Lansigan (2010) presented an approach using statistical hydrology and crop physiology to help distinguish the optimal planting schedules, with the aid of a physiological crop simulation model.

Improving land management system is another adaptation strategy that creates a suitable environment for crop growth (See Fig. 12.1). An example of this is employing agroforestry, organic farming, farm diversification, and Sloping Agricultural Land Technology (SALT). These production systems also combine adaptation and mitigation objectives. In particular, integrating trees in farms improves the carbon stocks in the soil and aboveground biomass, enhances microclimate, increases water retention, and enriches soil nutrients, among others. On top of these, the said production system also diversifies the farmer's income through the production of high value crops (Verchot et al. 2007).

Meanwhile, some agricultural adaptations rely on biotechnology to produce crops that are better adapted to the changing climate and extreme weather events, such as drought-, flood- or saline-resistant varieties. This may also involve plain selection of the best varieties that are tolerant to the above harsh elements. Nevertheless, these strategies rely on improving gene conservation and availability of deep gene banks (Rudinas et al. 2013; Cameron 2014). Best practice on this was demonstrated by a farmer-led network of people's organizations, non-government organizations and scientists, called MASIPAG, which has developed their own early-maturing rice varieties that can be harvested before the start of the typhoon season (Rudinas et al. 2013).

The above adaptations, which are mostly at the farm and watershed levels and focus on production support, are usually practiced in combination and depend on the characteristics of the agricultural area. However, these strategies should be accompanied by higher scale responses, as well as economic and institutional interventions, to truly build climate resilience in the agricultural sector. A comprehensive knowledge of the agricultural production areas, such as the locations of the

**Fig. 12.1** Transformation of farms in the Philippines through conservation farming village, a modality for enhancing transfer of agroforestry and other conservation farming technologies and practices (Photo Credit: PCAARRD-CFV Project)

rainfed and irrigated farms and types of ecosystems where they are situated, can aid decision-makers perform appropriate actions in times of calamity. Insurance mechanisms should be in place or expanded, preferentially covering the most vulnerable, as protection from production losses incurred due to extreme weather events (Roberts et al. 2009).

Market competition and high cost of inputs bring enormous burden to farmers. Moreover, any production support would be negated without the appropriate assistance in marketing the products at the right price (Acosta-Michlik and Espaldon 2008). Thus, ensuring a favorable economic environment (e.g., through management of rice imports) would level the playing field in favor of local farmers and aid them for better market competition. Other market instruments, such as paying the farmers by employing sustainable practices, risk sharing and subsidies, could also encourage adoption of the identified adaptation strategies (Cameron 2014).

Adaptation in water resources, on the other hand, involves tackling both the supply and demand sides (Jose and Cruz 1999). Supply adaptation consists of comprehensive watershed management and water allocation system and procedures. The former addresses the degradation in the country's critical watershed, due to diminishing forest cover, unsustainable agricultural practices, siltation, etc. It is important, as well, similar to the adaptation strategy above, that the hydrologic behavior of watersheds be understood with the aid of simulation models to design appropriate interventions, particularly in times of drought (Jaranilla-Sanchez et al. 2011). Water allocation system aims to manage water demand through needs priority. The demand adaptation, meanwhile, involves: (1) enhancement of irrigation efficiency; (2) introduction of low water use crops and efficient farming practices; (3) recycling (reuse) of water; (4) improvement of monitoring and forecasting systems for floods and droughts; and (5) use of water pricing policies and structures. At the institutional and governance level, the weak and fragmented management of the water sector due to the numerous government offices handling it, as clearly exposed by the series of storms and typhoons in 2009, should also be addressed to promote a more coordinated response for sustainable water supply (Climate Change Commission n.d.).

Application of the adaptation strategies for agriculture and water resources requires knowledge on barriers for their effective implementation. At the farm level, lack of money and information was found a hindrance in adopting technical measures, especially among subsistence or traditional farmers (Acosta-Michlick and Espaldon 2008). Other hurdles, as listed by Cameron (2014), include uncertainty about climate impacts, limited financial and human resources, limited coordination of different levels of governance and decision-making, different perceptions of risk, inadequate responses from political institutions, competing values, absence of leaders and champions, and limited tools to monitor effectiveness.

As can be noted above, adaptations for both agriculture and water sectors present areas for synergistic approaches to better manage climate change impacts. Integrated planning, through watershed, landscape approach or ecosystem-based approach, reconciles utilization, conservation and development strategies, and avoids disjointed or duplicative responses (Locatelli et al. 2008). Needless to say, this integrative adaptation should not only happen at the geographic realm, but also include economic and policy aspects, research and development, and education of concerned stakeholders.

## 12.6 Conclusions

The impacts of climate variability and extreme weather events, and therefore of climate change, on the agriculture and water sectors are more localized, and affect more the farmers, the immediate communities and the ecosystems. Adverse effects may also be felt at the provincial level or cut across municipalities and provinces, just like the case of food insecurity or widespread floods. However, as there are only selected areas in the country that are hit by the above natural hazards at one time, the magnitude of national level consequences, particularly on the economy, may be minimal and does not necessarily saddle economic growth. This, nonetheless, should not promote an attitude of complacency as historical records have shown increased frequency and intensity in extreme weather events, such as typhoons. Economists have also observed rising total damage cost and GDP ratio, which has even reached almost 3 % due to Tropical Storm Ketsana and Typhoon Parma in 2009. Moreover, apart from water shortages for irrigation and domestic supply, a critical national concern in times of drought is the reduced hydropower generation, which amplifies the threat of energy crisis in the country.

Meanwhile, there are a wide range of available adaptation strategies that can be considered in addressing the impacts of climate change, variability and extreme events. It is important to recognize the cause of vulnerability of a particular agricultural region or watershed so that an appropriate and complementary bundle of responses could be delivered, which does not only focus on production but takes into consideration the welfare of the people and the environment. Thus, it is crucial that integrated planning, through the employment of landscape, watershed, or ecosystem approaches, be employed to capture the synergies for both sectors. Adaptation should also be scaled up at the higher level, through economic and political reforms and support for research, development and extension, to create an enabling environment for truly building the resilience of the agriculture and water sectors to climate change. With the above actions, the vulnerabilities of the above sectors can be turned into one that does not only contribute to economic growth but that which promotes sustainable development.

## References


Acosta-Michlik L, Espaldon V (2008) Assessing vulnerability of selected farming communities in the Philippines based on behavioural model of agent's adaptation to global environmental change. Glob Environ Chang 18:554–563

Anderson MB (1995) Chapter 3. Vulnerability to disaster and sustainable development: a general framework for assessing vulnerability. In: Munasinghe M, Clarke C Disaster prevention for sustainable development: economic and policy issues. A report from the Yokohama World conference on natural disaster reduction, 23–27 May 1994. The International Decade for Natural Disaster Reduction (IDNDR) and The World Bank, Washington, DC

Ang A (2014) Typhoons and economic growth. Business Mirror, 24 July 2014. http://www.businessmirror.com.ph/index.php/en/news/opinion/35974-typhoons-and-economic-growth. Accessed 15 Aug 2014

Aquino AP, Ani PAB, Festejo MA (2013) An overview of policies and public sector investments in Philippine agriculture. http://ap.fftc.agnet.org/ap_db.php?id=65. Accessed 15 Aug 2014

Aquino AP, Ani PAB, Festejo MA (2014) Achieving sustainable and inclusive green growth: major agricultural policies in the Philippines. Paper presented at the International Workshop on Collection of Relevant Agricultural Policy Information and Its Policy Use, Council of Agriculture, Taipei, Taiwan, 23–27 June 2014

Asian Development Bank Institute (ADBI) Philippine Report (2013) Developing farming systems to mitigate climate change: the Philippine report. Submitted to Asian Development Bank Institute. www.adbi.org/files/2013.08.29.cpp.sess7.8country.ppr.philippines.pdf. Accessed 18 July 2014

Barbon WJ (2013) Typhoon Haiyan, the Philippine's tipping point to improve disaster risk reduction. International Institute of Rural Reconstruction, Across Asia, Regional Center for Asia, 2013 (2). http://www.iirr.org/images/uploads/Newsletter_2013_No_2.pdf. Accessed 18 July 2014

Benson C (1997) The economic impact of natural disasters in the Philippines. Working paper 99. Portland House, London
Benson C (2009) Mainstreaming disaster risk reduction into development: challenges and experience in the Philippines. Provention Consortium, Geneva
Bureau of Agricultural Statistics (BAS) (2013) Selected statistics on agriculture. BAS, Department of Agriculture, Quezon City, Philippines. http://www.bas.gov.ph. Accessed 7 July 2014
Cameron E (2014) Climate change: implications for agriculture. Key findings from the Intergovernmental Panel on Climate Change fifth assessment report. www.cisl.cam.ac.uk/ipcc. Accessed 18 July 2014
Climate Change and the Philippine Agriculture (n.d.) Research and development, extension agenda and programs 2011–2016
Climate Change Commission (n.d.) National climate change action plan 2011–2028
Combalicer EA, Im S (2012) Change anomalies of hydrologic responses to climate variability and land-use changes in Makiling Forest Reserve. J Environ Sci Manag, Special Issue (1–2012):1–13
Comiso JC, Blanche CA, Sarigumba TI, Espaldon MVO, Lansigan FP, Baguinon NT, Birosel RC, Coladilla JO, Cruz RVO, Florece LM, Guerrero RD III, Lasco RD, Perez RT, Pulhin JM, Tibig LV (2014) Changing Philippine climate: impacts on agriculture and natural resources. The University of the Philippines Press, Diliman
Evangelista R (2014) Strategies for inclusive development. Philippine Daily Inquirer, 29 June 2014. http://opinion.inquirer.net/76044/strategies-for-inclusive-development. Accessed 15 Aug 2014
Fazal SA, Wahab SA (2013) Economic impact of climate change on agricultural sector: a review. J Transform Entrep 1(1):39–49
Food and Agriculture Organization of the United Nations (2009) 2050: climate change will worsen the plight of the poor. http://www.fao.org/news/story/en/item/35831/icode/. Accessed 22 June 2014
Hilario F, de Guzman R, Ortega D, Hayman P, Alexander B (2009) El Niño southern oscillation in the Philippines: impacts, forecasts and risk management. Philipp J Dev 36(1):9–34. http://dirp3.pids.gov.ph/ris/pjd/pidspjd09-1enso.pdf. Accessed 15 Aug 2014
Israel DC, Briones RM (2012) Impacts of natural disasters on agriculture, food security, and natural resources and environment in the Philippines. Discussion paper series no. 2012-36. Philippine Institute for Development Studies (PIDS). http://dirp4.pids.gov.ph/ris/dps/pidsdps1236.pdf. Accessed 15 Aug. 2014
Jaranilla-Sanchez PA, Wang L, Koike T (2011) Modeling the hydrologic responses of the Pampanga River basin, Philippines: a quantitative approach for identifying droughts. Water Resour Res 47:W03514. doi:10.1029/2010WR009702
Jose AM (2002) El Niño-related droughts in the Philippines. Impacts and mitigation efforts. Asian Dis Manag News 8 (3), July–Sept 2002
Jose AM, Cruz NA (1999) Climate change impacts and responses in the Philippines: water resources. Clin Res 12:77–84
Juanillo E (2011) El Niño impacts on Philippine water resources: focus on Angat Dam. Abstract of paper presented at World Climate Research Programme (WCRP) Climate Research in Service to Society Conference, Denver, 24–28 October 2011
Kates RW (2011) What kind of a science is sustainability science? Proc Natl Acad Sci 108(49):19449–19450. doi:10.1073/pnas.1116097108
Lansigan FP (2010) Determining the seasonal climate-based optimal planting dates for rainfed rice in selected locations in the Philippines. 11th National Convention on Statistics (NCS), EDSA Shangri-La Hotel, 4–5 October 2010
Lansigan FP (2014) Implementation issues in weather index-based insurance in agricultural crop production in the Philippines. Paper presented at the SEARCA regional professorial chair lecture, SEARCA, College, Laguna, 28 Jan 2014

Lansigan FP, de los Santos WL, Coladilla JO (2000) Agronomic impacts of climate variability on rice production in the Philippines. Agric Ecosyst Environ 82(2000):129–137
Larano C (2013) Typhoon's economic impacts on Philippines likely to be minimal. Wall St J Southeast Asia, 30 November 2013. http://blogs.wsj.com/searealtime/2013/11/30/typhoons-economic-impact-on-philippines-likely-to-be-minimal/. Accessed 15 Aug 2014
Locatelli B, Herawati H, Brockhaus M, Idinoba M, Kanninen M (2008) Methods and tools for assessing the vulnerability of forests and people to climate change: an introduction. Working paper no. 43. Center for International Forestry Research, Bogor
Moss T (2014) Typhoon seen having small impact on Philippine economy. Wall St J World, 15 July 2014. http://online.wsj.com/articles/typhoon-rammasun-seen-having-small-impact-on-philippine-economy-1405419941. Accessed 15 Aug 2014
National Disaster Coordinating Council (NDCC) (2009) NDCC update: final report on tropical storm "Ondoy" (Ketsana) and Typhoon Pepeng (Parma). http://ndcc.gov.ph/attachments/092_NDCC%20Update%20Final%20Report%20re%20TS%20Ondoy%20and%20Pepeng.pdf. Accessed 15 Aug 2014
Philippine Council for Agriculture, Forestry and Natural Resources Research and Development – Department of Science and Technology (PCARRD-DOST) (2009) Philippine S&T agenda on climate change in the agriculture, forestry and natural resources sectors (2010–2016). PCARRD-DOST, Los Baños
Pulhin JM, Peras RJJ, Cruz RVO, Lasco RD, Pulhin FB, Tapia MA (2006) Vulnerability of communities to climate variability and extremes: Pantabangan-Carranglan watershed in the Philippines. AIACC working paper no. 44. International START Secretariat, Washington, DC, USA. Retrieved from URL: www.aiaccproject.org. Accessed 11 Apr 2011
Pulhin JM, Peras RJJ, Tapia MA (2010) Climate change adaptation in water and agricultural sectors in the Philippines: some issues and challenges. Asian J Environ Dis Manag 2(4):471–483
Pulhin JM, Tapia MA, Peras RJJ, Peria AS, Predo CD, Combalicer EA (2014) Adaptation to extreme events in Southeast Asia through sustainable land management systems. College of Forestry and Natural Resources, University of the Philippines Los Baños and World Agroforestry Centre (ICRAF) – Philippines, Los Baños
Renewable Facts (2011) Philippines: hydropower. http://www.renewablefacts.com/country/philippines/hydro. Accessed 15 Aug 2014
Roberts MG, Dawe D, Falcon WP, Naylor RL (2009) El Niño southern oscillation impacts on rise production in Luzon, the Philippines. Notes and Correspondence. J Appl Met Clim 48:1718–1724. doi:10.1175/2008JAMNC1628.1
Rubrico JGU (2010) Philippines' power at crisis point. Asia Times Online, April 10, 2010. http://www.atimes.com/atimes/Southeast_Asia/LD10Ae01.html. Accessed 15 Aug 2014
Rudinas JS, Godilano EC, Ilaga AG (2013) Implementing climate smart agriculture. Ridge-river-reef: the Philippine adaptation and mitigation initiative for agriculture. Paper presented to the ASEAN - FAO - GIZ regional expert forum on climate change: agriculture and food security in ASEAN. Organized jointly by the ASEAN Secretariat, the Food and Agriculture Organization of the United Nations (FAO) and German International Cooperation (GIZ), 2–3 May 2013, Nai Lert Hotel, Bangkok, Thailand
Senate Economic Planning Office (2013) Natural disasters at a glance. https://www.senate.gov.ph/publications/AAG%202013-04%20-%20Natural%20Disasters_final.pdf. Accessed 29 Sept 2014
Solomon S, Qin D, Manning M, Alley RB, Berntsen T, Bindoff NL, Chen Z, Chidthaisong A, Gregory JM, Hegerl JC, Heimann M, Hewitson B, Hoskins BJ, Joos F, Jouzel J, Kattsov V, Lohmann U, Matsuno T, Molina M, Nicholls N, Overpeck J, Raga G, Ramaswamy V, Ren J, Rusticucci M, Somerville R, Stocker TF, Whetton P, Wood RA, Wratt P (2007) Technical summary. In: Solomon S et al (eds) Climate change 2007: the physical science basis: contribution of Working Group I to the fourth assessment report of the Intergovernmental Panel on Climate Change. Cambridge University Press, Cambridge

Tapia MA, Pulhin JM, Peras RJJ (2014) Vulnerability and adaptation to climate change of selected community-based forest management areas in Oas, Albay, Philippines. Asia Life Sci 23(1):567–592

Tejada SQ, Tuddao VB Jr, Juanillo E, Brampio E (n.d.) Drought conditions and management strategies in the Philippines. http://www.ais.unwater.org/ais/pluginfile.php/597/mod_page/content/70/Philippines.pdf. Accessed 15 Aug 2014

Typhoon2000.com. The Philippines' first website on tropical cyclones. http://www.typhoon2000.ph. Accessed 22 June 2014

Valencia C (2012) Yearender: typhoons dent agri production in 2012. The Philippine Star, 28 December 2012, http://www.philstar.com/business/2012/12/28/890677/yearender-typhoons-dent-agri-production-2012

Verchot LV, Van Noordwijk M, Kandji S, Tomich T, Ong C, Albrecht A, Mackensen J, Bantilan C, Anupama KV, Palm C (2007) Climate change: linking adaptation and mitigation through agroforestry. Mitig Adapt Strat Glob Chang 12:901–918. doi:10.1007/s11027-007-9105-6

# Chapter 13
# Community Based Approaches of Sustainable Development and Disaster Risk Reduction

**Rajib Shaw**

**Abstract** Community based approaches of environment management, development and disaster risk reduction follow the basic principles of community ownership, leadership, and participation. Two key challenges are faced by most of the community-based practices: one, the sustainability issue, and the other up-scaling issue. Both of these issues have been widely discussed, and based on the current analysis of ten urban and rural cases, three specific key lessons are proposed: (1) to identify the right development needs of the community, and link the disaster risk issues to the development issues, (2) it is important to identify the right change agent, and (3) it is important to link the community initiative to local governance to enhance sustainability.

**Keywords** Community leadership • Community ownership • Sustainability • Up-scaling of community initiatives • Local governance

## 13.1 Introduction

Community based approaches have been accepted as one of the standard practices for either solving development issues or environmental issues or disaster risk reduction issues. Several publications have pointed out the importance and approaches of community based risk reduction (Shaw 2012). Two issues have often been described and analyzed as crucial challenges of community based risk reduction: the sustainability issue (how to sustain community initiative over a longer period of time) and up-scaling issue (how to disseminate the experiences of pilot community interventions to wider areas).

In this context, system integration is an important issue. The system can be defined in different aspects: in can be social system or governance/institutional system. In many cases, community based approaches are practiced over time, and have

R. Shaw (✉)
Graduate School of Global Environmental Studies, Kyoto University, Sakyo-ku, Kyoto, Japan
e-mail: shaw.rajib.5u@kyoto-u.ac.jp



J.I. Uitto, R. Shaw (eds.), *Sustainable Development and Disaster Risk Reduction*,
Disaster Risk Reduction, DOI 10.1007/978-4-431-55078-5_13

been part of the social system, norm, traditional practices. With time, the context gets changed, but the principles have remained unchanged. Thus, for social system integration, it is important to recognize the basic principles of community cohesion and community practices. The other aspect is the governance integration. In most cases, isolated efforts of the community practices, often started by external agencies have a natural death. To enhance sustainability, it is important to be recognized by the local governance system and make sure that the efforts are part of some sort of institutional mechanism. Recognition by the local government is an extremely important issue for sustainability as well as up-scaling.

The recent Sendai Framework of Disaster Risk Reduction (2015) states: "... *While the enabling, guiding and coordinating role of national and federal State Governments remain essential, it is necessary to empower local authorities and local communities to reduce disaster risk, including through resources, incentives and decision-making responsibilities, as appropriate*". The Sendai framework has emphasized that some of the non traditional stakeholders like scientific and technical communities and private sectors need to play increasingly pro-active roles in risk reduction at local and community level. The scientific information needs to be linked to the local decision-making, risk assessments, which need to be the base of the development practices of the local areas. Similarly, the private sectors can play a more important role to bring risk reduction as a part of their core business, and show more innovation in risk reduction, which will contribute to the local economy. Thus, community based approaches, which were so far the practices of the civil society, need more multi stakeholder involvement, business innovations and linkages to different development needs.

This chapter provides ten different examples of community based risk reduction, which are related to the different development issues in rural and urban areas. At the end, the chapter provides some basic analysis and draw common lessons from these examples.

## 13.2 Case Examples from Community Perspectives

### *13.2.1 Kobe Bokomi*

At the aftermath of the Kobe [The Great Hanshin Awaji] earthquake in Japan, the Kobe city government started a unique community based risk reduction scheme called BOKOMI ["Bousai Fukushi Community" in Japanese, meaning disaster prevention and social welfare community]. The key concept was to link the community based approaches to basic development needs of the community, and for most of the areas in Kobe, the need was social welfare, especially for the aged population. The basis of the community was elementary school district, where the school and community were linked together to provide collective solutions to their development needs. Disaster risk reduction was an additional issue, which was linked to the daily development needs, and thus the sustainability of the process was ensured. Through different approaches from the city government, and cooperation of diverse stakeholders, Bokomi was established in all 191 school districts by 2008 (Matsuoka et al. 2012). It took 13 years after the disaster, but ensured its sustainability through minimum support from the local government. This example is considered as one of the good examples of linking disaster risk reduction with development needs.

### 13.2.2 Mumbai ALM

The other example of community based risk reduction is ALM (Advanced Locality Management), which is practiced in Mumbai, in India. The Mumbai city government started the community solid waste management in 1998 with development of ALM system. The ALM calls for community formation at the neighborhood level, including both the middle income group and informal settlement, register it with the local government, and get small initial grant for waste management issues. There were around 800 ALM registered with the Mumbai city corporation, with different levels of activeness. During the 2005 historic flooding in Mumbai, the survey (Surjan et al. 2009) showed that the places where ALM were active, the responses were faster and flood levels were lower. The reason behind these were that the ALM took care of the drainage system, and used to collect solid waste from the drains, which otherwise clog the flow of water in most cases. Thus, a clear drainage system helped less flood level in mostly active ALM regions. Moreover, the community interactions through different activities helped them to gather information on the vulnerable households (like aged people, pregnant women, small child and physically handicapped people etc.), and rescue was faster in those cases. The key issue of the ALM and its success lies on the fact that the community interventions started with the daily development needs, like solid waste management, which helped during disaster event. The local government helped to develop the system, and it was sustained by local initiatives.

### 13.2.3 Bangladesh CPP

The famous Cyclone Preparedness Program (CPP) of Bangladesh is highly evaluated as a role model program for reducing rural risk and enhancing community resilience to disaster (Habiba and Shaw 2012). The CPP program, which was developed over 40 years after the major cyclone in 1970s in the coastal areas of Bangladesh, acted as a development initiative in the rural area. The CPP program developed the schools as cyclone shelter, and served the basic needs of the education in the rural areas of Bangladesh. It also developed a unique volunteer system, which helped people for evacuation during the emergency time. During the normal time, the volunteers played important roles in different development activities in the villages, be it the health issue, safer drinking water provision or even in different agriculture related practices. At different places rain water-harvesting system was developed an introduced by several NGOs, which solved the safe drinking water problems in the saline and arsenic hit coastal areas. The water also helped to develop vegetable garden and integrated fish pond, which also increased the economic status of the rural communities. Izumi and Shaw (2015) in their analysis has pointed out a new role of private sector in reducing rural risk [here daily safe drinking water provision], and enhance resilience to coastal hazards [like cyclone and salinity issues].

### 13.2.4 Reihoku Forest Management

Reihoku area is located in the border of Kochi and Ehime prefectures in Shikoku Island. The area has a high aged population, and almost 70 % of the area is covered by mountains (Fujita et al. 2011). Thus, one of the key issue in the local level is the mountain and forest management, which is done through formation of community based forest management groups, which have different members like community fire volunteers, local NPO [non profit organization], local timber industry and student volunteers from the local university. The key issue was to focus on community based forest co-management, bringing resources to the communities, enhancing local eco-tourism, organizing different community festivals, and thereby developing stronger community bonds. These bonds were found to be effective during the mountain disaster, especially landslide and flash flood, where the early warning system timely evacuation becomes important issues. The redundancy in the information sharing system is developed through these community based groups, which is found to be useful during rainy and typhoon season.

### 13.2.5 Ladakh Snow Water Harvesting

Norphel and Tashi (2015) have demonstrated a unique community based approach for snow water harvesting in Ladakh area, in the northern part of India. The unique approach makes some small physical interventions in the foothills so that the snow accumulates, and when it melts, can recharge the otherwise depleting ground water. The village community nominates a few people to look after the physical measure, and make regulatory measures for irrigation of water for a season. The members of the committee are authorized to take actions and solve legal disputes of water scarcity as and when required. The early spring water is essential for the farmers, and the artificial snow water harvesting system helps farmer to avail the water usage. A set of diversion canal, artificial glacier structure and construction of water reservoir are parts of the physical measures, which coupled with the community based management and decision making system serves the agriculture needs, and also helps in the community resilience to cope with extreme winter and snow fall.

### 13.2.6 Sri Lanka Eco-village Recovery Approach

After the 2004 Indian Ocean tsunami, Sarvodaya, a major NGO [non government organization], along with other partners have undertaken a unique eco-village approach for recovery of fisherman communities, who have been relocated from coastal to inland areas. Abe and Shaw (2015), from a series of survey, data collection and analysis between 2006 and 2012 showed that eco-village with strong environmental components [solid waste management, solar panel, rain water harvesting, home garden, environmental education etc.] helped to develop community bonding among the residents, who come from different affected villages after the tsunami. The key concept was to develop a new community in the resettled area, and environmental activities were the key binding forces of that. The survey also showed that through continuous environmental activities, the local residents had mind set change and developed their own responsibilities to take care of the local environment. This also helped to develop new community bonding among the residents.

### 13.2.7 Natori Community FM and Recovery

After the East Japan Earthquake and Tsunami [EJET], a city called Natori in Miyagi prefecture had developed emergency radio called Natori Radio [or "Natoraji", in short]. A detailed survey among the local residents [both in temporary housing and non-temporary housing] by Ideta et al. (2012) had pointed out that the number of listeners to radio has significantly increased after the disaster. The key reason was that the FM radio was able to provide lots of local information, which was crucial to the people in evacuation center to start their new lives. Local disaster information and local development needs were coupled together in the FM radio's information sharing scheme, and therefore, the local residents even opted for monetary contribution to ensure the sustainability of community FM from emergency FM. The key lesson was that the community radio can address daily development needs of the community, and thereby develops a trust and redundancy in the information flow. This, in turn, can play a key role during the emergency.

### 13.2.8 Kesennuma ESD and DRR Linkage

Another example provided by Oikawa (2014) from the Kesennuma city of Miyagi prefecture, which was also hit by the EJET had shown the importance of community based education approach. ESD (Education for Sustainable Development) has been practiced in several schools in Kesennuma, which was closely done with the local communities. The ESD had different components of environment, culture, local tradition, food and life etc., which were part of daily lives in the city and community. The community network developed through the ESD programs was found to be effective during the disaster, as well as post disaster recovery process. The key lessons were to incorporate the recovery lessons in the new dimension of ESD program, and the city has established a new community based school education program through these lessons.

### *13.2.9 Joint Mangroves Management in India*

A joint mangrove management program, which existed in Pichavaram area of the southern state of Tamil Nadu, India has been found to be effective to reduce the impact of Indian Ocean Tsunami of 2004. A survey by Ogino et al. (2010) showed that the villages which were protected by mangroves, had less damages, and have relatively higher perception on importance of mangroves conservation. A program of joint mangrove management was introduced by the local research institution, in close cooperation with local government departments, village council and local university. Different education programs were also developed along with the local schools. Eco-tourism program was developed with the village council. All these initiatives provided economic incentives, which enhanced the quality of life in the local villages, and the coastal buffer also reduced the impacts of coastal hazards, including tsunami and cyclones.

### *13.2.10 Indigenous Tank and Farming System in Sri Lanka*

Nianthi and Dharmasena (2009) had analyzed the indigenous tank system in the north central part of Sri Lanka, which has been in existence for more than 1,000 years. The tank system had a unique feature of ecosystem approach, where different areas are kept for different types of activities, some for cultivation, some for preservation for wetlands, some for drinking purposes etc. Since the area comes under the dry land areas, where water management is the key issue, this indigenous system had been developed and used over time to cope with usual drought and water scarcity in the area. This system of water usage is an unwritten rule among the local communities, and this community is termed as water community practices which strengthen the community bonding also.

## 13.3 Discussion

The ten above examples from different parts of the world have pointed out the link and role of communities in daily development issues, which, in turn help them in developing and enhancing resilience to disaster risks. Table 13.1 shows the key lessons from the ten examples, described above.

**Table 13.1** Summary of key lessons from case studies

| No. | Example | Urban/rural | Lessons |
|---|---|---|---|
| 1. | Kobe Bokomi | Urban | Risk reduction linked to social welfare |
| 2. | Mumbai ALM | Urban | Risk reduction linked to solid waste management in megacity |
| 3. | Bangladesh CPP | Rural | Risk reduction linked to education, health, water supply in coastal areas |
| 4. | Reihoku forest management | Rural | Risk reduction linked to forest management issues in mountain region |
| 5. | Ladakh snow water harvesting | Rural | Risk reduction linked to early spring agriculture in high altitude mountains |
| 6. | Sri Lanka eco-village recovery approach | Rural | Risk reduction linked to environmental issues for community relocation |
| 7. | Natori community FM | Semi urban | Risk reduction linked to daily local information sharing during recovery process |
| 8. | Kesennuma ESD DRR linkage | Semi urban | Risk reduction linked to ESD and community development |
| 9. | Joint mangrove management in India | Rural | Mangrove co-management linked to coastal resilience and life style change |
| 10. | Indigenous tank and farming system in Sri Lanka | Rural | Risk reduction linked to water management, agriculture and drought reduction |

As shown in Table 13.1, the risk reduction issues [either for pre-disaster preparedness aspects or post disaster recovery issues] need to be linked to development issues for its sustainability in the community-based approaches. Three aspects need to be considered for this.

Firstly, it is important to consider the right development needs at the local level. Based on the urban or rural context, development needs are different. It is important to find the right development needs around which the community based approaches can be formulated. The community needs to identify the appropriate pressing needs at the local level, which needs to be done in close cooperation with local governments.

Secondly, it is important to find the appropriate and effective "Change Agents" for the community based approaches. In some cases, the existing local community groups can be effective, like the women's group or youth group or other volunteer agencies. In some cases, new groups need to be formulated.

Thirdly, to enhance the sustainability of the approaches, it is required to link to the existing governance system, preferably at the local level. The city or municipality government needs to identify and recognize the community based approaches and the possible change agents, and develop the sustainability system so as to continue the activities in the long run. Therefore, a combination of right development needs, appropriate change agent and local governance system needs to be combined together for a successful community based disaster risk reduction and sustainable development system.

Finally, it needs to be kept in mind that communities are dynamic and change over time. Therefore, community practices need to be reviewed and updated based on the changes over time. However, the principle of community involvement remains unchanged, and can be applied to different contents and different time frame.

## References


Abe M, Shaw R (2015) Critical factors for sustainable post tsunami resettlement: cases from India and Sri Lanka. In: Shaw R (ed) Recovery from the Indian Ocean tsunami: a 10 years journey. Springer, Tokyo, pp 89–102

Fujita K, Takeuchi Y, Shaw R (2011) Voluntary self-help organization and fire volunteer for mountain disaster risk reduction in Reihoku Area, Japan. Asian J Environ Disaster Manag 3(3):317–337

Habiba U, Shaw R (2012) Bangladesh experiences of community based disaster risk reduction in disaster risk reduction. In: Community based disaster risk reduction. Emerald Publisher, Bradford, pp 91–111

Ideta A, Shaw R, Takeuchi Y (2012) Post disaster communication and role of FM radio: case of Natori, in East Japan earthquake and tsunami: evacuation, communication, education and volunteerism. Research Publishing, Singapore, pp 73–108

Izumi T, Shaw R (2015) A new approach of disaster management in Bangladesh: private sector involvement. Risk Hazards Crisis Public Policy. doi:10.1002/rhc3.12069

Matsuoka Y, Joerin J, Shaw R, Takeuchi Y (2012) Partnership between city government and community-based disaster prevention organizations in Kobe, Japan. In: Community based disaster risk reduction. Emerald Publisher, Bradford, pp 151–184

Nianthi R, Dharmasena PB (2009) Indigenous knowledge of farming practices and water management in the dry zone of Sri Lanka. In: Shaw R, Takeuchi Y, Sharma A (eds) Indigenous knowledge and disaster risk reduction, from practice to policy. Nova Publisher, New York, pp 329–348

Norphel C, Tashi P (2015) Snow water harvesting in the cold desert in Ladakh: an introduction to artificial glacier. In: Nibanupudi H, Shaw R (eds) Mountain hazards and disaster risk reduction. Springer, Tokyo, pp 199–201

Ogino Y, Takeuchi Y, Shaw R (2010) Community perspective of mangrove protection and its implication to coastal zone management. In: Shaw R, Krishnamurthy R (eds) Communities and coastal zone management. Research Publishers, Singapore, pp 113–126

Oikawa Y (2014) City level response: linking ESD and DRR in Kesennuma. In: Shaw R, Oikawa Y (eds) Education for sustainable development and disaster risk reduction. Springer, Tokyo, pp 155–176

SFDRR (2015) Sendai framework for disaster risk reduction, UN Document, Sendai, Japan

Shaw R (2012) Community based disaster risk reduction. Emerald Publisher, Bradford, 402 pages

Surjan A, Redkar S, Shaw R (2009) Community based urban risk reduction: case of Mumbai. In: Shaw R, Srinivas H, Sharma A (eds) Urban risk: an Asian perspective. Emerald Publication, Bingley, pp 339–354

# Chapter 14
# Education, Training, and Capacity Building for Sustainable Development

**Glenn Fernandez and Rajib Shaw**

**Abstract** The United Nations Decade of Education for Sustainable Development (DESD) ended in 2014. This chapter reviews the activities of the DESD in the area of Disaster Risk Reduction (DRR) education. Examples of education, training, and capacity building initiatives in formal and non-formal DRR education are presented. The role of higher education institution (HEI) consortia and other international networks in advancing DRR education is highlighted. The remaining challenges of utilizing DRR education as a tool to build a culture of disaster resilience will also be discussed to explore how DRR education can be enhanced and promoted more widely post-2014.

**Keywords** Disaster risk reduction education • Education for sustainable development • Training • Capacity-building

## 14.1 Introduction

The United Nations Decade of Education for Sustainable Development (DESD) began in 2005 and ended in 2014. DESD implementation was divided into two distinct phases (UNESCO 2014a). The first 3 years (2005–2008) were invested in defining and promoting Education for Sustainable Development (ESD), developing networks and partnerships, and putting monitoring and evaluation mechanisms in place. The second phase started with the 2009 World Conference on Education for

G. Fernandez (✉)
Environmental Education Laboratory, Graduate School of Global Environmental Studies, Kyoto University, Kyoto, Japan

Asian Disaster Preparedness Center, Bangkok, Thailand
e-mail: glenn.fernandez@gmail.com

R. Shaw
Graduate School of Global Environmental Studies, Kyoto University, Sakyo-ku, Kyoto, Japan
e-mail: shaw.rajib.5u@kyoto-u.ac.jp



J.I. Uitto, R. Shaw (eds.), *Sustainable Development and Disaster Risk Reduction*, Disaster Risk Reduction, DOI 10.1007/978-4-431-55078-5_14

Sustainable Development in Bonn, Germany. Based on the outcome document of the World Conference (the Bonn Declaration), UNESCO focused its work in ESD on three key issues: climate change, biodiversity, and disaster risk reduction (DRR) (UNESCO 2014a).

Tackling DRR in relation to ESD mostly involved promoting DRR education (UNESCO 2014a). For example, schools established DRR education programs to support students and their communities to reduce their vulnerability to disaster risks. DRR education "calls for systemic and systematic attention to the hazard combined with preparedness and resilience building so as to avoid the hazard morphing into disaster" (Selby and Kagawa 2012). DRR education refers to a wide range of formal and non-formal pedagogical tools that are used in preparing for disasters and in promoting disaster resilience knowledge at all age levels, including school-based initiatives; public information or awareness campaigns using mass media, social media, and special events; community learning through experiential activities like town-watching; adult education; life-long learning; understanding local indigenous knowledge; popular culture such as movies (e.g., Twister, Dante's Peak, Titanic, The Day After Tomorrow, Contagion, etc.) and other art forms; games; etc. (Benadusi 2014; Haigh et al. 2014; Shaw et al. 2011).

Benadusi (2014) gives a brief but comprehensive account of how the strategic use of education in DRR to enhance a culture of resilience was started. Previously, DRR education and training activities were technical in nature and were provided by experts in a top-down approach and almost exclusively catered to "fire fighters, police forces, civil defense volunteers, Red Cross personnel, relief workers, government officials, hazard managers, and technicians." It is only recently that DRR education at the grassroots or community level became widespread, upon the realization that an integrated approach involving capacity building from the bottom-up is what it will take to effectively reduce disaster risks.

The clearest endorsement of the relevance and significance of education in DRR appeared in the "Hyogo Framework for Action 2005–2015: Building the Resilience of Nations and Communities to Disasters" (HFA). HFA's Priority for Action 3 focuses on the use of knowledge, innovation, and education to build a culture of safety and resilience and states that "disasters can be substantially reduced if people are well informed and motivated to adopt a culture of disaster prevention and resilience, which in turn requires the collection, compilation, and dissemination of relevant knowledge and information on hazards, vulnerabilities, and capacities" (UNISDR 2005). HFA paved the way for DRR education to undergo an "explosive growth" and to become "a sort of universal element of current DRR strategies" (Benadusi 2014).

Shaw et al. (2011) classifies DRR education into three categories following the types of education in general: formal, non-formal, and informal (Fig. 14.1). Fundamentally, the main difference is that formal education and non-formal education are structured, while informal education is un-structured. Formal education takes place at an institutional facility (a school or learning center) and a certificate of graduation is awarded at the end of the course or program. Non-formal education may or may not be conducted at an institutional facility and certification, other than

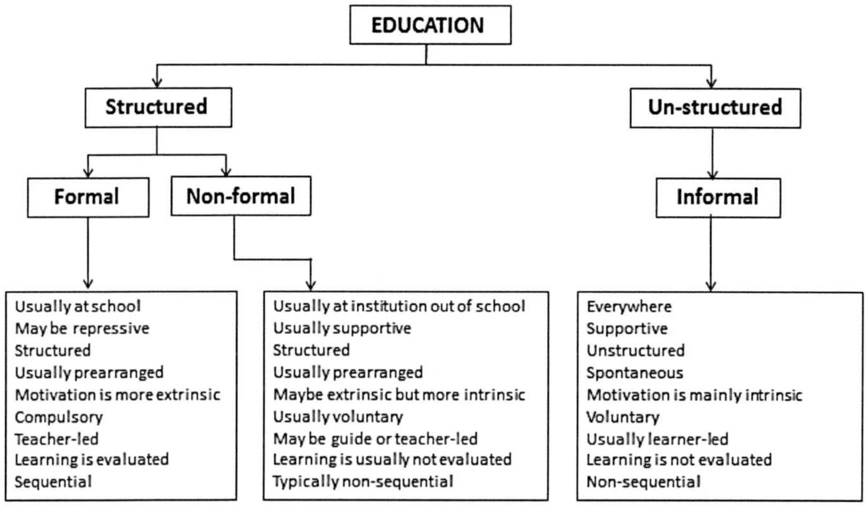


**Fig. 14.1** Three types of education (Source: Shaw et al. 2011)

a proof of attendance, is usually not awarded. Another difference is that formal education is compulsory and learning through this mode is evaluated but in non-formal education it is usually voluntary and not evaluated. Non-formal education is an alternative pathway to learning and can assist in filling the gaps in access to formal education systems (UNICEF 2012).

Informal education can happen spontaneously (e.g., during family conversation over dinner, while watching TV or reading the newspaper, etc.) and require no conscious effort on the individual to learn. Informal DRR education may either incidentally or purposefully come from reading posters, leaflets, or book on DRR; from using creative DRR education materials like toys and games, story books, comic books, computer games, etc.; from watching cultural art performance (musical instruments, songs, theater, puppetry, dance, etc.); from joining competitions (drawing and writing contests); or from getting involved in volunteer community activities (repairing school rooms before the start of classes, securing furnishings against earthquake shaking, digging channels to direct rainwater away from buildings, joining blood donation campaigns, etc.).

The next sections will discuss the formal and non-formal DRR education activities typically conducted during the DESD. Some specific examples are presented.

## 14.2 Formal Disaster Risk Reduction Education

Schools and universities have an important role to play in promoting safety against natural hazards and are places where young people can be easily reached (Baytiyeh and Naja 2014). Making DRR education part of the formal school curriculum fosters awareness and better understanding of the immediate environment in which students and their families live and work (Fernandez et al. 2014a). When students get DRR education at school, not only do they themselves benefit from it but also their household becomes better prepared as a result of the information gained (Tuswadi and Hayashi 2014).

DRR may be infused throughout the entire curriculum or integrated only into a limited number of carrier subjects, depending on implementation capacity (UNICEF 2012). Fernandez et al. (2014b) reports examples from Bicol Region in the Philippines, the region where the famous active Mayon Volcano is located. College-level and graduate school-level subjects that deal with certain aspects of disasters and disaster management are offered in a few universities such as the tertiary degree course in Disaster Risk Management in Camarines State Agricultural College and as an area of concentration for a masteral degree in Public Management in Bicol University. Starting from academic year 2008–2009, Central Bicol State University of Agriculture (CBSUA) has been offering its Master of Science in Disaster Risk Management (DRM). With the enactment of Republic Act 9729 (Climate Change Act of 2009) and Republic Act 9512 (National Environmental Awareness and Education Act of 2008), CBSUA intends to integrate climate change adaptation (CCA) concerns into the existing program and come out with a Master of Science in DRM and CCA program. Other examples of how formal DRR education is implemented are available in Selby and Kagawa (2012), Shaw et al. (2011), Shaw and Krishnamurthy (2009), CDP (2008), and other references.

But the integration of DRR into the education curriculum is not enough to bring about meaningful risk reduction (Tong et al. 2012). It should be accompanied by measures to address related issues such as structural and nonstructural school safety, legislative basis for DRR education, management mechanism, hiring adequate number of qualified and effective teachers, sufficient funding, collaboration with different partners, proper early warning system, and risk assessment, among others. Schools must prepare not only for the immediate issues of response for the safety of all their students during a disaster but also the longer term issues of recovery and school continuity to prevent students from being excluded from education for long periods of time (Boon et al. 2012). This means providing consideration also for alternative school locations, transportation, training for psychosocial counseling and support to deal with the aftermath of disasters; etc.

## 14.3 Non-formal Disaster Risk Reduction Education

There are multiple and flexible pathways to learning about DRR aside from formal DRR education. Many of the trainings and capacity building activities in DRR fall under non-formal DRR education, which provides opportunities to those who may be excluded from formal DRR education. Shaw et al. (2011) particularly promotes experience-based and action-oriented DRR education that clearly links the household, school, and community. Community resilience can be improved by incorporating education programs with activities that encourage community participation in problem solving (Bird et al. 2011).

Many organizations have focused on providing non-formal DRR education. For example, the Capacity for Disaster Reduction Initiative (CADRI) was launched in 2007 as a joint program of the United Nations Development Program, the United Nations Office for the Coordination of Humanitarian Affairs, and UNISDR (Zschocke et al. 2010). CADRI shares a range of training manuals and modules on all aspects of disaster risk management, such as thorough introductions to various hazard types and the appropriate response measures. According to Henstra (2010), people are more likely to take disaster preparedness and emergency response action if they have been previously educated about hazards and know what to do when they occur. Education and training are integral to capacity building in DRR as trained personnel respond much better to different disaster situations and will take proactive measures of mitigation and prevention (Haigh et al. 2014).

DRR trainings don't have to be face-to-face. They can now be provided online. The interactive mode of learning offered by advancements in information technology appears to be very promising when used in distance education programs (Marincioni 2007). Shiwaku and Fernandez (2011) had proposed some roles for DRR education websites in order to make them more useful and effective: sharing information in order for practitioners to learn from each other; providing information to promote implementation of DRR and DRR education activities; and collecting information useful and relevant to information users.

But across the globe DRR education and training goals and strategies vary widely, as illustrated in the case of health care providers reported by Slepski (2007). No standards or guidelines are clearly defined thus training and educational efforts lack standardization. Hagelsteen and Becker (2013) presents other challenges frequently faced in capacity development for DRR. Training activities are usually not institutionalized. In addition, because trainings are provided to individuals, with staff turnover, whatever little capacity that may be developed through training is lost. When it comes to evaluations of capacity development, they often assess output, not impact. For example, only the number of people who participated in trainings is recorded, without consideration on the mastery of the skills acquired or whether these skills were actually used after the training. To be able to monitor and evaluate the impact of trainings, baseline data and indicators are needed to measure progress.

Formal and non-formal education systems must be complementary to each other, working as one holistic system to provide quality education that meets the needs of all learners (UNICEF 2012). The integration of both formal and non-formal DRR education in curricular and extra-curricular programs is one way of ensuring that DRR messages reach every household and community and that learning can be sustained for future generations.

There are international networks doing excellent work in both ESD and DRR education. Most prominent among these networks is the United Nations University's Regional Centers of Expertise on Education for Sustainable Development (RCE) which were created to support the implementation of DESD (Mochizuki and Fadeeva 2008). As of May 2013, there are 117 acknowledged RCEs worldwide. The functions of RCEs include: creating a platform for dialogue among regional/local ESD stakeholders and for exchanging information, experience, and good practices in ESD; developing regional/local knowledge base; assisting in promoting vertical alignment of curricula from primary through university education and in linking formal and non-formal sectors of the education community (Tabucanon 2013).

One of the current RCE thematic areas is DRR. An RCE very active in this area is RCE Hyogo-Kobe. As a region that experienced the Great Hanshin-Awaji Earthquake in 1995, RCE Hyogo-Kobe positions DRR as an important component of ESD (UNU-IAS 2010). It aims at creating and disseminating a culture of disaster prevention and mitigation, rooted in people's daily lives, through collaboration with international organizations and research institutes in the region.

The secretariat of RCE Hyogo-Kobe is hosted by a higher education institution (HEI), Kobe University. The next section elaborates on the important role of HEIs in promoting both ESD and DRR, locally and globally.

## 14.4 The Role of Higher Education Institutions

The Sendai Framework for Disaster Risk Reduction 2015–2030 listed the roles and potential contributions of different stakeholder groups in DRR (UNISDR 2015). The role of academia, scientific and research entities and networks is to "focus on the disaster risk factors and scenarios, including emerging disaster risks, in the medium and long term; increase research for regional, national and local application; support action by local communities and authorities; and support the interface between policy and science for decision-making."

But although the vital role of higher education institutions (HEIs), such as colleges and universities, in helping society manage disaster risks are recognized, the potential of HEIs as active agents of change is not being fully realized (Virji et al. 2012). New and innovative educational approaches are needed to emphasize experiential learning that is problem-based and solution-focused. In addition, intra-regional and inter-regional university partnerships, as well as university collaborations with non-university partners, are needed to bring together the knowledge and priorities of the research community and the stakeholders they serve (Virji et al. 2012).

At the Third UN World Conference on Disaster Risk Reduction in Sendai, Japan, a public forum was organized on March 16, 2015 jointly by the Asian University Network of Environment and Disaster Risk Management (AUEDM), which was formed in 2010 and is composed of 30 universities from Asia, and by Partners Enhancing Resilience for People Exposed to Risks (Periperi U), an African consortium established in 2006 and now involving 11 universities, to discuss the roles of HEIs in advancing disaster risk knowledge and building local risk capacity (AUEDM and Periperi U 2015). Two major achievements of HEIs in the past decade have been (1) the increasing number of academic publications in DRR and (2) the formation and expansion of academic networks and consortia offering innovative, disaster risk-related courses.

Four HEI consortia gave presentations during the forum. They provided an overview of their membership as well as the direction their networks are heading towards in the future. The AUEDM presentation also featured its tie-up with another important network, the Asian Disaster Reduction and Response Network (ADRRN), a coalition of NGOs in Asia working on community-based developmental and DRR themes. The AUEDM-ADRRN linkage was presented as a knowledge-action partnership, bridging DRR research and practice.

Periperi U is a platform for university partnership to reduce disaster risks in Africa. This consortium of 11 HEIs has successfully generated accessible, sustainable, robust disaster risk academic programs across Africa. Periperi U's achievements include over 300 students graduating from disaster-related academic programs and courses in agriculture, engineering, urban planning, economics, environmental science, and public health, creating a home-grown surge in skilled human capital. Periperi U also has more than 50 disaster related short courses, reaching over 1500 professionals and practitioners in their local languages.

The academic network in the European Union is called ANDROID (Academic Network for Disaster Resilience to Optimize educational Development). ANDROID is an inter-disciplinary consortium of 67 partners from 31 countries and includes scientists from applied, human, social and natural disciplines. The network began in 2011 and during the first phase of its workplan has delivered several major activities and outputs: an inter-disciplinary doctoral school; a survey capturing and sharing innovative approaches to inter-disciplinary working; a survey of European education to map teaching and research programs in disaster resilience; a survey analyzing the capacity of European public administrators to address disaster risk; emerging research and teaching concerns in disaster resilience; and, open educational resources.

The fourth featured network was the APRU-IRIDeS Multi-Hazards Program, an international collaboration among the Asia-Pacific's 45 leading research universities to build safer and more disaster resilient societies through education, research, and partnerships. The International Research Institute of Disaster Science (IRIDeS) of Tohoku University provides secretariat services as the regional program hub that will harness the collective capabilities of the Association of Pacific Rim Universities (APRU) members for cutting-edge DRR research, focusing on better strategies to deal with low-frequency high impacts disasters. APRU has also established partnerships with other networks in the region, such as AUEDM, to further strengthen disaster science and research.

While the current outreach of the four consortia is still somewhat limited, they have accomplished major successes in the short time since they were established: formation of new disaster risk-related departments/centers; offering undergraduate and postgraduate DRR courses; and creating a new cadre of DRR professionals; and increased scientific research and publications advancing DRR scholarship. HEIs belonging to the consortia had actively contributed towards the development of the Sendai Framework for DRR. Hopefully they will also take lead roles in the implementation of the Sendai Framework. HEIs, especially when working in collaboration with other major stakeholders groups, can expedite knowledge dissemination to reach out to communities and local governments. As mentioned in the Sendai Framework, "in the context of increasing global interdependence, concerted international cooperation, an enabling international environment and means of implementation are needed to stimulate and contribute to developing the knowledge, capacities and motivation for disaster risk reduction at all levels, in particular for developing countries" (UNISDR 2015).

## 14.5 Prospects of Disaster Risk Reduction Education Post-DESD

As a follow-up to DESD, UNESCO has developed a Global Action Program (GAP) on ESD, which was endorsed by the UNESCO General Conference in 2013. The GAP has five priority action areas (Box 14.1) and aims at "scaling up action in ESD in order to accelerate overall progress towards sustainable development and marks an important contribution to the global post-2015 development agenda" (UNESCO 2014a). The GAP was officially launched in Nagoya, Japan during the World Conference on ESD in November 2014. The GAP mentions DRR only once, as one of the interrelated sustainable development issues to be addressed (UNESCO 2014b). DRR will continue to be integrated into the school curriculum, along with other critical issues, such as climate change, biodiversity, and sustainable consumption and production.

**Box 14.1: Five Priority Action Areas of GAP to Advance the ESD Agenda**

Priority Action Area 1: Advancing policy

Mainstream ESD into both education and sustainable development policies, to create an enabling environment for ESD and to bring about systemic change

Priority Action Area 2: Transforming learning and training environments: Integrate sustainability principles into education and training settings

Priority Action Area 3: Building capacities of educators and trainers

Increase the capacities of educators and trainers to more effectively deliver ESD

Priority Action Area 4: Empowering and mobilizing youth

Multiply ESD actions among youth

Priority Action Area 5: Accelerating sustainable solutions at local level

At community level, scale up ESD programs and multi-stakeholder ESD networks

Source: UNESCO (2014b)

The Sendai Framework for Disaster Risk Reduction has affirmed that effective disaster risk management contributes to sustainable development (UNISDR 2015). Priority 1 under the Sendai Framework is "Understanding disaster risk." DRR education will be essential in pursuing this priority. It is acknowledged that it is important to "build the knowledge of government officials at all levels, civil society, communities and volunteers, as well as the private sector, through sharing experiences, lessons learned, good practices and training and education on disaster risk reduction, including the use of existing training and education mechanisms and peer learning" and "promote the incorporation of disaster risk knowledge, including disaster prevention, mitigation, preparedness, response, recovery and rehabilitation, in formal and non-formal education, as well as in civic education at all levels, as well as in professional education and training," as well as "promote national strategies to strengthen public education and awareness in disaster risk reduction, including disaster risk information and knowledge, through campaigns, social media and community mobilization, taking into account specific audiences and their needs."

Both the GAP and the Sendai Framework will ensure that DRR education will continue to be an important undertaking in the future until 2030. Cole and Murphy (2014) noted that in order to promote a culture of safety and change human behavior, the DRR educational strategies must be long-term and on-going. However, DRR education should not be limited to schools, where education is provided, but should go beyond the school boundary and be linked to the communities and households (Oktari et al. 2015). The role of household and community participation is crucial for the enhancement as well as the sustainability of DRR education.

The importance of sufficient capacity in the area of DRR education cannot be overemphasized (UNICEF 2012). Partnerships, collaboration, outsourcing, and networking are crucial for getting the needed capacities. Multiple stakeholders must get involved in promoting and enhancing DRR education. McBean and Rodgers (2010) noted a Chair's summary at a UNISDR Global Platform on DRR meeting in 2007 that states that "a core challenge in disaster risk reduction is to scale up proven practices," highlighting the importance of capacity building of all types. It is widely acknowledged that education plays a pivotal role in reducing disaster risks and achieving human security (Shaw et al. 2011). According to Petal (2008), the goal of developing disaster-resilient communities is heavily dependent on the success of DRR education. DRR education should be woven into our daily life. It is important that people not only see DRR information but also actually use it. This way, DRR education can save and sustain life and give people confidence to face the future.

**Acknowledgement** We thank the Ministry of Education, Culture, Sports, Science, and Technology (MEXT) of Japan for the generous scholarship grant that made it possible for the first author to pursue his MA and PhD degrees at Kyoto University. We also acknowledge the financial support provided by the Global Center of Excellence Program on Sustainability/Survivability Science for a Resilient Society Adaptable to Extreme Weather Conditions (GCOE-ARS) of Kyoto University for the first author's research activities.

## References

AUEDM (Asian University Network of Environment and Disaster Risk Management), Partners Enhancing Resilience for People Exposed to Risks (Periperi U) (2015) Role of Higher Education Institutions (HEIs): advancing disaster risk knowledge and building local risk capacity. http://www.wcdrr.org/wcdrr-data/uploads/84/WCDRR_%20Seminar%20on%20Role%20of%20HEIs.pdf

Baytiyeh H, Naja M (2014) Can education reduce middle eastern fatalistic attitude regarding earthquake disasters? Disaster Prev Manag 23(4):343–355

Benadusi M (2014) Pedagogies of the unknown: unpacking 'Culture' in disaster risk reduction education. J Conting Crisis Manag 22(3):174–183

Bird D, Chagué-Goff C, Gero A (2011) Human response to extreme events: a review of three post-tsunami disaster case studies. Aust Geogr 42(3):225–239

Boon H, Pagliano P, Brown L, Tsey K (2012) An assessment of policies guiding school emergency disaster management for students with disabilities in Australia. J Policy Pract Intellect Disabil 9(1):17–26

CDP (Center for Disaster Preparedness) (2008) Mainstreaming disaster risk reduction in the education sector in the Philippines. Asian Disaster Preparedness Center, Bangkok

Cole J, Murphy B (2014) Rural hazard risk communication and public education: strategic and tactical best practices. Int J Disaster Risk Reduction 10:292–304

Fernandez G, Shaw R, Abe M (2014a) Lessons from the recovery of the education sector after the Indian ocean tsunami. In: Shaw R (ed) Recovery from Indian Ocean tsunami: ten years journey. Springer, Tokyo

Fernandez G, Thi TTM, Shaw R (2014b) Climate change education: recent trends and future prospects. In: Shaw R, Oikawa Y (eds) Education for sustainable development and disaster risk reduction. Springer, Tokyo, pp 53–74

Hagelsteen M, Becker P (2013) Challenging disparities in capacity development for disaster risk reduction. Int J Disaster Risk Reduction 3:4–13

Haigh R, Amaratunga D, Thayaparan M (2014) ANDROID: an inter-disciplinary academic network that promotes co-operation and innovation among European Higher Education to increase society's resilience to disasters. Procedia Econ Financ 18:857–864

Henstra D (2010) Evaluating local government emergency management programs: what framework should public managers adopt? Public Adm Rev 70(2):236–246

Marincioni F (2007) Information technologies and the sharing of disaster knowledge: the critical role of professional culture. Disasters 31(4):459–476

McBean G, Rodgers C (2010) Climate hazards and disasters: the need for capacity building. Wiley Interdiscip Rev Clim Chang 1(6):871–884

Mochizuki Y, Fadeeva Z (2008) Regional Centres of Expertise on Education for Sustainable Development (RCEs): an overview. Int J Sustain High Educ 9(4):369–381

Oktari R, Shiwaku K, Munadi K, Syamsidik, Shaw R (2015) A conceptual model of a school – community collaborative network in enhancing coastal community resilience in Banda Aceh, Indonesia. Int J Disaster Risk Reduction. doi:10.1016/j.ijdrr.2015.02.006

Petal M (2008) Concept note: formal and informal education for disaster risk reduction. http://www.riskred.org/activities/ddredislamabad.pdf

Selby D, Kagawa F (2012) Disaster risk reduction in school curricula: case studies from thirty countries. http://www.unicef.org/education/files/DRRinCurricula-Mapping30countriesFINAL.pdf

Shaw R, Krishnamurthy RR (eds) (2009) Disaster management: global challenges and local solutions. Universities Press, Hyderabad

Shaw R, Shiwaku K, Takeuchi Y (eds) (2011) Disaster education. Emerald Group Publishing, Bingley

Shiwaku K, Fernandez G (2011) Innovative approaches in disaster education. In: Shaw R, Shiwaku K, Takeuchi Y (eds) Disaster education. Emerald Group Publishing, Bingley, pp 115–136

Slepski L (2007) Emergency preparedness and professional competency among health care providers during hurricanes Katrina and Rita: pilot study results. Disaster Manag Response 5(4):99–110

Tabucanon M (2013) Engaging the youth in education for sustainable development. http://aeeid.asean.org/wp-content/uploads/2013/12/Engaging-the-Youth-in-Education-for-Sustainable-Development-Mario-T.-Tabuchanon.pdf

Tong TMT, Shaw R, Takeuchi Y (2012) Climate disaster resilience of the education sector in Thua Thien Hue Province, Central Vietnam. Nat Hazards 63(2):685–709

Tuswadi, Hayashi T (2014) Disaster prevention education in Merapi Volcano Area Primary Schools: focusing on students' perception and teachers' performance. In: Procedia environmental sciences, the 4th international conference on sustainable future for human security SUSTAIN 2013, vol 20, pp 668–677

UNESCO (2014a) Shaping the future we want: UN decade of education for sustainable development (2005–2014) final report. UNESCO, Paris

UNESCO (2014b) UNESCO roadmap for implementing the global action programme on education for sustainable development. UNESCO, Paris

UNICEF (United Nations Children's Fund) (2012) Climate change adaptation and disaster risk reduction in the education sector: resource manual. UNICEF, New York

UNISDR (United Nations International Strategy for Disaster Reduction) (2005) Hyogo framework for action 2005–2015: building the resilience of nations and communities to disasters. http://www.unisdr.org/2005/wcdr/intergover/official-doc/L-docs/Hyogo-framework-for-action-english.pdf

UNISDR (United Nations Office for Disaster Risk Reduction) (2015) Sendai framework for disaster risk reduction 2015–2030. http://www.wcdrr.org/uploads/Sendai_Framework_for_Disaster_Risk_Reduction_2015-2030.pdf

UNU-IAS (United Nations University – Institute of Advance Studies) (2010) Five years of regional centres of expertise on ESD. UNU-IAS, Yokohama

Virji H, Padgham J, Seipt C (2012) Capacity building to support knowledge systems for resilient development – approaches, actions, and needs. Curr Opin Environ Sustain 4(1):115–121

Zschocke T, De León JC, Beniest J (2010) Enriching the description of learning resources on disaster risk reduction in the agricultural domain: an ontological approach. In: Sánchez-Alonso S, Athanasiadis I (eds) Metadata and semantic research. Springer, Berlin, pp 320–330

# Chapter 15
# Missing the Forests for the Trees? Assessing the Use of Impact Evaluations in Forestry Programmes

**Jyotsna Puri and Bharat Dhody**

**Abstract** In this chapter, we examine how impact evaluations can contribute to measuring and understanding the overall effectiveness, efficiency and sustainability of forestry programmes. In most cases we find that impact evaluations have used quasi-experimental methods rather than experimental methods to identify and measure change that can be causally attributed to forestry programmes. We conclude that in measuring the change that be attributed to these programmes, impact evaluation methods help to measure the overall effect, deal with sources of potential bias and mitigate confounding factors while undertaking these measurements. Impact evaluations also hold enormous potential because they are able to leverage the potential held by big and open data. However caution must also be exercised in using these methods. Impact evaluation methodologies must also incorporate causal pathways and methods of implementation research if they are to be relevant to policy and programme managers.

**Keywords** Forests • Impact evaluation • Experimental methods • Quasi-experimental methods • Adaptation • Sustainable development • Big data

## 15.1 Introduction

Forests contribute to greater resilience and reduced vulnerability of ecosystems (UNEP 2011). They provide important ecosystem services, influence micro weather systems, are an important carbon sink that in the long run mitigates the risk of climate change and constitute an important part of an ecosystem themselves (World Bank 2015; Wunder et al. 2014). Therefore knowing what policies and programmes are important in preserving the health of forestry systems is an important question for anyone wanting to discuss a sustainable economy.

J. Puri (✉) • B. Dhody
International Initiative for Impact Evaluation (3ie), New Delhi, India
e-mail: jpuri@3ieimpact.org



J.I. Uitto, R. Shaw (eds.), *Sustainable Development and Disaster Risk Reduction*, Disaster Risk Reduction, DOI 10.1007/978-4-431-55078-5_15

The chapter lays out a paradigm for evaluating forestry projects, policies and programmes. It examines how experimental methods and quasi-experimental methods can be used to understand the effectiveness of the forestry sector and discusses their limitations and possible solutions. We then make some recommendations for methodology.

Overall we find that, first, not surprisingly there have been few studies that have used robust attribution methods to assess the impacts of programmes on how well forests are adapting and are sustainable; second, we find that there are big contributions that impact evaluation methodologies can make to the field. However there are also many limitations in traditional experimental and quasi-experimental methods that can limit the understanding of impacts in these multi-intervention and multi-sectoral contexts. Finally, we present possibilities in methodology and data that represent an important way forward.

The chapter is laid out as follows. 15.2 lays out the objectives including a brief exposition of the forestry sector. 15.3 lays out the definition of attribution methods and discusses briefly what impact evaluations are and why they can be powerful for evidence informed policy. 15.4 discusses the main contributions that impact evaluation methodologies and studies can make to this sector. 15.5 discusses some limitations of current approaches. 15.6 discusses some possible solutions and opportunities and 15.7 concludes.

## 15.2 Objectives

The objective of the chapter is to examine the experience of impact evaluations in the field of forestry. While examining these methods we examine the strengths and limitations of impact evaluation methodologies used in this field. Impact evaluation evidence can be potentially used to inform the extent to which forestry programmes are relevant, effective, replicable and scale-able. They thus respond to the needs of programme managers who are being increasingly asked to provide high quality evidence of whether their programmes are working or not, why and for whom and how much. They also respond to the needs of policy makers to help them assess the effectiveness of forestry programmes in general and to assess in a more robust manner the trade-offs they are required to consistently make (UNEP 2011). In examining impact evaluation studies we also take a step back to see objectively if these are indeed meeting the needs of the sector.

Additionally we also discuss how current impact evaluation practices and methodologies can be extended to attend to the needs of the forestry sector in particular, and climate change discipline in general, to better assess questions of timing, threshold effects, internal validity, complexity and external validity. Since this field is rapidly evolving, the chapter examines how current methods of impact evaluations may be modified to cater to the different needs of the sector. Indeed an overall and practical consideration is cost – impact evaluations are expensive and clearly it is neither required nor possible to undertake impact evaluations of all programmes in this area, given that the average cost of impact evaluations is $450,000 (Puri et al. 2015).

The questions that this study answers are the following:

- What can impact evaluations of forestry programmes tell us?
- What additional value do impact evaluation studies for evaluating forestry programmes provide over traditional process evaluations?
- What are the limitations of current impact evaluation approaches and how can they be made more relevant to the forestry sector?

In this chapter we use a shortlist of impact evaluations that examine forestry programmes, policies or projects to illustrate some of the contributions and limitations of impact evaluations (see Annex II). The list of these studies is not exhaustive. Indeed it is purposive. However they are all impact evaluation studies – they all use experimental or quasi-experimental methods to measure changes caused by forestry programmes.

## 15.3 Impact Evaluations and Definitions

In recent times the use of experimental and quasi-experimental methods to understand changes that can be attributed to programmes and policies has become popular also see Bowler et al. (2010).

The reason impact evaluations have gained a lot of traction in recent times is because 'identification strategies' allow us to clearly measure the amount of change that can be attributed to the programme, while dealing with confounding factors and potential endogeneity, that may have bias results (see for example Gertler et al. 2011). They thus make it possible to measure this effect but also therefore truly compare programmes across different settings (see Table 15.1). Impact evaluations use either experimental or quasi-experimental methods to identify and measure these effects and also deal with selection bias and programme placement bias (programmes are placed in areas where they are likely to be most successful, which means that what you are measuring when measuring effectiveness is the conflated effect of the programme *and* the context. Since context is not something programme managers typically have control over, this gives biased measures of the success of the programme or its failure).[1] In Annex I, we define some important terms.

[1]This is the sense in which we use the term 'impact evaluations'. Other definitions also exist. Different definitions of impact evaluations emphasize different aspects of the causal chain. The OECD-DAC defines it as the 'The positive and negative, primary and secondary long term effects produced by a development intervention, directly or indirectly, intended or unintended.' The World Bank defines it as 'assessing changes in the well-being of individuals, households, communities or firms, that can be attributed to a particular programme, policy or work'. We use the meaning of impact evaluations as the one used by 3ie: 'Rigorous impact evaluation studies are analyses that measure the net change in outcomes for a particular group of people that can be attributed to a specific programme using the best methodology available, feasible and appropriate to the evaluation question that is being investigated and to the specific context.'

**Table 15.1** Types of identification methods that may be used in impact evaluations

| Methods | Description | Pros | Cons |
|---|---|---|---|
| ***I. Experimental design all of which require randomized assignment*** | | | |
| (a) The basic randomized control trial (RCT) | A sample of eligible subjects (areas, people, communities) are randomly assigned into those who receive the programme or intervention and those who do not. Impact is the difference in outcomes between the two groups (i.e. between the treatment group and the comparison group). | Straight forward estimation (difference in statistical averages) | Requires a comparison group; |
| | | It is commonly argued that these designs do not require baseline data but these are usually desired so that balance between different arms can be checked. | Requires check of balance (i.e. whether randomization was successful). If randomization is not successful, then the results are not valid |
| | | | Usually requires large sample sizes. |
| (b) Factorial designs | These should be considered as a variation to the basic RCT (see above where randomized assignment is a one time event.) In this design the comparison group gets the standard package that everyone is getting. The treatment arms get the standard package *plus* small changes or additions that in turn need to be tested and compared with the comparison group. So it requires at least four groups: the comparison arm, treatment group A, treatment group B and a treatment group A+B. | Different new interventions can be tested. | The maintained assumption is that there is little doubt about the efficacy and effectiveness of the 'standard package' This method is used to assess the effectiveness of additional and possibly 'innovative' methods. It requires careful planning and a large eligible sample. |
| | | Cost effectiveness comparisons can be done. | |
| | | Is easily done in contexts where interventions are new. | |
| (c) Pipeline or phased designs | The treatment or policy is implemented in phases where units are assigned to different phases *randomly*. | It overcomes the traditional resistance to random assignment that the comparison group is being denied the benefits of a policy or an intervention. | It requires that the units are assigned to different phases in a random manner. The phased design and the random assignment usually require a keen understanding and close engagement by the project team that is implementing the programme. |

| | | | |
|---|---|---|---|
| ***II. Quasi-experimental designs*** | | | |
| Regression discontinuity | A cut-off determines who is eligible to participate. Outcomes of beneficiaries and non-beneficiaries close to the cut-off line are compared. | All quasi-experimental designs require baseline data. Additional data on attributes is also required. | The credibility of the technique hinges strongly on establishing that the discontinuity is random. |
| | | | Requires data on many variables that are unrelated to the programme. |
| | | | Precision of impact is limited to those participants most close to the allocation cut-off. |
| Matching (including propensity score matching.) | Programme beneficiaries are compared to a group of non-beneficiaries. The match is constructed by finding people whose observable characteristics are similar to those of the people in the treatment group | Can be designed after the programme has been implemented and rolled out. However requires high quality baseline data that is relevant to the programme on many variables. | Requires a comparison group and credibility hinges strongly on establishing that there is a good match and balance between matched arms/groups. |
| | | Can use previously existing datasets such as censuses, DHS, etc. *if* they are high quality to establish baseline and balance in level and trends. | Requires data on "matching variables" and knowledge of important covariates. The maintained assumption is that a match on 'observed' characteristics is also accounting for any differences in *un*observed characteristics. |
| Instrumental variables | Participation in a programme can be predicted by an incidental factor, or "instrumental" variable, that is uncorrelated with the outcome (other than by predicting participation) | Can be undertaken after the programme has been rolled out. | Requires strong assumption that the instrument affects the outcome only through one specific channel, which is usually hard to establish. |
| | | The counterfactual is determined by the program | |

## 15.4 What Do Impact Evaluation Studies Show?

In this section, we discuss the contributions that impact evaluations of forestry programmes have been able to make in assessing their effectiveness.

**Measure change**: Impact evaluations help to understand the direction and measure the magnitude of change of forestry programmes. Thus for example Somanathan et al. (2005) show that in India forest cover increased by 12–16 % as a result of community management. Similarly other studies have measured the effect caused by drivers of deforestation. Andersson et al. (2011) examine the effect of safety net transfers on forestry cover in Ethiopia and find that there is an increase in livestock but no effect on tree plantation and Alix-Garcia et al. (2013) examine the effect of cash transfers on forestry management in Mexico and find a positive effect. Bensch and Peters (2011) examine the effect of improved cookstoves in Senegal and find that charcoal use reduced by 25 % depending on the extent of use of these stoves. They infer the effect on deforestation.

In many cases, these are likely to have repercussions for policy. Pfaff et al. (2008) find that in Costa Rica, the programme for payment of ecosystem services (PSA) had little effect on the 1997–2000 forest clearing. They find that despite a large amount of resources devoted to this programme, PSA prevented deforestation in the first few years of contracting by only 0.21 % in the land that was enrolled. The main reason for the small change is that the overall national deforestation rate was reducing during this period and the incremental contribution made by this programme was minimal. In the same country, Andam et al. (2007) measure the effect of protected areas and conclude that between 1967 and 1997, protected areas helped to reduce deforestation by 10 %.

**Deal with bias**: In many cases non-impact evaluation studies are unable to disentangle the effect of inherent characteristics of areas/people receiving forestry programmes on one side and the effect caused by the programmes themselves (irrespective of context). Consequently they erroneously attribute all changes in forest cover to the forest programme or policy (e.g. programmes that manage forests differently or offer incentives to reduce deforestation). This is the programme placement bias problem. In other cases, people that select themselves into being beneficiaries of programmes have inherently different characteristics from people that don't. This is the selection bias problem. Both these need to be accounted for, if we want to measure results of a forestry programme in an unbiased way. So Cropper et al. (2001), and Nelson and Chomitz (2009), account for the fact that protected areas that aim to protect forests, are likely to be placed in areas that have low agricultural productivity and profitability. All increases in forest cover in these areas, compared to other areas, cannot be then rightly attributed to the effectiveness of protected areas. This is because many of these areas would have remained forested even in the absence of these policies.[2] In these cases, once programme placement

[2] See Cropper et al. (2001).

bias and selection bias is accounted for, the change in outcomes is usually quite small.

Similarly for PES (payment for ecosystem services) programmes, the effectiveness of these systems is likely to be jeopardized by the potential for adverse self-selection and poor administrative targeting which may result in low effectiveness of land being under forests (Alix-Garcia et al. 2013).

**Assess effectiveness of targeting**: Impact evaluations can also help understand the effectiveness of targeting, i.e. whether those that are most likely to clear forests are the ones that are being targeted by forestry programmes and policies. Therefore Alix-Garcia et al. (2013) examine whether the PES programme in Mexico was targeting the most vulnerable areas. They find that the country wide programme, with a budget of more than US$5 million, was quite successful in targeting eligible households. Assessments such as these also help to inform how programmes should be designed and some of the pitfalls to be aware of when designing PES programmes for example. In Costa Rica Pfaff et al. (2008) find that the PSA programme did not target locations that were most likely to change land use. Hence there were very small changes in forest cover *caused* by the PES programme.

**Estimating impacts on sub-groups**: Impact evaluations also can help to address questions of equity and heterogeneous impacts. Somanathan et al. (2005) show that after accounting for potential selection bias and placement bias, community managed forests in India performed better in raising crown cover by 12–16 % compared to unmanaged commons, but only for forests comprising of broadleaf trees but not pine trees. (Understanding the effects on sub-groups however requires that sample sizes are so selected to be representative for the sub-groups of interest.)

**Compare different forestry programmes**: Many studies examine programmes that engage communities and compare their effectiveness with the status quo such as government managed systems or unmanaged systems (Somanathan et al. 2005; Tachibana and Adhikari 2009; Scullion et al. 2011; Edmonds 2002).[3] In India Somanathan et al. (2005) show that in community managed forests performed no worse and perhaps better than state managed forests. Tachibana and Adhikari (2009) examine this question in Nepal and show that community co-managed forests are more successful in helping deforested areas recover forests than forests that are solely managed by communities. They also find that *co*-managed forests are able to especially protect forests better where large visible extraction is taking place (such as of timber). Similarly Cropper et al. (2001) find that after accounting for selection biases, protected areas as a whole are less effective in protecting forests than specially designated wildlife sanctuaries most likely because the latter have more resources devoted to them.

**Unintended consequences and spill-overs**: Impact evaluations can also help to measure unintended effects, spill-over effects (see for example Andam et al. (2007) and Arriagada et al. (2012)) as well as strength of these neighbourhood effects and

---

[3] Three of the four studies (one is ongoing) show that community based management systems did reduce forest clearing (see Table 15.2). The time periods over which these interventions are also examined are very short.

peer effects (Bravo-Ureta et al. 2011; Chibwana et al. 2013). Bravo-Ureta et al. (2011) show that plots *close* to households that received natural resource management training were as likely to benefit from resource management techniques as those much further away concluding that the evidence for peer effects in training programmes is very sparse.

**Assess Trade-offs**: In the context of forestry projects, it's clearly important to know whether a change in policy is reducing deforestation. Almost as importantly, it's also important to know whether the changes planned for and foreseen by the interventions and the policy changes would in fact have occurred *without* the intervention. Knowing whether and how *much* these effects were is important. Thus Sims (2008) shows that in Thailand, protected areas have prevented forest clearing that otherwise would have occurred: sub-districts with more land in protected areas have between 9 and 32 % more forest cover than typical sub-districts. These programmes have also reduced land available for agriculture. However, social costs have been minimal: consumption levels in sub-districts with more land in national parks was higher by 2 and 7 % respectively and poverty levels 4 and 12 % lower than for comparison sub-districts. On the other hand, inequality measures were higher on average for communities near national parks, indicating a disproportionate share of these gains went to higher income households.[4] In another case, Alix-Garcia et al. (2013) examine the effect ofcash transfers under the aegis of Opportunidades and find that forests were affected detrimentally as a consequence of a cash transfer programme. They use an RCT to measure the magnitude of this effect. This is important because programme managers and policy makers can make choices once they know the magnitudes of change. Similarly in the Ghana (Burwen and Levine 2012) and Senegal (Bensch and Peters (2011) studies of fuel wood use, the studies *measured* the impact on fuel wood and charcoal.

## 15.5 What Do Impact Evaluation Studies of Forest Adaptation Projects *Not* Show and Challenges

Although impact evaluation studies in forestry have been successful in helping to measure various policy and programmatically relevant topics, there are several areas that impact evaluations have fallen short of.

First, it is also important to see what kinds of forestry programmes these impact evaluations are *not* evaluating. Most forestry programmes are **complex, multi-intervention, multi-sectoral** programmes with livelihoods, health, agriculture and income poverty as their primary outcomes. These sorts of complex programmes that have multiple arms, have large possibilities for additional programmes being tagged on, are implemented by different agencies on the ground and are usually scaled up in a slow, organic manner, are not traditionally evaluated by impact evaluations.

[4]The most probable mechanism for the positive income effects of national parks is increased income from tourism.

Additionally, if programmes have not built in an impact evaluation plan into them from the start, these are usually not impact evaluated either. Last but not least, in all the programmes listed here, researchers use clever and innovative ways to construct counterfactuals or comparison groups. Although with randomized experiments, this is easier to do, programmes also use matching methods to 'construct' these counterfactuals. Other programmes and policies that do not traditionally get impact evaluated are: large national or regional policies, programmes that aim to change institutions, and programmes that are very small (and have no clear possible counterfactual).

Secondly, forestry projects are also frequently projects with high '**causal density**'. As defined by Woolcock (2013), this means that forestry programmes typically have different interventions that start at different times, and are implemented with different intensities. This means that it is not always clear what the 'treatment' group receives and what the comparison group receives. Many studies choose to deal with this by making the treatment a 'package' so that *variations* within the package are ignored. If a package for instance has two types of interventions, frequently an impact evaluation assumes that the treatment group has received both interventions equally (while the comparison group doesn't usually get any part of the package). In vector terminology, an impact evaluation this can be represented as testing (1,1) vs (0,0) where '1' represents the full intervention being implemented and '0' means the intervention is not implemented. The '( )' term represents a full 'package'. But frequently real-world implementation means that equal intensities and amounts across different treatment subjects is not possible. Furthermore,in many cases it is clear that the relevant and more useful impact evaluation question is to evaluate what would happen with different 'dosages' of the same intervention i.e. (1,1) vs. (0.5,1) vs. (1,0.5) vs. (0,0) vs. (0.5,0.5) for instance. (This route has many repercussions for design, roll-out of programmes and sampling and costs of evaluations which we don't discuss here.)

Another gap we found in impact evaluations was the lack of **implementation research**. Impact evaluations presume that the programmes they evaluate have high 'efficacy' i.e. if implemented correctly and completely, will work on the ground, as the programmes were envisioned in the laboratory.This is not borne out in the real world and we maintain that it will be very useful to additionallyexamine what is required to *implement* programmes better: an important question for programme managers and also for policy makers. None of the impact evaluation studies that we reviewed had an implementation research component. Understanding for example what possible methods of community based management programmes are most effective (in the relative sense), what makes forest officials more efficient, what types, frequencies and magnitudes of payments for ecosystem services makes them most effective in ensuring better current and future forest cover are all important questions that most impact evaluation studies shy away from answering.

Fourthly, there is the question of **assumed trends**. In many cases the timelines of programmes that are evaluatedare different and keep changing. They are also different at different points in time. Impact evaluations presume that past trends in comparison and treatment areas will be mirrored during the period of the impact evaluation. As Woolcock (2013) points out, this is incorrect.

The fifth point relates to **data and capacity**. Impact evaluation studies require a lot of data: they require high quality relevant baseline and endline quantitative and qualitative data not just for programme areas/beneficiaries but also for non-programme areas/beneficiaries. Additionally, they also require good and timely implementation data. This has also constrained the application of impact evaluation methodologies. Indeed in most cases unless impact evaluations are planned for, at the inception of the project, it is hard for these to be done at the end.We also found in our assessment that some impact evaluations used randomized control trials as their 'identification' strategy (or methods that allow them to measure the strength of the causal relationship). These included studies by Burwine and Levine (2012) and Hafashimana et al. (forthcoming). But most studies used quasi-experimental methods (Arriagada et al. 2012; Edmonds 2002; Sims 2008; Cropper et al. 2001; Andersson et al. 2011). The technical expertise required to specify, estimate, analyse and understand quasi-experimental methods is, arguably, greater than those for randomized control trials. Indeed this might also account for the small number of impact evaluations in the area. A recent systematic review (Samii et al. 2014) indeed found only 12 studies in developing countries that could reliably be measuring changes in outcomes of forestry programmes (after accounting for inclusion and exclusion criteria).

Last but not least, impact evaluations can be robust but they can also provide **contradictory results** for the effectiveness of programmes. Therefore in Thailand, Sims (2008) and Cropper et al. (2001) differ in their conclusions about the effectiveness of protected areas; in Costa Rica, Andam et al. (2007) and Pfaff et al. (2008) differ in their conclusions about the effectiveness of deforestation related policies. Although clearly not the objective, one of the consequences of using impact evaluation studies is that they usually require a registration of protocol, clear pre-analysis plans and in many cases now, replicability of results, the robustness of these studies and therefore their implications for policies and programmes can be assessed very easily. Systematic reviews (see for example Waddington et al. 2014) also usually incorporate meta-analysis of data (if it is possible) to see what the overall result from a collection of studies is likely to be.[5]

## 15.6 A Discussion of Some Ways Forward

Overall there are two types of programmes that can directly or indirectly increase forest cover. The first is through development projects where the development programmes affect the intensive or extensive frontier of forests indirectly – by affecting

[5]Although this requires that several fairly stringent conditions are fulfilled – for example the intervention needs to be the same, the outcome needs to be the same and the assumption that the different datasets are coming from the *same* underlying statistical population which has the same underlying distribution, can be a strong one.

the pressure on livelihoods where the substitution effect is hypothesized to be greater than the income effect (see for example Puri 2006).

The second type of programmes are those that aim to affect forests directly, usually through the way they are managed. Following Samii et al. (2014) these can be divided into the following categories: (i) community forestry management or joint forestry management programmes; (ii) protected area programmes; and (iii) payment for ecosystem services programmes.

We argue that this understanding of possible **causal pathways** is critical in any impact evaluation. Causal pathways (also called theories of change or impact pathways) help to identify nodes that are critical for realizing changes in outcomes or impact indicators but also help identify possible unintended consequences and spillover effects. Discussion of causal pathways can also allay some doubts that have been raised about the uses and limitations of impact evaluation methodologies (see Stern et al. 2012).

The second critical limitation affecting the use of impact evaluations is the extensive demands on data. We argue that the methodologies used in impact evaluations allow researchers to be creative in understanding many questions of bias, placement and measure effectiveness in a robust manner. For example, Nelson and Chomitz (2009) assess the impact of tropical protected areas on forest fires, which they argue is the best available global proxy for deforestation at a fine spatial scale. Using forest fires as a proxy for depleted forest cover, they conclude that in Latin America and the Caribbean, protected areas reduce forest loss – by 4.3 %. Cropper et al. (2001), Sims (2008), Andam et al. (2007), Arriagada et al. (2012), Scullion et al. (2011), Edmonds (2002), Somanathan et al (2005), Pfaff et al. (2008), Alix-Garcia et al. (2014), all use **GIS data in different ways and in combination with survey data** to understand attributable impact. In Table 15.2 below we discuss the various data sources that can be used in different ways to understand and measure attributable impact.

## 15.7 Conclusions

There are several good and other not so good reasons for why impact evaluations are not routinely undertaken in the forestry sector. Capacity and costs are the two most cited reasons for the limited number of impact evaluations in this area.[6] Both these problems can be dealt with. Clearly impact evaluations should not be done in all cases but in some cases their use is critical.

Woolcock (2013) in his excellent exposition on exploring various dimensions of external validity lays out three domains which are likely to influence the external validity of results. He categorizes these domains as, first, causal density ('the extent to which an intervention or its constituent elements are complex'), second, implementation capability (the extent to which any other organization can faithfully

[6] Personal conversations.

**Table 15.2** Sources of big and small data for understanding attributable impact in forestry studies

| No. | Type | Helps to inform | Other notes |
|---|---|---|---|
| 1. | Satellite images | Forest cover, change in forest cover, density of forests, crops grown, land use, land cover. | Landsat high resolution images. |
| | | | Need interpretation and ground truthing. |
| 2. | Aerial photographs | Forest cover, surveillance, cartography and drawing maps, | Can be combined easily with GPS data. |
| | | | Is unobtrusive and depending on resolution can be used easily in GIS. |
| 3 | Other remote sensing images | Forest fires, deforestation, land use, conversion, conservation, altitudes, elevations, topographic maps | Helps to collect data on inaccessible or dangerous objects. |
| | | | Replaces expensive on the ground data and ensures that areas are not disturbed in the bargain. |
| 4. | Other maps | Soil, roads, access, use, boundaries of properties including villages and states and other administrative units, altitude, population density, climate maps, ethnicity, migration patterns etc. Maps showing indices such as disaggregated poverty have also become common. | Include topographic maps, soil maps. Usually require other methods for construction such as remote sensing but also careful census type data collection (e.g. for property rights and boundaries.) |
| 5. | Social and household surveys | Livelihoods, behaviour reasons, behaviour patterns, eligibility for programmes, perceptions, socio economic indicators, physical/economic/social access, use, income, assets and impacts on welfare | Requires careful qualitative work beforehand and afterwards to interpret. It also requires a lot of training for data collectors, careful piloting of instruments to ensure questions are conveying what is being asked. Also requires careful data entry and algorithmic checks which can, if done on mobile phones or PDAs be done on the survey itself. May also require data cleaning. Cross-sectional and repeated time series or panel data have attributes that are specifically useful for understanding changes over time. It is important to understand attrition in this data and reasons for it. |

| | | | |
|---|---|---|---|
| 6. | Administrative data | Eligibility, socio-economic processes, administrative processes, forest management processes, laws and changes in legislation. | Typically is broad brushstrokes data that can be used for sub-populations and illustrates main changes or attributes without explaining motivations or behaviour change of populations |
| 7. | Individual, structured or semi-structured interviews | Help to explain perceptions, beliefs, customs, reasons for behaviour change, determinants of actions, social status, processes and exchange. Also help to explain unobservable selection bias. As well as explain participation and non-participation. | Usually required before, during and after most quantitative data collection so quantitative data can be anchored, collected well and interpreted clearly. Also required for questionnaire design and sampling. |
| 8. | Case study | Qualitative understanding of local drivers and dynamics including processes. | Case studies are especially important to understand processes and behaviours and get deeper insights into what quantitative data may be telling us. |
| 9. | GPS data | Can help determine locations for cities, markets, hospitals, schools as well as boundaries for properties and areas. | Can be combined into a GIS with other data and helps to combine aerial data with on the ground data. GPS data can provide spatial coordinates that may be combined with mosaics of satellite images to then make these useful with other types of data layers as well. |
| 10. | Management information systems | This data can help capture delivery, implementation fidelity, extent to which project targets have been met | These usually are used with process data to understand implementation fidelity. |

implement the type of intervention or programme), and finally, reasoned expectation (the extent to which claims of actual or potential are understood within the context of an evidence based theory of change, which also in turn specifies what can be achieved by when). Using this typology, it is clear that the external validity of impact evaluations is limited when one considers that forestry projects are usually complex and have many intervention arms.

One way to deal with these is to critically incorporate discussions of causal pathways into impact evaluations.

It is not easy to randomize or generate counterfactuals in this area. In our review we recognize that few impact evaluations have usedrandomized assignment. Impact evaluations that use quasi-experimental methods such as regression discontinuity, matching, switching regressions, propensity score matching techniques all require technical expertise that is much more academic than randomized control trials.

Another question that is important to answer is: are impact evaluations necessary and sufficient to answer the big questions? Not unlike in other fields, forestry programmes are complex, are implemented with multiple arms, at different times with different intensities. There are clearly many possible confounding factors that can affect the ultimate outcomes that programmes in this area seek to influence. With caution, we argue that impact evaluations are in fact necessary to answer some of the questions, especially if we want to measure effects robustly that take into account the various sources of bias and are still able to measure the magnitude of these effects.

Impact evaluations require a large amount of data, which is highly disaggregated and has layers that are able to deal with endogeneity and account for confounding effects. Spatially explicit data that use satellite imagery, aerial photographs and other GIS can clearly be combined with traditional sources of data such as household and individual surveys to make these possible. Clearly it is important to be creative here.

We recommend multidisciplinary teams that are able to measure the different variables around the implicit causal chain. We also recommend the use of different sources of data. High resolution data has become available and can be used with real time socio-economic data to understand and also account for socio-economic effects.

Impact evaluation should clearly not be undertaken in all cases. We recommend that **impact evaluations should be undertaken in three cases.**[7] First, there are programmes and policies that are innovative where there is no previous evidence that they even work. Much like pharmaceutical medical phased trials that typically undergo four phases of efficacy testing (see for example Meinert and Tonascia 1986) these are really efficacy trials that just need the proof of concept. So new media campaigns, awareness raising campaigns, programmes that introduce new technologies or new technical ways to save forests, should be tested in this type of impact evaluations. This is where implementation control is critical and these are

[7] This is not new. Several other agencies have adopted this nomenclature. Innovations for Poverty Action (IPA) also uses a similar nomenclature.

laboratory to field experiments. Second, there are programmes where the original programme design has been tweaked to fit the context or have been implemented in a slightly different way. These external replication programmes can and should be impact evaluated because not only does it tell us whether the tweak in the design has the same effectiveness as originally envisioned but also because, impact evaluations can help throw up good lessons for implementation. The third type of projects and policies that this should be done for are large scale programmes where accountability to stakeholders such as donors or the people funding the programme (ordinary people through their taxes) want to know just how much difference the programme is making (so Oportunidades is an example or the Mexico PES programme) and whether it's really important to know if there is value for money.

**Acknowledgements** The authors thank Raag Bhatia, research assistant, 3ie for excellent support.

## Annex I: Definition of Important Terms for Impact Evaluations

| | |
|---|---|
| Intervention | Is used here interchangeably with the programme or the policy implemented or planned to increase the resilience and reduce the vulnerability of forests. |
| Treatment group | The stakeholders that receive or are beneficiaries of the programme or the intervention. Can be individuals, households, plots of land, communities, villages, districts etc. |
| Comparison group | Is the group that is typically compared with the treatment group and (at least for some time) does not get the treatment. |
| Identification design | Are methods that can help *identify* and help to attribute changes in measured effects to a programme/policy/project. Usually these require that implicit or explicit counterfactuals (also called comparisons) to understand what would not have occurred had the programme not occurred. |
| Unit of assignment | Is the level or unit at which a programme is implemented. |
| Unit of measurement | Is the unit for which measurement is undertaken and the units for which the measurement of the effects is important. |
| Mixed methods | Is the collection of methods that are interdisciplinary and use qualitative and quantitative methods in an integrated manner, informing each other and supporting and assisting each other to provide and supplement each other to provide a more wholistic understanding and measure of the effects of a policy, programme or project. |

## Annex II: List of Studies, Locations, Interventions and Identification Methods Reviewed in this Chapter

| No. | Location of study (author) | The main intervention | Intended outcome | Identification method and data used |
|---|---|---|---|---|
| 1. | Ethiopia (Andersson et al. 2011) | A productive safety net (food for work) | Changes in livestock and tree holdings | Regression with propensity score matching. |
| 2. | Mexico (Alix-Garcia et al. 2013) | Conditional cash transfer programme | Deforestation | Method: Regression discontinuity along with IV discontinuity. |
| 3. | Ghana (Burwen and Levine 2012) | Distribution and use of improved cookstoves | Fuel use | A randomized trial. |
| 4. | Thailand (Cropper et al. 2001) | Road building and protected areas | Deforestation in protected areas | Instrumental variables |
| 5. | Senegal (Bensch and Peters 2011) | Improved cookstoves | Demand for charcoal | propensity score weighted regression approach. |
| 6. | India (Somanathan et al. 2005) | Managed forests | Deforestation measured by crown cover. | Difference in difference |
| 7. | Nepal (Tachibana and Adhikari 2009) | Community co-management of forests | Deforestation. | A switching regression model |

| | | | | |
|---|---|---|---|---|
| 8. | Tanzania (Scullion et al. 2011) | Management of forests | Governance, forest conditions and local livelihoods | Quasi experimental methods |
| 9. | Nepal (Edmonds 2002) | Management of forests | Forest cover | Instrumental variables and regression discontinuity approach |
| 10. | Costa Rica (Arriagada et al. 2012) | Protected areas | Avoided deforestation | Mahalanobis weighting with propensity score matching. |
| 11. | Thailand (Sims 2008) | Wildlife Sanctuaries and National Parks | Forest clearing | Quasi-experimental matching techniques |
| 12. | Developing countries (Nelson and Chomitz 2009) | Tropical protected areas. | Deforestation fires which proxy deforestation | Differences in differences with matching |
| 13. | Uganda (Hafashimana et al. forthcoming) | Payment for ecosystem services | | The study is on-going (randomized assignment) |
| 14. | Mexico (Alix-Garcia et al. 2014) | Payment for ecosystem services | Forest cover and socio-economic outcomes. | Difference in difference with matching |
| 15. | Costa Rica (Arriagada et al. 2012) | Payment for ecosystem services | Forest cover | Difference in difference with matching |
| 16. | Costa Rica (Andam et al. 2007) | Payment for ecosystem services | Forest clearing | Difference in difference with matching |

## References


Alix-Garcia J, Mcintosh C, Sims KRE, Welch JR (2013) The ecological footprint of poverty alleviation: evidence from Mexico's oportunidades program. Rev Econ Stat 95:417–435

Alix-Garcia J, Aronson G, Radeloff V, Ramirez-Reyes C, Shapiro E, Sims K, Yañez-Pagans P (2014) Environmental and socioeconomic impacts of Mexico's payments for ecosystem services program

Andam KS, Ferraro PJ, Pfaff ASP, Sanchez-Azofeifa GA (2007) Protected areas and avoided deforestation: a statistical evaluation. Final report. Global Environment Facility Evaluation Office, Washington, DC

Andersson C, Mekonnen A, Stage J (2011) Impacts of the productive safety net program in Ethiopia on livestock and tree holdings of rural households. J Dev Econ, Elsevier 94(1):119–126

Arriagada RA, Ferraro PJ, Sills EO, Pattanayak SK, Cordero-Sancho S (2012) Do payments for environmental services affect forest cover? A farm-level evaluation from Costa Rica. Land Econ 88:382–399

Bensch G, Peters J (2011) Alleviating deforestation pressures? Impacts of improved stove dissemination on Charcoal consumption in urban Senegal. Land Econ 89(4):676–698

Bowler D, Buyung-Ali L, Healey JR, Jones JPG, Knight T, Pullin AS (2010) The evidence base of community forest management as a mechanism for supplying global environmental benefits and improving local welfare. Environmental Evidence: www.environmentalevidence.org/SR48.html

Bravo-Ureta BE, Almeida AN, Solís D, Inestroza A (2011) The economic impact of Marena's investments on sustainable agricultural systems in Honduras. J Agric Econ 62(2):429–448

Burwen J, Levine DI (2012) A rapid assessment randomized-controlled trial of improved cookstoves in rural Ghana. Energy Sustain Dev 16:328–338

Chibwana C, Jumbe CBL, Shively G (2013) Agricultural subsidies and forest clearing in Malawi. Environ Conserv 40(01):60–70

Cropper M, Puri J, Griffiths C (2001) How the location of roads and protected areas affects deforestation in North Thailand

Edmonds EV (2002) Government-initiated community resource management and local resource extraction from Nepal's forests. J Dev Econ 68:89–115

Gertler PJ, Martinez S, Premand P, Rawlings LB, Vermeersch CM (2011) Impact evaluation in practice. World Bank, Washington, DC

Hafashimana D, Jayachandran S, Stanton C, de Laat J, Kalenscher T (forthcoming) Testing the effectiveness of payments for ecosystem services to enhance conservation in productive landscapes in Uganda; a prospective randomized evaluation. The International Initiative for Impact Evaluation (3ie)

List JA, Levitt SD (2005) What do laboratory experiments tell us about the real world. NBER working paper

Nelson A, Chomitz KM (2009) Protected area effectiveness in reducing tropical deforestation: a global analysis of the impact of protection status. Eval Brief 7:31

Pfaff A, Robalino JA, Sanchez-Azofeifa GA (2008) Payments for environmental services: empirical analysis for Costa Rica. Terry Sanford Institute of Public Policy, Duke University, Durham

Puri J (2006) Factors affecting agricultural expansion in forest reserves of Thailand: the role of population and roads. Ph.D. dissertation, University of Maryland

Puri J, Khosla A, Oldenbeuving M (2015) Grants for real-world impact evaluations: What are we learning? A statistical overview and a process analysis of 3ie Open Window Grants. 3ie working paper 23. International Initiative for Impact Evaluation (3ie) (forthcoming), New Delhi

Samii C, Lisiecki M, Kulkarni P, Paler L, Chavis L (2014) Effects of Decentralized Forest Management (DFM) on deforestation and poverty in low and middle income countries: a systematic review. Campbell Syst Rev 2014:10

Scullion J, Thomas CW, Vogt KA, Perez-Maqueo O, Logsdon MG (2011) Evaluating the environmental impact of payments for ecosystem services in Coatepec (Mexico) using remote sensing and on-site interviews. Environ Conserv 38:426–434

Sims K (2008) Evaluating the local socio-economic impacts of protected areas: a system level comparison group approach. Global environment facility impact evaluation information document, Washington, DC

Somanathan E, Prabhakar R, Mehta BS (2005) Does decentralization work? Forest conservation in the Himalayas. Planning unit discussion paper (05-04), Indian Statistical Institute, Planning Unit, New Delhi Discussion Papers, Indian Statistical Institute, New Delhi

Stern E, Stame N, Mayne J, Forss K, Davies R, Befani B (2012) Broadening the range of designs and methods for impact evaluations. Report of a study commissioned by the Department for International Development, working paper 38. DFID report

Tachibana T, Adhikari S (2009) Does community-based management improve natural resource condition? Evidence from the forests in Nepal. Land Econ 85:107–131

UNEP (2011) Forests in a green economy, a synthesis. United Nations Environment Programme

Waddington H, Snilstveit B, Hombrados J, Vojtkova M, Phillips D, Davies P, White H (2014) Farmer field schools for improving farming practices and farmer outcomes: a systematic review. Campbell Syst Rev 2014:6. doi:10.4073/Csr.2014.6

Woolcock M (2013) Using case studies to explore the external validity of 'Complex' development interventions, Wider working paper no. 2013/096. UNU/WIDER, Helsinki

World Bank (2015) World development report 2015: mind, society, and behavior. World Bank, Washington, DC. doi:10.1596/978-1-4648-0342-0. License: Creative Commons Attribution CC BY 3.0 IGO

Wunder S, Angelsen A, Belcher B (2014) Forests, livelihoods, and conservation: broadening the empirical base. World Dev 64:S1–S11

# Chapter 16
# Integration of Indigenous Knowledge into Disaster Risk Reduction and Management (DRRM) Policies for Sustainable Development: The Case of the Agta in Casiguran, Philippines

**Jesusa Grace J. Molina and Andreas Neef**

**Abstract** Due to a combination of physical, socio-economic and political factors, the Agta, an indigenous group in Casiguran, Philippines, are highly susceptible to the threat of natural hazards, especially typhoons, floods, storm surges and landslides. Despite their evident vulnerabilities, the Agta possess valuable indigenous knowledge, generated through practical and long-standing experiences, culture and local resources, which they utilise in coping and in ensuring their safety from the detrimental impacts of disasters. However, the decision-making and planning processes of the local government in the area of disaster risk reduction and management (DRRM) remain insensitive to Agta's knowledge and context, putting them in a more precarious condition and compromising the sustainability of their livelihoods. Employing qualitative and participatory methods, such as semi-structured interviews, policy and document analysis, participant observation and validation workshop, it is argued that there is a need for integrating Agta's indigenous knowledge into the existing DRRM policies and plans of the local government in respect of the rights to sustainable development and survival of the former and in response to the legal obligation of the latter. A sustainable development framework that calls for a process of harmonising indigenous knowledge and science-based information in DRRM towards vulnerability reduction and disaster resilience guided the investigation. While the local government recognises the importance of indigenous knowledge in DRRM, integration with science only happens at the individual level and is not applied in formal settings such as planning and decision-making processes of the municipality. The study recommends mechanisms to ensure Agta's inclusion in the local government's DRRM decision-making, planning, and policy formulation processes such as effective implementation of national laws on DRRM and indigenous peoples; active representation in DRRM council and committees at the municipal and village scales; documentation, validation and integration of

J.G.J. Molina • A. Neef (✉)
Development Studies, School of Social Sciences, Faculty of Arts,
University of Auckland, Auckland 1142, New Zealand
e-mail: a.neef@auckland.ac.nz



J.I. Uitto, R. Shaw (eds.), *Sustainable Development and Disaster Risk Reduction*,
Disaster Risk Reduction, DOI 10.1007/978-4-431-55078-5_16

indigenous knowledge in different sectors such as education, health and livelihood; organising work; and capacity building initiatives that will realise Agta's rights to sustainable development and disaster safety.

**Keywords** Indigenous knowledge • Sustainable development • Vulnerability • Natural hazards • Disaster risk reduction • Philippines

## 16.1 Introduction: The Value of Integrating Indigenous Knowledge in DRRM Policies and Plans of Local Governments for Sustainable Development

When disasters strike, the local people themselves carry out self-initiated response to ensure their safety and survival (Pelling and Wisner 2009; Neef and Shaw 2013; Yila et al. 2013). Acting as frontliners, the locals capitalise on their own capacities, which highlights the role of indigenous knowledge as an important resource for protecting their lives, livelihoods and properties from natural hazards towards sustainable development. Due to their long-time and intimate interaction with the environment, they have acquired a wealth of experience and practices comprising their indigenous knowledge that serves as a major lifeline in the absence of prompt action from concerned authorities and other external actors (Sillitoe 1998; Mercer et al. 2009). Indigenous peoples (IP) are among those who possess a rich collection of indigenous knowledge that has historically enabled them to cope and endure the detrimental impacts of disasters (Bankoff et al. 2004; Scott et al. 2013). A testimony to this is the case of the Agta, an ethnic group in the Philippines that managed to survive strong typhoons along with their secondary hazards for generations by depending on their local practices, environmental resources and networks for disaster preparedness, response and recovery. This study focuses on the Agta of Casiguran, who are among the early IP groups that settled in the Philippine coasts and mountains (Headland 2004).

Different success stories on the value of indigenous knowledge in disaster risk reduction and management (DRRM) have been documented by international scholars and practitioners especially in the Asia Pacific Region (Dekens 2007; McAdoo et al. 2009; Shaw et al. 2009; Mercer et al. 2010). Furthermore, in pursuit of institutionalising the use of indigenous knowledge together with science-based information in DRRM decision-making and planning processes at all levels, a set of indicators and relevant frameworks have been crafted as well (Baumwoll 2008; Mercer et al. 2010; Hiwasaki et al. 2014). The paradigm shift from a hazard-focused orientation to a multi-dimensional disaster approach, which began in the 1990s, paved the way for indigenous knowledge to gain greater recognition as a form of capacity that can contribute to vulnerability reduction, resilience building and sustainable development (Yodmani 2001; Gaillard and Mercer 2013). The integration of indigenous knowledge with science and expert knowledge provides an avenue for local people themselves to take on the role of development agents rather than acting as mere subjects of intervention. With rapid globalisation and climate change influencing the occurrence and impacts of hazards and risks, the active and empowering participation of IP is crucial not only for disaster safety but also for capacity enhancement and realisation of ownership towards effective and sustainable development.

However, despite the obvious value of integrating indigenous knowledge with science-dominated DRRM policies and plans at all levels, its implementation still remains a challenge to many countries across the globe (Shaw et al. 2008). The Philippines is among those countries that struggle in integrating IPs' indigenous knowledge in DRRM processes and systems, especially at the local government level. Focus is given at this level since local governments are legally mandated as duty bearers to provide the first line response in times of disasters when community people are no longer capable of responding on their own (United Nations Development Programme 2004). Despite the passage of the Philippine DRRM Act of 2010, which transformed the traditional reactive nature of the country's response to disaster risks to a more proactive approach (Fernandez et al. 2012), many local government units (LGUs) are still driven by a top-down approach in which indigenous knowledge remains untapped, leaving IP and other vulnerable groups in a perilous condition as they lack representation in defining not only their resilience roadmap but also their greater development agenda.

In 2014, the Philippines was identified as the second most at risk country in the world (Alliance Development Works and United Nations University-Institute for Environment and Human Security 2014). With the evident impacts of climate change along with anthropogenic activities, the escalating frequency and magnitude of hazards and risks especially typhoons and floods in the country poses an intensified threat to the safety and development of the most vulnerable sectors such as the IP. Recent typhoon disasters have resulted in large-scale destruction of livelihoods, assets, resources and environment which contributed to a significant loss in the country's development gains, further subjecting the underprivileged and marginalised communities to the chains of poverty (Collins 2009; Abon et al. 2011; Chiu 2013). This scenario is a clear manifestation of impeding sustainable development since it compromises not only the subsistence and survival of the current populace but also the future of the next generations.

Guided by a sustainable development framework, this chapter specifically explores how the indigenous knowledge of Agta is integrated into DRRM policies and plans of the local government of Casiguran. Section 16.2 describes the basis for the selection of the study site along with the research framework, methods and tools used. Section 16.3 presents the results and is divided into two parts: Agta's indigenous knowledge in DRRM and Agta's participation in village and municipal DRRM planning and policy-making processes. The need for integrating indigenous knowledge and scientific information in DRRM decision-making processes towards sustainable development is discussed in Sect. 16.4. Section 16.5 offers some concluding remarks.

## 16.2 Methodology and Study Concept

### *16.2.1 Study Site*

The selection of study site was made in consideration of two factors: (1) number of Agta households living in the community and (2) the area's vulnerability to natural hazards. *Sitio*[1] Dipontian in Cozo, a coastal community in Casiguran, Aurora (Fig. 16.1) was chosen since a significant population of Agta reside in the area and it was

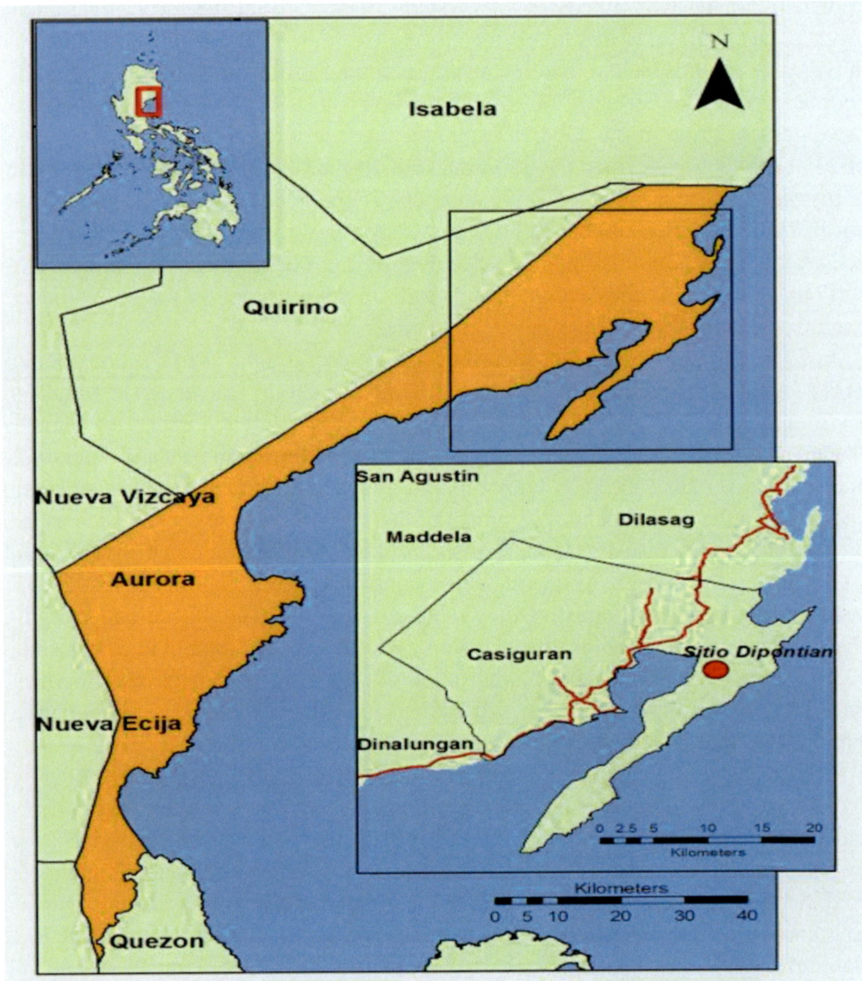


**Fig. 16.1** Map of Aurora province showing the study site

[1] *Sitio* is the Filipino term for zone, a subdivision of a village.

severely affected by super typhoon Utor, locally known as *Labuyo,* in August 2013 that triggered floods, storm surges and strong winds (Cabreza and Orejas 2013). As a result of the disaster, all the Agta's houses were severely damaged, livelihoods were destroyed, environmental resources were depleted and belongings were swept away leaving them with almost nothing.

The municipality of Casiguran can be found in the East-Central side of Luzon Island (Provincial Government of Aurora 2012). It is one of eight municipalities comprising Aurora, a province located in the eastern seaboard making it prone to varying hydro-meteorological hazards such as typhoons, floods and storm surges especially during the rainy season from November to February (Philippine Atmospheric, Geophysical and Astronomical Services Administration 2011). Since Casiguran is surrounded by the Sierra Madre mountain range, it is also susceptible to the risk of landslides, flashfloods and earthquakes. With the Agta's ecosystem-dependent and subsistence-oriented livelihoods, the impact of these hazards – aggravated by climate change and rapid urbanisation (Scott et al. 2013) – is considered a significant threat to their survival and sustainable development since they have to constantly deal with economic shocks, social distress and environmental losses.

## 16.2.2 Conceptual Framework

The conduct of this study was guided by a sustainable development framework (Fig. 16.2). Given the unprecedented magnitude of disasters leading to depleted resources, ill economy and greater poverty at present time, there is an increasing

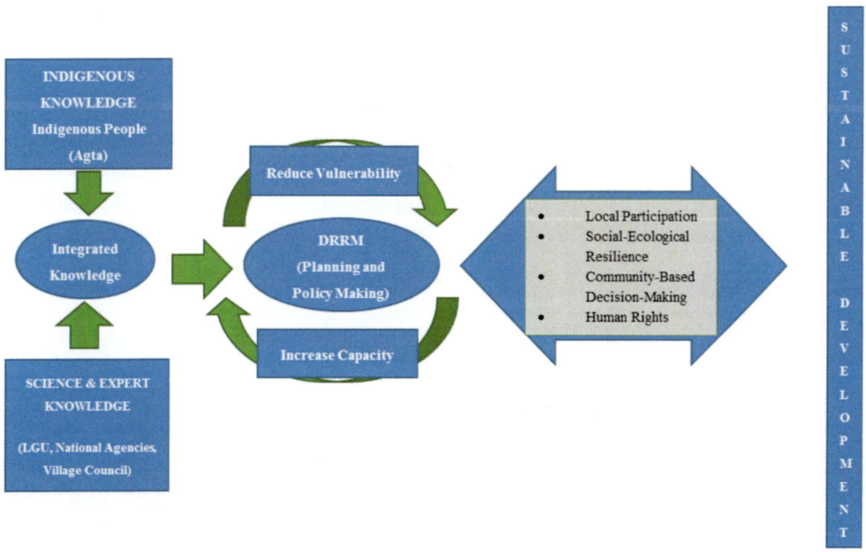


**Fig. 16.2** Integrating indigenous knowledge and science in DRRM towards sustainable development

recognition among stakeholders, especially governments, on the role of DRRM as a foundation for successful sustainable development (Pelling et al. 2014. In many cases, the large-scale damage to livelihoods and assets caused by disasters result in development setbacks, exposing the most vulnerable such as indigenous peoples to the vicious cycle of poverty. This reality testifies to the close interaction of disasters and development which adversely affects developing countries the most (Collins 2013).

The framework emphasises that the integration of indigenous knowledge with science and expert knowledge in the DRRM agenda is fundamental for vulnerability reduction since its realisation strongly adheres to the relevant principles of sustainable development stipulated in the Rio Declaration on Environment and Development (United Nations 1992). These principles include local participation, social-ecological resilience, community-based decision-making and human rights (Boyer-Villemaire et al. 2014; United Nations 1992).

### *16.2.3 Methods*

This study was conducted utilising a qualitative approach as demonstrated in the various face-to-face interviews and participatory methods used for data collection in the field from May to August 2014. Semi-structured interviews were carried out with 41 individuals from the national, municipal and village levels. The respondents comprised Agta, non-Agta residents of Cozo, LGU officials and staff of Casiguran, national government officers and civil society organisation (CSO) representatives working in the field of DRRM. Most interviews lasted for an hour, and a combination of English and Filipino languages was used during interactions. Although a set of guiding questions was prepared, open discussion was still facilitated through probing.

To experience the Agta's way of life and have a broader knowledge of their physical and socio-economic condition, participant observation through living in their village and engagement in daily activities was carried out during the fieldwork period. The integration process provided a channel to understand the Agta's environment, livelihood activities and cultural practices and beliefs, which influence their level of vulnerability. This understanding played an important role in developing a better grasp of the problems raised by this study. The close interaction with the members of Agta households supplemented the information from interviews and documents while building rapport with them at the same time.

Another primary method utilised for data collection was the gathering of DRRM-related documents and policies for analysis. DRRM-related documents such as plans, maps, profiles and ordinance were secured from relevant offices at the municipal and village levels. Through analysing and scrutinising these documents, an improved understanding of the social, cultural, institutional and structural processes and factors influencing the LGU's DRRM approach and the Agta's participation in decision- making and planning activities was gained.

As part of the concluding activity, two validation workshops were organised to present the collected information and generate feedback from the respondents at the

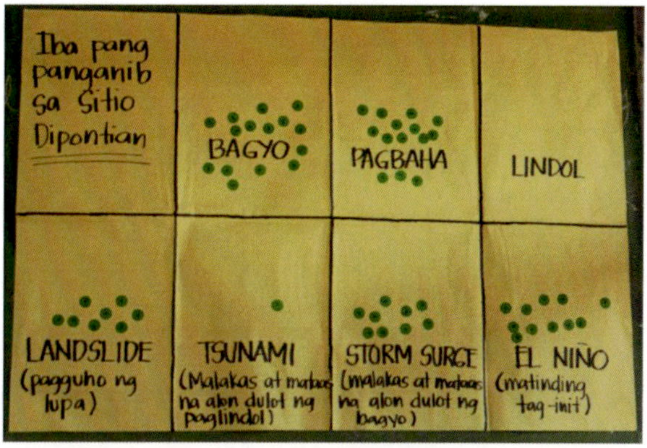


**Fig. 16.3** Agta's hazard ranking output from the workshop (Photo by J. Molina)

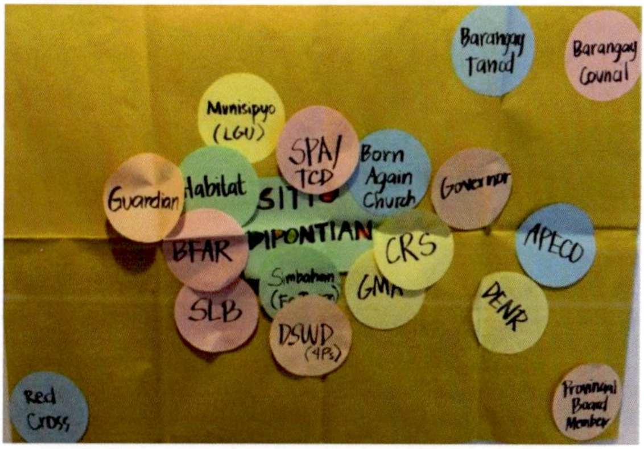


**Fig. 16.4** Mapping of Agta's DRRM partners using Social Venn diagram tool (Photo by J. Molina)

municipal and village levels. The workshop at the municipal level, which was attended by the LGU department heads and staff, was conducted using a focus group approach. On the other hand, at the village level, where Agta comprise the participants, participatory and interactive tools like ranking (Fig. 16.3) and mapping (Fig. 16.4) were used to encourage active participation and create a non-threatening atmosphere. The validation activity provided a significant avenue in confirming the accuracy of information and identify gaps that needed to be addressed.

## 16.3 Results

### *16.3.1 Agta's Indigenous Knowledge in DRRM*

The high exposure of the Agta of *Sitio* Dipontian to natural hazards has impelled them to depend and hone their traditional knowledge and practices not only for personal safety but also for the protection of their ecosystem-based livelihoods over the years. Despite the presence of modern technology, the Agta have managed to sustain the practice of indigenous knowledge and relay it to the younger generation in their *sitio*. With the aid of local resources, the Agta's indigenous knowledge is employed in early warning, construction of emergency shelters through digging holes in the ground and making roofs out of coconut leaves, building multi-purpose structures and stockpiling. In addition to the physical component of indigenous knowledge, the Agta also practice *bayanihan* (spirit of working together) culture which exhibits their social resourcefulness as they respond to recognised risks and vulnerabilities for survival.

Among the identified uses of indigenous knowledge, its application in the aspect of hazard prediction and early warning is the most common and well-known. Dependence on indigenous indicators allows them to initiate preparedness measures, such as securing food supply; safekeeping of belongings; and constructing emergency shelters for protection and survival. Globally, scholars and practitioners have documented good practices confirming the value of indigenous knowledge in hazard prone countries such as Nepal (Dekens 2009); India (Acharya 2011); Bangladesh (Paul and Routray 2013); Indonesia and Timor Leste (Hiwasaki et al. 2014).

Observations relating to animal behaviour, atmospheric changes and the natural environment were identified by the Agta as early warning indicators that enable them to prepare for either an approaching typhoon or strong rains (Table 16.1). These two hazards were mentioned frequently in interviews as they cause the greatest and most frequent impacts to their livelihoods, health and housing. The Agta have recalled four strongest typhoons that resulted to massive damage and

**Table 16.1** Indigenous knowledge for hazard prediction of Agta in Sitio Dipontian

| | Observations |
|---|---|
| Animal behaviour | Ants transport food to an elevated area |
| | Some seashells crawl upward |
| | Birds (i.e. Philippine hornbill) squeak and become playful |
| | Monkeys hide in the forests |
| Atmospheric changes | Red sunrays in the West |
| | Cloud movement and colour |
| | Clouds appear in clusters and move fast |
| | Clouds are thick in appearance |
| | Skies get dark |
| | Wind movement |
| | Strong and warm winds from the East |
| | *Habagat* (southwest monsoon) |
| | Winds locally known as *timog* and *balat* |
| Natural environment | Silence in the surroundings |
| | Sea level rises |
| | Presence of huge waves |
| | The sea looks shallow |

Source: Own fieldwork 2014

threatened their subsistence: Pitang (1970), Harurot (2003), Queenie (2006) and Labuyo (2013).

The Agta respondents who frequently go to the sea and mountains for a living reported that when unusual behaviours are witnessed in animals such as ants crawling upward transporting food; seashells moving in groups to reach higher elevation; certain bird species such as *kalaw* (Philippine hornbill) becoming restless; and monkeys hiding in the forests, they expect that a heavy rain or typhoon may hit them within a few days. Atmosphere-related observations such as strong and warm winds from the East, the presence of southwest monsoon and dark and thick clouds moving fast warn them of the possibility of torrential rains and typhoons as well. Similar hazards are expected when sun rays in the West appear reddish in colour, unusual silence is observed in the community's environment, the sea level rises and big waves are spotted. This is verified in the excerpt below as shared by one of the respondents:

> When our community experienced a typhoon before, we noticed that there was silence in the surroundings, even in the village centre. You cannot hear anything unlike in normal days where people play loud music. When this was observed, we were alerted. Then, after a few days, the weather got bad. The clouds moved fast and the sky was dark. (Male Agta, Sitio Dipontian, interview, June 2014)

The reliability of Agta's indigenous knowledge over generations earned it a reputation as a credible source of information, which enabled them to cope with disasters and risks in the absence of regular and timely dissemination of warning from the Barangay[2] Council or the LGU, as mentioned in the interviews and workshop. Even non-Agta respondents such as the LGU officials, representatives from national government agencies and CSOs and village residents of Cozo expressed affirmation on the relevance of indigenous knowledge up to these days, since they too depend on them for early warning and as a support to scientific forecasts from concerned authorities and media. However, despite the perceived value of indigenous knowledge, it was also shared that the reliability of indigenous knowledge is presently challenged especially by climate change and science as demonstrated in the following quote:

> I grew up here in Casiguran and I also depend on environmental observations for warning such as the appearance of red sunrays, which signals the arrival of a typhoon. However, there are times that this knowledge is inaccurate, especially nowadays. You know why? It is because of climate change. (LGU Department Head, Casiguran, interview, June 2014)

With the phenomenon of a changing climate becoming observable globally, the risk of stronger typhoons and other associated hazards has escalated in recent years and has posed an alarming threat especially to the safety and development of most vulnerable groups (Kelman et al. 2012). The changing behaviour of hazards and risks at the same time exposes indigenous knowledge to uncertainty which challenges its existence (Mercer et al. 2009). Although the Agta possess indigenous knowledge such as those for hazard prediction, the unprecedented magnitude and varying frequency of recent disasters bring in unimaginable and large-scale impacts

[2] *Barangay* is the native Filipino term for village, the smallest political unit in the Philippines.

that exceed their experience and capacity. Of the Agta's existing indigenous knowledge described above, their housing practices are challenged by stronger typhoons, which have become more prominent in recent years, causing heavy rains and destructive winds. The escalating intensity and worsening magnitude of typhoon risks threaten the reliability of indigenous materials that the Agta use for building emergency shelters at present. Furthermore, the evident irresponsiveness of local governments to their legal mandate as first line of defence as attested by the Agta aggravates their heightening vulnerability and results in greater damage as confirmed in the following excerpt:

> None from the Barangay Council informs us. We are informed of an approaching typhoon when we go to the municipal proper and sell orchids. In case we are not able to come home, we remain unprepared. This happened during the time of typhoon Labuyo. We did not receive any warning. (Female Agta, Sitio Dipontian, interview, June 2014)

Moreover, while local government actors have acknowledged the value of indigenous knowledge, its use remains at the individual level and is not translated in the DRRM work of either the village or municipality. All respondents from the LGU, particularly the members of the Municipal Disaster Risk Reduction and Management Council (MDRRMC) have confirmed that DRRM planning and decision-making processes exclusively utilise scientific and technical information produced by planners, engineers and scientists. After doing an analysis of the collected documents, the 2014 Municipal Disaster Risk Reduction and Management Plan (MDRRMP) showed the dominance of science in the LGU's approach as it depended mainly on Geographic Information System (GIS) generated multi-hazard maps in identifying high-, medium- and low-risk villages to various hydro-meteorological and geologic hazards. Although a municipal-wide profiling with Global Positioning System (GPS) tagging was carried out by the LGU to determine hazards and at-risk households together with the Agta, the participation of the latter was limited to being respondents, and a validation of information gathered was not conducted. The absence of such a dialogue inhibited knowledge exchange that could have been a helpful channel to make the information more precise and responsive to their needs.

## 16.4 Agta's Participation in Village and Municipal DRRM Planning and Policy-Making Processes

The circumstances mentioned in the previous section are testimony of the technocratic approach that drives local governments in crafting DRRM policies and plans, giving high regard to science while relegating the value of indigenous knowledge and IP's participation. The invisibility of indigenous knowledge in the DRRM agenda of village and municipal governments reflects the weak democratisation of political processes as community people are compelled to embrace imposed solutions, which is a deliberate violation of the latter's right to self-determination leading to their exclusion in the sustainable development agenda.

The marginalised position and docility of the Agta in the DRRM agenda is rooted in their weak level of participation in the general village and municipal development planning activities and policy-making processes, which have been the predominant scenario for decades. Excerpts from interviews with the Agta demonstrate that the Municipal Government and Barangay Council usually design programmes or projects on their behalf and make the final decision:

> We actually had a request to the Municipal Government when I became the representative of our sitio in the barangay meeting. We would like to have an evacuation centre in our sitio. However, it was not realised. What the Mayor did was he pushed for the establishment of a health centre here. It is yet to be constructed though. (Male Agta, Sitio Dipontian, interview, June 2014)
>
> The Barangay Council does not consult us in planning for disasters and other development activities. When decisions are made, they just consult those who are close to them. You know, patronage politics. That's how things work. (Mother Agta, Sitio Dipontian, interview, June 2014)

Although there are instances where Agta are invited to participate in the meetings and assemblies through their Chieftain, despite physical presence, they remain passive during discussions which inhibit them from actively contributing inputs into the conceptualisation of development interventions relating to livelihoods, health, education and DRRM. Furthermore, it is a common scenario that a staff member from the service centre of the National Commission on Indigenous Peoples (NCIP), the country's focal agency for IP affairs, or a representative from the Barangay Council attends on their behalf if the meeting is convened at the municipal level. One of the contributing factors why the Agta exhibit such a form of participation is the LGU's dismissive attitude towards them as reflected in the narratives of a government employee:

> The Agta's lifestyle is different compared to others which explains why most of them are impoverished. There are times when you provide financial assistance, the Agta just use it for drinking alcohol. They usually go home drunk. I always tell them to build and protect the reputation of their tribe given their poor condition. They are not like other indigenous peoples such as the Igorot who are very hardworking. The Agta are really different. (LGU department head, Casiguran, interview, June 2014)

The Agta are commonly perceived as lazy, dependent, spoiled and alcoholic individuals, a clear reflection of the discrimination they suffer, which inhibits them from exercising participation in defining socio-economic and environmental plans and programmes for sustainable development. Aside from the negative perception of them, geographical isolation, when meetings are held in the municipal proper, as well as work and home responsibilities, especially of those who fish and hunt in the mountains, serve as barriers to their involvement. This shows the significant influence of economic condition, an underlying determinant of vulnerability (Bankoff et al. 2004), on participation as revealed in the Agta's unsecured livelihoods. Furthermore, the absence of Agta's involvement is influenced by the local governments' focus on enhancing response capacities and lack of training in facilitating participatory approach and tools in DRRM as shared in the following excerpt:

> At the LGU level, we really need trainings related to disaster preparedness. We had trainings in the past such as Water Search and Rescue (WASAR) and Basic Life Support (BLS) but I find it inadequate. These trainings should be updated regularly. In fact, I have not received any trainings yet since I assumed my position a year and 4 months ago. (LGU department head, Casiguran, interview, June 2014)

Despite the passage of proactive laws in the Philippines for advancing DRRM and IP welfare, the dominant centralised approach of local governments reinforces the Agta's vulnerability since it promotes a culture of dependency that inhibits the latter from harnessing their potentials as able community members. The lack of consultations and exclusivity of decision-making and planning processes in the hands of the LGU contributes to the perpetuation of a "one-size-fits-all" scheme making the policies and programmes including DRRM insensitive and unreceptive to Agta's needs. This is further aggravated by the template-driven DRRM planning strategy of the LGU where they simply follow outlines provided by the national office, which are not context-specific. Such an approach fails to provide a holistic picture of the underlying causes of the Agta's vulnerability which are linked to varying physical, socio-economic, political and environmental factors. This reality is confirmed by interviews with 14 of the 20 Agta respondents who consistently mentioned that there is no regular committee or representative from either the Barangay Council or Municipal Government that visits and talks to them to get to know their specific needs and concerns.

The Agta's overlooked position in DRRM processes of local governments has resulted in the former's distrust and hesitance to approach the latter, making collaboration towards sustainable strategies intangible. This being the case, the Agta are more eager to engage and work with the non-government organisations (NGOs) operating in Casiguran as shared in the excerpt below:

> We go there and our representative whom we expect to speak up and share our needs and situation is not acknowledged. Maybe the Barangay Council or the Municipal Government takes us for granted. Because of this treatment, the indigenous peoples do not have the drive to participate anymore given that the same thing always happens. We feel left out. Instead of approaching the Barangay Council or the Municipal Government, we prefer to seek help from NGOs. (Male Agta, Sitio Dipontian, interview, June 2014)

NGOs, particularly humanitarian and church organisations, are perceived to provide a better avenue for Agta's engagement in planning and policy-making processes given the participatory approach they employ when it comes to project and programme implementation. Such a strategy provides an avenue for the Agta to exercise an empowered participation through their indigenous knowledge. Although majority of the NGOs' programmes are focused on response and rehabilitation, a proactive approach through collaborative engagement and dialogues with the Agta was ensured throughout the process, starting from damage assessment until the actual implementation and this is evidenced by the quote below:

> Before we implemented our shelter assistance, we went to the IP community and consulted them on the materials to be used for house construction. Since the Agta are not familiar with Hardiflex plywood, we then asked them about the most appropriate substitute for it. They recommended sawali (interwoven splits of bamboo for walling) and we agreed to their sug-

gestion since it is locally available and has been a part of their traditional practice. (NGO staff, Makati City, interview, June 2014)

The inclusive approach employed by the NGOs is a practice that remains a challenge to local governments. Utilising participatory mechanisms is essential in achieving sustainable strategies not just in DRRM but in any development work, given that the Agta are able to play a significant role in the process. Instead of acting as helpless beneficiaries, the Agta are able to express their voice which is crucial not only in exercising their political entitlement but also economic and socio-cultural rights since they are able to define the kind of development that is acceptable and sensitive to their needs.

## 16.5 Discussion: The Need for Integration of Indigenous Knowledge and Scientific Information in DRRM Decision-Making Processes Towards Sustainable Development

The high vulnerability of indigenous peoples and their ecosystems (Carling et al. 2013) and the cross-cutting relationship of DRRM, governance and development (Collins 2013; Pelling et al. 2014), as shown in the case of the Agta of Sitio Dipontian, underscores the need for mainstreaming indigenous knowledge in science-dominated DRRM planning and policy-making strategies of governments at all levels. This is emphasised in the works of DRRM scholars and practitioners who also advocate and support such an integration (Shaw et al. 2009; Mercer et al. 2010; Gaillard and Mercer 2013; Hiwasaki et al. 2014). Although the local government of Casiguran is still challenged to realise such a form of integration, efforts are made in the locality towards promoting the mainstreaming of indigenous knowledge into the education sector. In particular, the Tribal Centre for Development (TCD), an NGO that manages a school for the Agta, has included Indigenous Knowledge System and Practices (IKSP) subject in their academic curriculum. The inclusion of IKSP serves as an effective mechanism for the younger generation of Agta to revive the traditional practices of their tribe in livelihood and housing construction for survival. Integration of indigenous knowledge in education is a fundamental channel to disseminate its value, improve the current indigenous practices and thus ensure their sustainability.

The reliability of indigenous knowledge over generations forms a basis for its inclusion in DRRM processes such as risk assessment, planning and implementation since it can be an effective tool for determining the vulnerabilities and capacities to certain hazards of particular communities (Cadag and Gaillard 2012). Giving voice to the locals through validating their indigenous knowledge is fundamental in fostering ownership towards effective and sustainable DRRM decision-making since they themselves are able to identify their own needs.

The role of indigenous knowledge as an important source of risk information for the Agta has proven its significance in securing personal safety and economic subsistence, which are both fundamental for achieving sustainable development. Indigenous knowledge can play a pivotal role for the Agta in realising their right to survival as it can be utilised to secure their livelihood, particularly those related to agriculture. Utilising hazard indicators, which are cost-efficient and familiar to the Agta, would allow appropriate planning for farming and hunting activities and security of harvest towards hunger prevention and poverty reduction. A good practice showcasing the effective use of indigenous indicators similar to that of the Agta in the agricultural sector has been documented in Northern Philippines (Galacgac and Balisacan 2009) and India (Acharya 2011). With the aid of observing animal behaviour, plants and atmospheric conditions, farmers are able to prepare and protect their crops to prevent losses. In Casiguran's case, the technical and financial support of the LGU is necessary for the enhancement and optimal utilisation of indigenous knowledge along with science to equip the Agta with better farming skills and techniques towards an increased and sustainable yield. As Schipper and Pelling (2006) argued, DRRM takes into consideration the protection of livelihoods from external pressures and shocks while pursuing the agenda of strengthening them at the same time.

Capitalising on indigenous knowledge is also advantageous to local governments with limited financial resources as it requires minimal cost in comparison to employing modern mechanisms alone. Given that disasters greatly affect impoverished communities like that of Sitio Dipontian, reconciling and integrating indigenous knowledge with science offers a viable option for the local governments that have scant resources towards progressively realising their constituents' rights to safety and sustainable development (Scott et al. 2013).

Despite the proven value of indigenous knowledge, it is continuously confronted by challenges relating to the top-down governance and science-dominated approach of the LGU, climate change, culture change, modernisation, assimilation pressures and resource limitations at present. Although this is the case, the Philippines need not to start from scratch in pursuing the integration of indigenous knowledge with science in DRRM policy-making and planning processes. Along with the affirmation of respondents from all levels on the value of integrating both knowledge forms, preliminary efforts in developing an integration framework and enactment of laws have also been carried out, which should be used to the best advantage to promote its institutionalisation in DRRM processes and systems.

Besides the identified gaps, the study also sought viable means to promote the value of indigenous knowledge in DRRM policy-making and planning processes towards effective and sustainable development. The identified strategies by Agta and non-Agta respondents in the interviews and workshop include popularisation; capacity building activities; representation of Agta to DRRM/ Legislative Council at all levels; integration of indigenous knowledge in different sectors; conduct of validation; and organising work (Table 16.2).

The achievement of these suggested strategies would offer effective avenues to engage the Agta and local governments in collaboration, despite cultural differences

**Table 16.2** Suggested ways to promote indigenous knowledge in DRRM processes

| Strategy | Description |
|---|---|
| Popularisation | Use of various information, education and communication (IEC) channels as follows: invite Agta elderly to share indigenous knowledge in cultural presentations and community celebrations (i.e. fiesta, celebration of IP month); documentation and publication; social media; story telling |
| Capacity building activities | Conduct of DRRM trainings and Indigenous Peoples Rights Act orientation for the Agta, LGU and other stakeholders working in DRRM |
| Representation of Agta | Selection of an Agta representative to the DRRM/Legislative Council at all levels |
| Integration in different sectors | Integration of indigenous knowledge in education through the academic curriculum; health through various services and facilities; and livelihood through income-generating programmes |
| Validation | Conduct of further research to confirm the scientific basis and enhance the value of indigenous knowledge |
| Organising work | Organising work for knowledge exchange and networking among stakeholders |

and varying capacities, and recognise the value of each other's knowledge for pursuing not only disaster risk reduction but also an equitable, holistic and sustainable development.

## 16.6 Conclusion

Findings from this study show that, in general, the Agta and the local government of Casiguran have acted separately in response to DRRM, the former depending on indigenous knowledge and experience, while the latter capitalises on science and technology. Despite the proven value of Agta's indigenous knowledge, it is disregarded in formal settings as shown by the limited collaboration between them and the LGU, leading to poorly integrated mechanisms for vulnerability identification and uncoordinated strategies for planning and policy-making in DRRM and greater sustainable development agenda. Our results support the argument of Mercer et al. (2009) that indigenous knowledge is subject to oppression in manifold ways because of the marginalisation, discrimination, powerlessness and violence experienced by its bearers. The study confirms that dominant unequal power relations, discriminatory practices and external pressures brought by modernisation are among the contributing factors why the local governments and Agta continue to work in isolation subjecting the latter and their knowledge to an uninterrupted marginalised state, which worsens their underdevelopment since the underlying physical, economic, social and political vulnerabilities are left unaddressed.

While the Philippine DRRM system is faced by challenges to realise the integration of indigenous knowledge and science, the study concludes that it does offer a viable environment to move forward. Although the country's DRRM practitioners believe that the Philippines is still in an infancy stage of implementing such integration, there has been a good beginning for local and national decision makers since they can now build on the early efforts such as the passage of the Philippine DRRM Act of 2010 recognising the value of indigenous knowledge (DRRNet Philippines 2011) and the existing integration framework tested in selected IP communities in the country (Hiwasaki et al. 2014).

Relevant policy implications of this study emphasise the necessity for local and national governments to provide an enabling environment for the effective implementation of proactive DRRM legislations such as the one mentioned above; formal adoption of the existing integration framework; and representation of indigenous peoples at all levels. Furthermore, validation and dissemination of indigenous knowledge need to be conscientiously carried out together with the locals and experts (Hiwasaki et al. 2014). Although the research was able to gather the Agta and the local government representatives to conduct a validation, experts from the science community were not involved because of time and resource restraints. Holding a multi-stakeholder validation of indigenous knowledge is important for its recognition; to maximise its value; and achieve its successful application and institutionalisation towards disaster resilience and sustainable development.

**Acknowledgements** The authors are grateful to the New Zealand Agency for International Development (NZAid); The University of Auckland Human Participants Ethics Committee; Simbahang Lingkod ng Bayan; and *Nuestra Señora de la Salvacion* Parish that have contributed to the successful implementation of the research in one way or another. The authors are also thankful for the warmth and cooperation of the Municipal Government of Casiguran, the Village Council of Cozo and, most especially, for the active participation of the Agta of Sitio Dipontian.

## References

Abon C, David C, Pellejera N (2011) Reconstructing the tropical storm Ketsana flood event in Marikina River, Philippines. Hydrol Earth Syst Sci 15(4):1283–1289

Acharya S (2011) Presage biology: lessons from nature in weather forecasting. Indian J Tradit Knowl 10(1):114–124

Alliance Development Works, United Nations University- Institute for Environment and Human Security (2014) World risk report. Alliance Development Works & UNU-EHS, Bonn

Bankoff G, Frerks G, Hilhorst D (2004) Mapping vulnerability: disasters, development, and people. Earthscan, London

Baumwoll J (2008) The value of indigenous knowledge for disaster risk reduction. Webster University, Webster Groves

Boyer-Villemaire U, Benavente J, Cooper JAG, Bernatchez P (2014) Analysis of power distribution and participation in sustainable natural hazard risk governance: a call for active participation. Environ Hazards 13(1):38–57

Cabreza V, Orejas T (2013) Typhoons test Aurora town's resilience. Online article. URL: http://newsinfo.inquirer.net/470473/typhoons-test-aurora-towns-resilience. Last accessed Jan 2015

Cadag JRD, Gaillard J (2012) Integrating knowledge and actions in disaster risk reduction: the contribution of participatory mapping. Area 44(1):100–109

Carling J, Wessendorf K, Feiring B (2013) Indigenous peoples and climate change. A human rights challenge. The Danish Institute for Human Rights, Copenhagen
Chiu Y (2013) Typhoon Haiyan: Philippines faces long road to recovery. Lancet 382(9906):1691–1692
Collins AE (2009) Disaster and development. Routledge, London
Collins AE (2013) Linking disaster and development: further challenges and opportunities. Environ Hazard 12(1):1–4. doi:10.1080/17477891.2013.779137
Dekens J (2007) Disaster preparedness: a literature review. International Center for Integrated Mountain Development, Patan
Dekens J (2009) Local knowledge on flood preparedness in Eastern Terai of Nepal. In: Shaw R, Sharma A, Takeuchi Y (eds) Indigenous knowledge and disaster risk reduction: from practice to policy. Nova Science Publishers, New York, pp 147–166
Disaster Risk Reduction Network Philippines (2011) Primer on the disaster risk reduction and management act of 2010. Christian Aid, Quezon City
Fernandez G, Uy N, Shaw R (2012) Community-based disaster risk management experience of the Philippines. In: Shaw R (ed) Community, environment and disaster risk management, vol 10. Emerald Publishers, Bingley, pp 205–231
Gaillard JC, Mercer J (2013) From knowledge to action: bridging gaps in disaster risk reduction. Prog Hum Geogr 37(1):93–114
Galacgac ES, Balisacan CM (2009) Traditional weather forecasting for sustainable agroforestry practices in Ilocos Norte province, Philippines. For Ecol Manage 257(10):2044–2053
Headland TN (2004) Basketballs for bows & arrows: deforestation and Agta culture change. Cult Surviv Q 28(2):41–44
Hiwasaki L, Luna E, Shaw R (2014) Process for integrating local and indigenous knowledge with science for hydro-meteorological disaster risk reduction and climate change adaptation in coastal and small island communities. Int J Disaster Risk Reduction 10:15–27
Kelman I, Mercer J, Gaillard JC (2012) Indigenous knowledge and disaster risk reduction. Geography 97:12–21
McAdoo BG, Moore A, Baumwoll J (2009) Indigenous knowledge and the near field population response during the 2007 Solomon Islands tsunami. Nat Hazards 48(1):73–82
Mercer J, Kelman I, Dekens J (2009) Integrating indigenous and scientific knowledge for disaster risk reduction. In: Shaw R, Sharma A, Takeuchi Y (eds) Indigenous knowledge and disaster risk reduction: from practice to policy. Nova Science Publishers, New York, pp 115–131
Mercer J, Kelman I, Taranis L, Suchet S (2010) Framework for integrating indigenous and scientific knowledge for disaster risk reduction. Disasters 34(1):214–239
Neef A, Shaw R (2013) Local responses to natural disasters: issues and challenges. In: Neef A, Shaw R (eds) Risks and conflicts: local responses to natural disasters. Emerald Publishers, Bingley, pp 1–8
Paul SK, Routray JK (2013) An analysis of the causes of non-responses to cyclone warnings and the use of indigenous knowledge for cyclone forecasting in Bangladesh. In: Filho WL (ed) Climate change and disaster risk management. Springer, Heidelberg, pp 15–39
Pelling M, Wisner B (2009) Disaster risk reduction: cases from urban Africa. Earthscan, London
Pelling M, Basher R, Birkmann J, Cutter S, Desai B, Fakhruddin SHM, Ferrugini F, Mitchell T, Oliver-Smith T, Rees J, Takeuchi K (2014) Disaster risk reduction and sustainable development. In: Sassa K (ed) Landslide science for a safer Geoenvironment, vol 1. Springer, Cham, pp 211–216
Philippine Atmospheric Geophysical and Astronomical Services Administration (2011) Climate change in the Philippines. PAGASA, Quezon City
Provincial Government of Aurora (2012) General information: land and climate. URL: http://www.aurora.ph/geninfo.html. Last accessed 10 Jan 2015
Schipper L, Pelling M (2006) Disaster risk, climate change and international development: scope for, and challenges to, integration. Disasters 30(1):19–38
Scott J, Llamas-Cabello D, Bittner P (2013) Engaging indigenous peoples in disaster risk reduction. A white paper prepared for the United Nations Permanent Forum on Indigenous issues. Center for Public Service Communications, Claiborne

Shaw R, Uy N, Baumwoll J (eds) (2008) Indigenous knowledge for disaster risk reduction: good practices and lessons learned from experiences in the Asia-Pacific region. United Nations International Strategy for Disaster Reduction, Bangkok

Shaw R, Sharma A, Takeuchi Y (2009) Introduction: indigenous knowledge and disaster risk reduction. In: Shaw R, Sharma A, Takeuchi Y (eds) Indigenous knowledge and disaster risk reduction: from practice to policy. Nova Science Publishers, New York, pp 1–13

Sillitoe P (1998) The development of indigenous knowledge: a new applied anthropology 1. Curr Anthropol 39(2):223–252

United Nations (1992) Rio Declaration on Environment and Development. URL: http://www.unep.org/Documents.Multilingual/Default.asp?documentid=78&articleid=1163. Last accessed 3 Jan 2015

United Nations Development Program (2004) A global report reducing disaster risk: a challenge for development. UNDP Bureau of Crisis Prevention and Recovery, New York

Yila O, Weber E, Neef A (2013) The role of social capital in post-flood response and recovery among downstream communities of the Ba River, Western Viti Levu, Fiji Islands. In: Neef A, Shaw R (eds) Risks and conflicts: local responses to natural disasters. Emerald Publishers, Bingley, pp 79–107

Yodmani S (2001) Disaster risk management and vulnerability reduction: protecting the poor. The Asian Disaster Preparedness Center, Bangkok

# Chapter 17
# Usefulness of a Sustainability Literacy Test

**Thi Kinh Kieu, Glenn Fernandez, and Rajib Shaw**

**Abstract** This chapter will trace the history, development, and purpose of the Sustainability Literacy Test (SLT) being promoted by several universities worldwide to ensure that they are producing sustainability literate graduates. A comparison between SLT and similar pioneering tests will be made to offer insights on lessons learned from past experiences and provide suggestions for improving SLT. In addition, this chapter will present initial feedback from Kyoto University students, who were among the first batch of students in Asia to take the global pilot version of the test in 2014, on how they found SLT and what recommendations they could share to make SLT more useful from the perspective of test-takers.

**Keywords** Sustainable development • Sustainability literacy test • Higher education institutions

## 17.1 Introduction: What Is Sustainability Literacy?

According to Kanj and Mitic (2009), "to function well in the twenty-first century a person must possess a wide range of abilities and competencies, in essence many 'literacies'." This smorgasbord of literacies includes but is not limited to: civic literacy, computer literacy, consumer literacy, cultural literacy, energy literacy, environmental literacy, financial literacy, geographic literacy, health literacy, historical literacy, investment literacy, legal literacy, mathematical literacy, media literacy, moral literacy, political literacy, scientific literacy, technological literacy, and

T.K. Kieu (✉)
Environmental Education Laboratory, Graduate School of Global Environmental Studies, Kyoto University, Kyoto, Japan
e-mail: kieukinh@gmail.com

G. Fernandez
Environmental Education Laboratory, Graduate School of Global Environmental Studies, Kyoto University, Kyoto, Japan

Asian Disaster Preparedness Center, Bangkok, Thailand

R. Shaw
Graduate School of Global Environmental Studies, Kyoto University, Sakyo-ku, Kyoto, Japan



J.I. Uitto, R. Shaw (eds.), *Sustainable Development and Disaster Risk Reduction*, Disaster Risk Reduction, DOI 10.1007/978-4-431-55078-5_17

workplace literacy. Literacy is defined as basic knowledge in a subject or field (Snavely and Cooper 1997).

One the emerging literacies is sustainability literacy. It follows in the footsteps of environmental literacy and ecological literacy (El Ansari and Stibbe 2009; Lugg 2007). The UK Government's sustainable development strategy, 'Securing the Future,' was pioneering in calling on all education sectors to "embrace sustainable development and promote the concept of sustainability literacy among their students" (HEA 2006). 'Securing the Future' stipulates the need to make sustainability literacy a core competency for professional graduates (DEFRA 2005).

According to Stibbe and Luna (2009), building a more sustainable self, community, society and world requires more than knowledge about sustainability – it requires sustainability literacy, which refers to "the skills, attitudes, competencies, dispositions, and values that are necessary for surviving and thriving in the declining conditions of the world in ways which mitigate that decline as far as possible". Sustainability literacy helps compel individuals to become deeply committed to building a sustainable future (Carteron and Decamps 2014). A sustainable literate person recognizes the necessity for shifting to a more sustainable way of doing things, has adequate knowledge and skills to decide and act in a way that favors sustainable development, and is able to appreciate and reinforce other people's decisions and actions that favor sustainable development (Adderley 2007).

## 17.2 History and Development of the Sustainability Literacy Test

The last decade has witnessed a growing public awareness of sustainability and higher education institutions (HEIs) have also joined the bandwagon (Yuan and Zuo 2013). HEIs have significantly contributed in the generation of knowledge and in shaping social and scientific paradigms. Through their teaching and research activities, universities and colleges are expected to fashion a more sustainable future. The Rio+20 United Nations Conference on Sustainable Development in 2012 highlighted the role of HEIs in increasing awareness of sustainability challenges for graduates. Given the objectives of Rio+20, HEIs have a special responsibility to provide leadership on education for sustainable development (ESD) which aims at enabling every graduate student to acquire the values, competencies, skills, and knowledge for a sustainability literate society.

In the Declaration on Higher Education Sustainable Initiative launched at Rio+20, Chancellors, Presidents, Rectors, Deans, and leaders of HEIs and related organizations, acknowledged the responsibility that they bear in the international pursuit of sustainable development. They committed to (1) teach sustainable development concepts; (2) encourage research on sustainable development issues; (3) green their campuses by reducing their environmental footprint; adopting sustainable procurement practices; providing sustainable mobility options for students

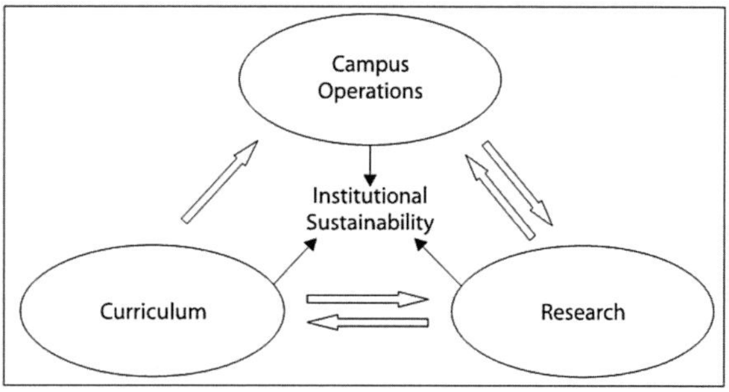


**Fig. 17.1** A whole-of-university approach to sustainability (Source: Mcmillin and Dyball 2009)

and faculty; adopting effective programs for waste minimization, recycling, and reuse; and encouraging more sustainable lifestyles; (4) support sustainability efforts in the communities in which they reside; and (5) engage with and share results through international frameworks.

Since the UN Decade of Education for Sustainable Development was established by UNESCO in 2004, HEIs have advanced sustainability principles on their campuses through a variety of activities in all dimensions in a whole-of-university approach (Fig. 17.1), including in governance, campus operations, education, research, and outreach (Mcmillin and Dyball 2009; Yarime and Tanaka 2012; Savelyeva and McKenna 2011).

After a certain period, HEIs developed the Sustainability Assessment Tool (SAT) to identify obstacles to the current approaches towards sustainability. A question is raised on how HEIs can assess and report on their global performance? In this context, the Sustainability Literacy Test (SLT) was suggested as "a tool for the various initiatives on sustainability lead by HEIs to assess and verify the sustainability literacy of their students when they graduate" (http://www.sustainabilitytest.org). SLT "assesses the minimum level knowledge in economic, social, and environmental responsibility for higher education students, applicable all over the world, in any kind of HEI, in any country, studying any kind of tertiary-level course (Bachelors, Masters, MBAs, PhD)." The objectives of SLT are to: (1) get feedback to teach and enhance the quality of students' knowledge on sustainability; (2) enhance sustainability literacy worldwide; (3) create a benchmark for ESD (with statistics and worldwide survey); and (4) serve as a potential recruitment tool for employers (NGOs, government institutions, private companies).

In October 2013 a draft version of the Sustainability Literacy Test (version 0) was launched in France and between January and October 2014 a pilot version (version 1) was launched worldwide (Carteron and Decamps 2014). More than 24,500 students from 30 countries had taken the Sustainability Literacy Test versions 0 and 1 combined (Carteron and Decamps 2014).

The scope of SLT focuses on two key areas: (1) questions about the current challenges facing society and the planet, e.g., general knowledge on social, environmental, and economic issues; basic understanding of the Earth system, e.g., water and carbon cycles, greenhouse effect, etc. and (2) questions on an organization's

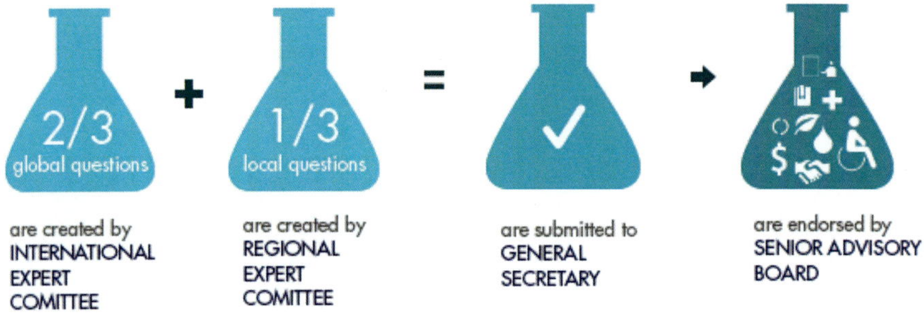


**Fig. 17.2** Structure of the sustainability literacy test (Source: http://www.sustainabilitytest.org)

responsibility, e.g., questions about organizational practices for integrating social responsibility in their activities and questions on the responsibility of individuals as employees and citizens.

A Multiple Choice Question (MCQ) format was chosen to make the test easier to use and implement worldwide. Fifty MCQs are randomly selected among a wide range of questions from a question bank. Out of these 50 questions, 30 are related to supra/international level issues (e.g., global warming) and 20 are linked to national/regional issues (e.g., local regulations and laws, culture, and practices) (Fig. 17.2). Completing this web-based test usually takes 30 min.

The supra/international level questions are used to be able to compare knowledge from one region of the globe to another and to allow institutions/students to benchmark at a worldwide level. The national/regional level questions are used to ensure that SLT remains relevant. Questions in the Sustainability Literacy Test were suggested by the Regional/National Expert Committee (RNEC) and International Expert Committee (IEC) members, in most cases after consulting the broader community of educators. The questions were reviewed and revised by a reviewing committee (part of the General Secretariat), and then, after a series of revisions, questions were posted on a secure platform for comments and validation by members of the Senior Advisory Board.

Carteron and Decamps (2014) presented a 1-year report on the implementation of the Sustainability Literacy Test on the occasion of the World Conference on Education for Sustainable Development in Nagoya, Japan in November 2014. They shared the achievements of SLT and its next steps going forward.

## 17.3 Experiences of Similar Tests by Pioneering Universities

The idea behind sustainability literacy testing is not new. Sustainability literacy is a topic of increasing interest among a growing number of higher education sustainability faculty and staff. Several groups have been working on sustainability literacy tests for quite some time. In the early stage of DESD, tests were utilized for assessing the sustainability knowledge of students before and after one course, frequently in the manner of pre-, sometimes mid-, and post-test checks on content (Erdogan and Tuncer 2009). Hence the number of students tested was limited to the size of the class taking a course. Since the adoption of the Sustainability Tracking Assessment and Rating System (STARS) by over 200 HEIs, there had been attempts to evaluate the understanding of basic knowledge on sustainability among the student population (AASHE 2014). In this chapter, several case studies which will be analyzed based on these two main approaches: course-based sustainability literacy tests and university-scale sustainability literacy tests. Here we are briefly presenting the experiences of Middle East Technical University Ankara (Turkey), Liverpool John Moores University (UK), University of Maryland (USA), and Ohio State University (USA).

The convergence of a number of trends and events in recent years led to the emerging transformation in university curriculum to develop a responsible citizenry capable of applying ecological, economic, and socio-cultural knowledge to solve current and future global problems. Sustainability topics are now included in formal or non-formal education in the form of short-term courses, lectures, or even training programs which are related to specific topics connected to the specialization of the students. This new educational shift leads to research on how to evaluate the efficiency of lecture design, contents, pedagogical approaches, and other dimensions of sustainability education at each university (Erdogan and Tuncer 2009; Connell et al. 2012; Cotgrave and Kokkarinen 2011).

A set of questions is frequently used to assess the improvements in the sustainability literacy of participating students. In the Middle East Technical University Ankara (Turkey) for instance, a set of seven open-ended questions was utilized to evaluate the change in sustainability outlook of 68 university students from the Faculty of Education, after taking a course titled "Education and Awareness for Sustainability" (Erdogan and Tuncer 2009). The questions focused on future employment, consumption decisions, lifestyle choices of the students, and how they contribute to the improvement of communities in which they live. The students' response on the whole presented the positive change in their sustainable ways of living.

Liverpool John Moores University (UK) similarly designed a course to promote sustainability literacy among undergraduate construction students (Cotgrave and Kokkarinen 2011). In order to test the efficiency of the new course, questions were provided to students via email pre-, mid-, and post-course. Majority of questions asked the students to use a rating scale to compare the motivation of students during the course via their knowledge, awareness, attitudes, and learning styles as well as to identify the difference in terms of mode of study, program of study, and student age. Additionally, each student was required to write an essay after finishing the course. The results showed that 216 final-year students' attitudes consistently increased at every stage of the course, particularly students from a construction management major. Although students did not claim that their knowledge was higher than before, the results from the test and essays implied their holistic thinking in the selection of construction materials, which was not only based on the quality and price but also on health, safety, and environmental considerations.

As can been seen from the two case studies above, sustainability themes with strong linkage to the students' future employment were designed to nurture an autonomous citizenry for diverse fields. Testing the sustainability literacy of students at all stages of the education process has been pondered as an effective tool to evaluate and modify ESD courses. Using open-ended questions or essays allow professional educators to deeply understand to what extent students perceive and improve their competencies to construct a sustainable future. Nonetheless, to confirm the change in sustainability literacy requires researchers to follow the students over the long term, at least for several semesters.

When HEIs have become incrementally more aware of ESD, there is a demand to ensure a certain level of sustainability understanding among university students regardless of their majors. The University of Maryland (USA) was the first HEI to conduct a university-wide sustainability literacy test, named "Sustainability Quick Quiz" (Horvath et al. 2013). The test which included 15 close-ended questions and 1 open-ended question was hosted on the Survey Monkey website. The main contents of the test included: (i) the meaning of sustainability; (ii) how to live sustainably; and (iii) the relationship between humans and Nature. There were 1,442 students who took the test, out of a random sample of 9,170 students registered in the 2011 spring semester (one-fourth of student population): 68 % of those who took the test were undergraduate students and 32 % were graduate students.

The mean raw score for all assessment respondents was 23 points or a mean sustainability score of 74.9 %; the mode for all respondents was 83.8 %; the median sustainability score was 77.4 %; the range was between 16 % and 100 %, with a standard deviation of 15.66. The data showed that graduate students (master and doctoral levels) scored significantly higher (mean sustainability score = 77 %) than undergraduate students (mean sustainability score = 74 %). Another meaningful finding was that the students who took three or more sustainability-themed courses had more sustainability knowledge than students who took zero, one, or two courses. There were several challenges identified after conducting the Sustainability Quick Quiz such as the low rate of participation (only 16 %) via online survey and lack of participation of key members (administrators, faculty members, and campus sustainability staff). Only those who were interested in sustainability took the test so the survey could not capture the whole picture of the sustainability literacy of all the students. Moreover, the questions were not able to assess students' awareness, sensitivity, knowledge, level of concern, and level of responsibility.

In Ohio State University (USA), their Sustainability Literacy Test included 16 multiple choice questions across three domains of sustainability knowledge: six environmental sustainability questions, five social sustainability questions, and five economic sustainability questions. The test was sent through email to over 10,000 enrolled undergraduate students and about 1,930 responded (Zwickle et al. 2014). The results are shown in Table 17.1:

The overall average score was 11.08 out of 16. For the three separate domains, there was no remarkable difference between environmental and social domains but the students obtained the lowest score in the economic domain. In terms of academic level, students at higher levels achieved higher scores (Fig. 17.3).

**Table 17.1** Results of the sustainability literacy test of Ohio State University

| Result | Environmental | Social | Economic | Total |
|---|---|---|---|---|
| Mean raw score | 4.39/6 | 3.55/5 | 3.03/5 | 11.08/16 |
| Mean sustainability score | 73 % | 71 % | 61 % | 69 % |
| Standard deviation | 1.48 | 1.23 | 1.27 | 3.21 |

Source: Zwickle et al. (2014)

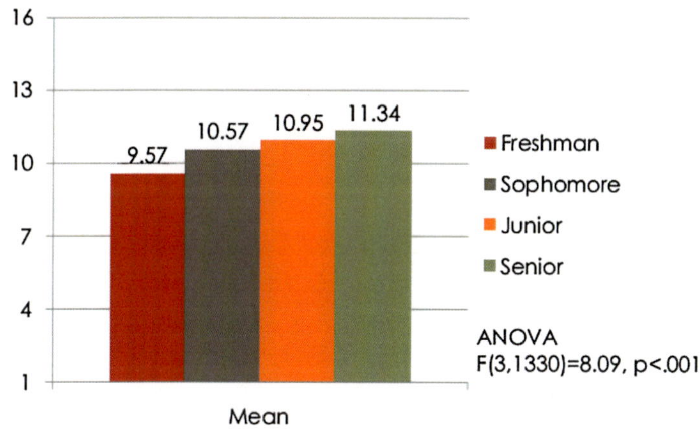


**Fig. 17.3** Sustainability literacy test results by academic level (Source: Zwickle et al. 2014)

Based on their experience, Ohio State University came to the realization that developing an assessment and methodology that provides meaningful results can be difficult; analyzing and interpreting the results can be time-consuming; buy-in from administration may not be there; students may already be experiencing survey fatigue; and no central assessment tool currently exists.

Table 17.2 summarizes the comparison among the sustainability literacy tests of pioneering universities and the international SLT launched in 2014.

## 17.4 Students' Early Reactions to the Sustainability Literacy Test

Since the global pilot version of SLT was launched in 2014, 261 HEIs had registered to join the network and conducted the test (Carteron and Decamps 2014) (Table 17.3).

Kyoto University was the first HEI in Japan to join the network and conduct the pilot SLT in July 2014. In order to understand the reaction of the students after taking the test, a survey was conducted. A total of 43 students from 7 faculties (more than half of all the test-takers from Japan) voluntarily joined the survey: 16 of them were Japanese and the rest came from ten different countries. Most of survey

**Table 17.2** Comparison among different sustainability literacy tests

| | Middle East Technical University Ankara | Liverpool John Moores University | University of Maryland | Ohio State University | International SLT |
|---|---|---|---|---|---|
| Year | 2004–2005 academic year | 2010 | 2011 | 2013 | 2014 |
| Time | – | – | – | – | 30 min |
| No. of items | 7 | | 16 | 16 | 50 |
| Maximum score | – | – | 31 points | 16 points | 50 points |
| Languages | Turkish | English only | English only | English only | Six languages |
| Mode | Survey | E-mail | Survey monkey | E-mail | Online |
| Types of questions | Open-ended | Open ended + Essay | 15 close ended, one open ended | Multiple choice | Multiple choice |
| Target students | Faculty of Education students who took the “Education and Awareness for Sustainability” course | Undergraduate construction students | 9,170 undergraduate and graduate students | 10,478 undergraduate students | All higher education students |
| No. of students tested | 85 | 216 | 1,442 (68 % undergraduate, 32 % graduate) | 1,930 | 24,555 students from 30 countries (as of 2014) |

**Table 17.3** Universities registered and tests completed as of October 24, 2014

| Countries where the test is already customized | Registered universities (V1) | Students who have completed the test |
|---|---|---|
| Argentina | 1 | 176 |
| Brazil | 20 | 2229 |
| Canada - Quebec | 3 | 348 |
| China | 2 | 132 |
| China (Hong Kong) | 8 | 824 |
| Costa Rica | 9 | 75 |
| Egypt | 1 | 88 |
| France | 86 | 8776 |
| India | 7 | 38 |
| Ireland | 1 | 91 |
| Italy | 6 | 675 |
| Japan | 3 | 76 |
| Peru | 1 | 600 |
| South Africa | 4 | 73 |
| Spain | 7 | 56 |
| United Kingdom | 19 | 1926 |
| USA | 58 | 1553 |
| | 236 | 17736 |
| Countries without customized questions | | |
| Australia | 2 | 4 |
| Belgium | 2 | 77 |
| Burkina Faso | 1 | |
| Canada | 5 | |
| Ecuador | 1 | |
| Finland | 1 | |
| Germany | 2 | |
| Haiti | 1 | |
| Israel | 1 | |
| Kenya | 4 | |
| Malaysia | 1 | 708 |
| New Zealand | 1 | |
| Sweden | 1 | 20 |
| Switzerland | 1 | 7 |
| Taiwan | 1 | 1 |
| Dominican republic | 0 | |
| Senegal | 0 | |
| | 25 | 817 |
| TOTAL | 261 | 18553 |

Source: Carteron and Decamps (2014)

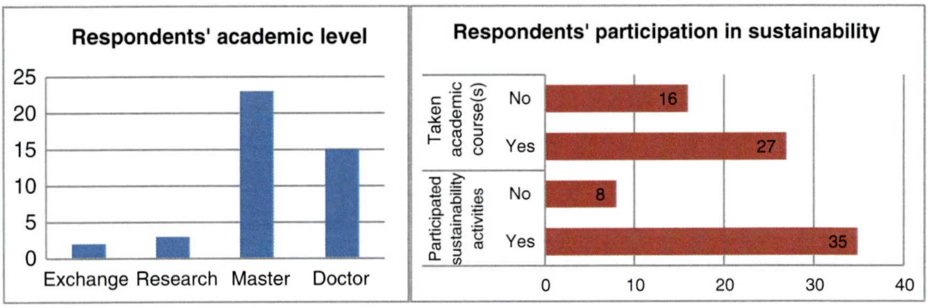


Fig. 17.4 Survey respondents' background information

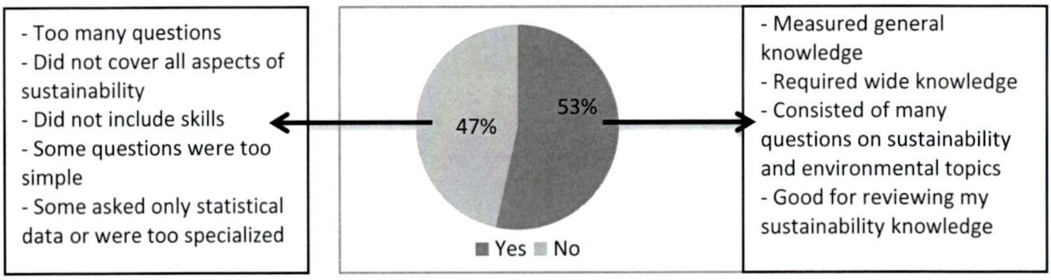


Fig. 17.5 Students' reactions to the questions in SLT

respondents had experienced involvement in sustainability activities by taking academic courses or joining outdoor events such as the Green Festival, planting trees, recycling wastes, or participating in the Sustainability Day in campus (there were a total of 21 diverse activities listed by the survey respondents) (Fig. 17.4). Indeed, sustainability extra-curricular activities have become more popular among higher education students. It was expected that with students' sustainability experiences, they would be able to provide relevant feedback after finishing the SLT.

When asked if this test accurately measured their knowledge on sustainability and environmental topics, the students' answers are presented in Fig. 17.5:

The number of respondents who agreed and disagreed with the claim that SLT accurately measured their knowledge on sustainability is not significantly different (53 % agreed and 47 % disagreed). The reasons for their arguments are indicated in Fig. 17.5. These reasons can somehow be initially considered as the weaknesses and strengths of the SLT.

Despite the disagreement of the respondents on whether SLT measure their sustainability knowledge or not, the replies reflect some positive effects of the SLT on the student, as shown in Fig. 17.6. Only five students said that the test was not useful at all while the rest concluded that the test helped them to improve their knowledge and encouraged them to learn more and get involved in sustainability activities more often.

Obviously to create a test relevant worldwide is not an easy task. The pilot version of SLT is certainly not perfect. The test also does not claim to be able to evaluate the ability of students and graduates to contribute to a sustainable world, which

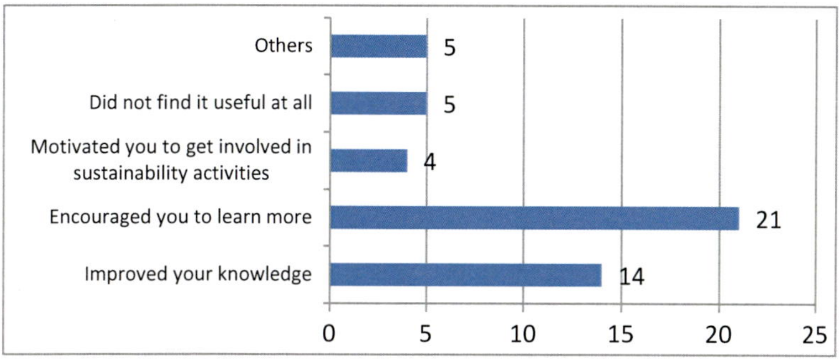


**Fig. 17.6** The usefulness of the SLT to the students

is more important than their sustainability knowledge. The survey respondents provided several suggestions on how to improve the SLT. In terms of contents, most of the respondents recommended that the questions should be qualitative rather than quantitative. Quantitative, trivia-like questions should not form the majority of the test questions. According to the survey respondents, application of the concept of sustainability and skills were more important than memorizing statistics and dates hence questions involving practical problem-solving should be added. In order to inspire students to learn about sustainability, the SLT should be conducted regularly and the website should enable students to learn through the process at their own pace and not during a certain period only (which may conflict with other school activities). The format of the test should be improved to facilitate learning. At the end of each question, there should be an explanation related to the answer of the respondents, why it is correct or not. Furthermore, at the end of the test, it is necessary to interpret what the final score of the students mean.

Improving the quality of SLT will require testing and updating over a long term. SLT should also be complemented with other assessment tools on values and competencies necessary to create systemic changes for a sustainable future. This is because to be truly sustainability literate, students must be able to combine knowledge from the environmental, economic, and social domains and put this knowledge into practice.

## 17.5 Lessons Learned and Suggestions for Improvement

Jackson (2014) has also noted that one of the limitations of SLT is that it only tests students' knowledge. Nevertheless, SLT has a great potential because it provides visibility to sustainability in higher education communities and reveals the gap between "minimal knowledge" and what students are learning, which should prompts educators to reflect on when and where the missing information should be taught, whether in formal or in nonformal education (Jackson 2014). SLT's transnational nature is its great advantage: sustainable development implies a notion of global vision (Jolly et al. 2014). With SLT, the test is open to anyone anywhere in the world and will allow a comparison between countries. If we want to collectively find solutions to different sustainability issues, we need at the very least a shared core of basic knowledge (Carteron and Decamps 2014).

But due to its inherent limitations, SLT must not be used as the single means to assess sustainability literacy. Students' educational backgrounds and interests vary widely. Answering questions randomly taken from a pool of questions sounds like a lottery. Students will likely get a wide range of scores if they have the opportunity to take the test several times. SLT should be complemented with other assessment tools on values and competencies necessary to create systemic changes for a sustainable future (Carteron and Decamps 2014). In conjunction with evaluation of the sustainability literacy, HEIs should promote campus sustainability initiative in several categories such as education and research, operations, planning, administration, and innovation. Under the combination of this hard soft approach, students are expected to reach more visual experiences towards sustainability (Nakamura et al. 2014).

The team behind SLT is aware that the test will never guarantee that students will behave responsibly (Carteron and Decamps 2014). Anyone can have good knowledge about crucial social and environmental issues but still decide not to act. Without a sense of personal connection to sustainability issues, the knowledge and skill sets gained by students may not lead to positive actions either in the workplace or in private life (Murray et al. 2007). Sustainability literacy requires practical skills for transitioning away from consumerist societies to communities capable of fulfilling human needs with minimal use of energy and resources (Stibbe and Luna 2009). But exploration and assessment of sustainability literacy is something that may never be complete since the changing conditions of the world will continuously require new and different skills, i.e., an evolving sustainability literacy (El Ansari and Stibbe 2009).

**Acknowledgement** We acknowledge Mr. Takayuki Nakamura, President of Fukushima National College of Technology and the Kyoto University Facility Management Department staffs for involving us in the sustainability literacy test. We thank the Ministry of Education, Culture, Sports, Science, and Technology (MEXT) of Japan for the generous scholarship grant that made it possible for the first and second authors to pursue their PhD degrees at Kyoto University. We also acknowledge the financial support provided by the Global Center of Excellence Program on Sustainability/Survivability Science for a Resilient Society Adaptable to Extreme Weather Conditions (GCOE-ARS) of Kyoto University for the first and second author's research activities.

## References

AASHE (2014) STARS technical manual version 2.0. http://www.aashe.org/files/documents/STARS/2.0/stars_2.0_technical_manual_-_administrative_update_two.pdf

Adderley P. (2007) Developing sustainability literacy cross-discipline resources. http://www.heacademy.ac.uk/assets/documents/sustainability/paul_adderley.ppt

Carteron JC, Decamps A (2014) The sustainability literacy test: one year report of the sustainability literacy test. Presented on the occasion of the UNESCO World Conference on Education for Sustainable Development, Nagoya (Japan), November 2014. http://www.comite21.org/docs/actualites-adherents/2014/sustainability-literacy-test-report---dec-2014.pdf

Connell KYH, Remington SM, Armstrong CM (2012) Assessing systems thinking skills in two undergraduate sustainability courses: a comparison of teaching strategies, vol 3. http://www.jsedimensions.org/wordpress/wp-content/uploads/2012/03/HillerConnellRemingtonArmstrongJSE2012.pdf

Cotgrave A, Kokkarinen N (2011) Promoting sustainability literacy in construction students: implementation and testing of a curriculum design model. Struct Surv 29(3):197–212

DEFRA (Department for Environment, Food & Rural Affairs, UK) (2005) Securing the future: delivering UK sustainable development strategy. https://www.gov.uk/government/publications/securing-the-future-delivering-uk-sustainable-development-strategy

El Ansari W, Stibbe A (2009) Public health and the environment: what skills for sustainability literacy – and why? Sustainability 1:425–440

Erdogan M, Tuncer G (2009) Evaluation of a course: education and awareness for sustainability. Int J Environ Sci Educ 4(2):133–146

HEA (Higher Education Academy) (2006) Sustainable development in higher education: current practice and future developments. The Higher Education Academy, York

Horvath N, Stewart M, Shea M (2013). Toward instruments of assessing sustainability knowledge: assessment development, process, and results from a pilot survey at the University of Maryland. J Sustain Educ 5. http://www.jsedimensions.org/wordpress/wp-content/uploads/2013/06/HorvathSustainability-Literacy-Assessment-Final-Version-PDF-REady-1.pdf

Jackson L (2014) Hong Kong, global citizenship and the UN sustainability literacy test. http://conference.ntu.edu.sg/asaihl/Documents/PPTs/1_1_1%20Liz%20Jackson.pdf

Jolly AM, Leger C, Cadorel JY (2014) Sustainable development in engineering universities: how accreditation agencies can help them in this demarche? The case of French HEIs. http://www.sefi.be/conference-2014/0008.pdf

Kanj M, Mitic W (2009) Promoting health and development: closing the implementation gap. http://www.who.int/healthpromotion/conferences/7gchp/Track1_Inner.pdf

Lugg A (2007) Developing sustainability-literate citizens through outdoor learning: possibilities for outdoor education in higher education. J Adventure Educ Outdoor Learn 7(2):97–112

Mcmillin J, Dyball R (2009) Developing a whole-of-university approach to educating for sustainability: linking curriculum, research and sustainable campus operations. J Educ Sustain Dev 3(1):55–64

Murray PE, Brown N, Murray S (2007) Deconstructing sustainability literacy: the cornerstone of education for sustainability? The role of values. Int J Environ Cult Econ Soc Sustain 2(7):83–92

Nakamura T, Fujisawa M, Shaw R (2014) Towards the establishment of sustainable campus. In: Shaw R, Yukihiko O (eds) Education for sustainable development and disaster risk reduction. Springer, Tokyo, pp 177–187

Savelyeva T, McKenna JR (2011) Campus sustainability: emerging curricula models in higher education. Int J Sustain High Educ 12(1):55–66

Snavely L, Cooper N (1997) The information literacy debate. J Acad Librariansh 23(1):9–14

Stibbe A, Luna H (2009) Introduction. In: Stibbe A (ed) The handbook of sustainability literacy: skills for a changing world. Green Books, Dartington

Yarime M, Tanaka Y (2012) The issues and methodologies in sustainability assessment tools for higher education institutions: a review of recent trends and future challenges. J Educ Sustain Dev 6(1):63–77

Yuan X, Zuo J (2013) A critical assessment of the higher education for sustainable development from students' perspectives – a Chinese study. J Clean Prod 48:108–115

Zwickle AM, Koontz TM, Slagle K, Bruskotter JT (2014) Assessing sustainability knowledge of a student population. Int J Sustain High Educ 15:375–389

# Chapter 18
# Sustainable Development and Disaster Risk Reduction in Post-2015

**Nitin Srivastava, Glenn Fernandez, Rajarshi DasGupta, Akhilesh Surjan, and Rajib Shaw**

**Abstract** Disaster risk reduction needs to be mainstreamed with development, and it has been increasingly identified at the global level. This chapter explores the direction of inclusion of disaster risk reduction (DRR) and resilience in sustainable development goals post-2015. The Rio+20 Conference also stressed the same principle in its outcome document. However, investment for DRR, enhancement of disaster knowledge and access to such information and building conducive international environment still pose challenges in post-2015 scenario.

**Keywords** Sustainable development goals • Disaster risk reduction • Post-2015 • Sendai Framework

N. Srivastava (✉)
Graduate School of Global Environmental Studies, Kyoto University, Kyoto, Japan
e-mail: imnitin@gmail.com

G. Fernandez
Environmental Education Laboratory, Graduate School of Global Environmental Studies, Kyoto University, Kyoto, Japan

Asian Disaster Preparedness Center, Bangkok, Thailand

R. DasGupta
International Environment and Disaster Management Research Group, Graduate School of Global Environmental Studies, Kyoto University, Yoshida Honmachi, Sakyo-ku, Kyoto 606-8501, Japan
e-mail: rajarshidg1@yahoo.co.in

A. Surjan
Faculty of Engineering, Health, Science and the Environment, Charles Darwin University, Darwin, Australia

R. Shaw
Graduate School of Global Environmental Studies, Kyoto University, Sakyo-ku, Kyoto, Japan
e-mail: shaw.rajib.5u@kyoto-u.ac.jp



J.I. Uitto, R. Shaw (eds.), *Sustainable Development and Disaster Risk Reduction*, Disaster Risk Reduction, DOI 10.1007/978-4-431-55078-5_18

## 18.1 Introduction

Disaster derails development and lack of development curtails progress. Therefore, disaster-considerate development is one of the most cost-effective investments in poverty reduction for any country. A great deal has been documented on why it is important to mainstream disaster risk reduction (DRR) with the development goals. It includes all important issues like food security, health, education, training and capacity building, ecosystem and coastal management, gender balance and environmental security and complex issues like evaluation of climate change and subsequent adaptation and managing trans-boundary water resources under this adaptation. These issues have been discussed in detail in the earlier chapters in this book.

In early 2000s, 85 % of people exposed to earthquakes, cyclones, floods and droughts live in developing countries (UNDP 2004). Development is at the core of the future plans for any developing country. Direct disaster damage costs alone have shot up from US$ 75.5 billion in the 1960s to nearly a trillion dollars in the past 10 years (CRED 2009). These monumental figures do not include the indirect damages to the communities and their sources of livelihood and their future development prospects.

Disaster risk reduction is also critical for ensuring one of the most basic human rights – freedom from hunger. Disaster risk reduction needs to adapt to climate change, manage growth and stop environmental degradation so that disasters cease to threaten more lives and livelihoods than ever before. In summary, disaster risk reduction protects development investments and helps societies to accumulate wealth in spite of hazards. For example, China spent US$ 3.15 billion on reducing the impact of floods, thereby averting losses estimated at US$ 12 billion (DFID 2004).

Davies et al. (2009) advocates 'DRR and adaptation are advocated as cost-effective means of preventing future negative impacts on development investments.' A similar opinion was echoed by a report by IPU (2010) which emphasizes on all eight Millennium Development Goals (MDGs). MDGs have achieved considerably in bringing forth DRR in conjunction with development. Similarly, the path the sustainable development goals (SDGs) would take in the post-2015 era would be instrumental in shaping the future of the world in general and the countries in particular.

This chapter discusses the pre- and post-2015 scenario in terms of inclusion of disaster risk reduction measures in development agenda.

## 18.2 Disaster Risk Reduction and Millennium Development Goals (MDGs)

The initiatives and strategies adopted prior to the year 2015 need to be understood to comprehend the direction of the post-2015 development agenda and to what extent it has included disaster risk reduction in its objectives.

In September 2000, the UN Millennium Summit gathered in New York, and a total of 189 world leaders met and adopted the UN Millennium Declaration. Under

'Protecting our common environment' the declaration says 'adopt in all our environmental action a new ethic of conservation and stewardship and, as first steps, resolves…to intensify cooperation to reduce the number and effects of natural and man-made disasters' [paragraph 23] (UNISDR 2005a). Thereafter, strategies linked to ISDR for moving ahead on this goal were outlined in the road map towards the implementation of the United Nations Millennium Declaration (Secretary-General Report to GA A/56/326) (UNISDR 2005b), which included:

- Developing early warning systems, vulnerability mapping, technological transfer and training
- Supporting interdisciplinary and inter-sectoral partnerships, improved scientific research on the causes of natural disasters and better international cooperation to reduce the impact of climate variables, such as El Niño and La Niña
- Encouraging governments to address the problems created by megacities, the location of settlements in high-risk areas and other manmade determinants of disasters
- Encouraging governments to incorporate disaster risk reduction into national planning processes, including building codes

The road map also established the eight targets, known as the *Millennium Development Goals* (MDGs), providing guidance for achieving development for the international community, national governments and, in particular, the United Nations. All of these targets focused on development issues while touching upon domains 'which are closely linked to vulnerability to natural hazards, such as eradicating extreme poverty and hunger, achieving universal primary education, promoting gender equality, ensuring environmental stability and partnerships for development'. For example, the goal of improving the lives of thousands of slum dwellers around the world living in high-risk areas involves poverty eradication, proper land use planning and the improved understanding of vulnerability to disasters in densely populated areas.

The UN Millennium Project on MDGs entitled 'Investing in Development: A Practical Plan to Achieve the Millennium Development Goals' laid out practical plans for achieving MDGs by 2015. It recommended fivefold strategy (UN Millennium Project 2005) for reducing losses from disasters:

1. Strategies to reduce disaster losses need to be mainstreamed
2. Infrastructure investment to incorporate DRR
3. Social safety nets for the vulnerable, particularly through government provisions
4. Early warning capacities and information campaigns supported by governments
5. Pre crisis emergency and contingency planning

To what extent MDGs were successful in achieving disaster risk reduction and to what extent they were designed to achieve so is open to scrutiny, as MDGs come to an end of their tenure in September 2015. However, there is still time to devise and design the post-2015 development agenda for what we could not achieve as a global community.

## 18.3 Post-2015: Disaster Risk Reduction and Sustainable Development Goals (SDGs)

The United Nations is in the process of defining a post-2015 development agenda scheduled to be launched at a summit in September 2015, which is the target date for realizing the MDGs. It is currently being detailed and refined through informal consultations of the UN General Assembly. There have been numerous inputs to the agenda, notably a set of sustainable development goals (SDGs) proposed by an open working group of the General Assembly, the report of an intergovernmental committee of experts on sustainable development financing, General Assembly dialogues on technology facilitation and many others.

### *18.3.1 Sustainable Development Goals*

In the Rio+20 Conference in June 2012, the member states agreed to launch a process to develop a set of sustainable development goals (SDGs), which will build upon the Millennium Development Goals and converge with the post-2015 development agenda. It was decided to establish an 'inclusive and transparent intergovernmental process open to all stakeholders, with a view to developing global sustainable development goals to be agreed by the General Assembly'. In the Rio+20 outcome document, *The Future We Want*, member states agreed that sustainable development goals (SDGs) must:

- Be based on Agenda 21 and the Johannesburg Plan of Implementation
- Fully respect all the Rio Principles
- Be consistent with international law
- Build upon commitments already made
- Contribute to the full implementation of the outcomes of all major summits in the economic, social and environmental fields
- Focus on priority areas for the achievement of sustainable development, being guided by the outcome document
- Address and incorporate in a balanced way all three dimensions of sustainable development and their interlinkages
- Be coherent with and integrated into the United Nations Development Agenda beyond 2015
- Not divert focus or effort from the achievement of the Millennium Development Goals
- Include active involvement of all relevant stakeholders, as appropriate, in the process

It was further agreed that SDGs must be action oriented, concise, easy to communicate, limited in number, aspirational, global in nature, universally applicable to all countries while taking into account 'different national realities, capacities and levels of development and respecting national policies and priorities' (OWG 2014). The outcome document mandated the creation of an inter governmental Open Working Group (OWG) that will submit a report to the 68th session of the General Assembly containing a proposal for sustainable development goals for consideration and appropriate action. It also resolved to establish an inclusive and transparent intergovernmental process on SDGs that is open to all stakeholders with a view to developing global sustainable development goals to be agreed by the United Nations General Assembly. In August 2014, OWG came up with a report with 17 Goals and 169 Targets under these goals, as shown in Table 18.1.

Disaster risk reduction has been considered in parts in SDGs and still is not a major pillar for the development agenda. Table 18.2 shows that Goals 1, 2, 11 and 13 have verbatim references to the word 'disasters' in their targets. In addition, Table 18.3 shows other disaster risk reduction issues which have been addressed in other targets, as per a publication by UNISDR (2015) and UNSC (2015). There is some promise in addressing these issues. However, there is still a long way to go in addressing the challenges associated with mainstreaming DRR in SDGs.

**Table 18.1** Sustainable development goals in OWG report, 2014

| | |
|---|---|
| Goal 1 | End poverty in all its forms everywhere |
| Goal 2 | End hunger, achieve food security and improved nutrition and promote sustainable agriculture |
| Goal 3 | Ensure healthy lives and promote well-being for all at all ages |
| Goal 4 | Ensure inclusive and equitable quality education and promote lifelong learning opportunities for all |
| Goal 5 | Achieve gender equality and empower all women and girls |
| Goal 6 | Ensure availability and sustainable management of water and sanitation for all |
| Goal 7 | Ensure access to affordable, reliable, sustainable and modern energy for all |
| Goal 8 | Promote sustained, inclusive and sustainable economic growth, full and productive employment and decent work for all |
| Goal 9 | Build resilient infrastructure, promote inclusive and sustainable industrialization and foster innovation |
| Goal 10 | Reduce inequality within and among countries |
| Goal 11 | Make cities and human settlements inclusive, safe, resilient and sustainable |
| Goal 12 | Ensure sustainable consumption and production patterns |
| Goal 13 | Take urgent action to combat climate change and its impacts |
| Goal 14 | Conserve and sustainably use the oceans, seas and marine resources for sustainable development |
| Goal 15 | Protect, restore and promote sustainable use of terrestrial ecosystems, sustainably manage forests, combat desertification and halt and reverse land degradation and halt biodiversity loss |
| Goal 16 | Promote peaceful and inclusive societies for sustainable development, provide access to justice for all and build effective, accountable and inclusive institutions at all levels |
| Goal 17 | Strengthen the means of implementation and revitalize the global partnership for sustainable development |

**Table 18.2** Verbatim references of 'disaster' in SDGs' targets in OWG report, 2014

| Goals | Targets |
|---|---|
| 1 | **1.5** By 2030, build the resilience of the poor and those in vulnerable situations and reduce their exposure and vulnerability to climate-related extreme events and other economic, social and environmental shocks and disasters |
| 2 | **2.4** By 2030, ensure sustainable food production systems and implement resilient agricultural practices that increase productivity and production; that help maintain ecosystems; that strengthen capacity for adaptation to climate change, extreme weather, drought, flooding and other disasters; and that progressively improve land and soil quality |
| 11 | **11.5** By 2030, significantly reduce the number of deaths and the number of people affected and decrease by [x] per cent the economic losses relative to gross domestic product caused by disasters, including water-related disasters, with a focus on protecting the poor and people in vulnerable situations. **11.b** By 2020, increase by [x] per cent the number of cities and human settlements adopting and implementing integrated policies and plans towards inclusion, resource efficiency, mitigation and adaptation to climate change, resilience to disasters, develop and implement, in line with the forthcoming Hyogo Framework, holistic disaster risk management at all levels |
| 13 | **13.1** Strengthen resilience and adaptive capacity to climate-related hazards and natural disasters in all countries |

**Table 18.3** UNISDR (2015) and UNSC (2015) analysis on involvement of DRR in SDGs

| | |
|---|---|
| 3 | **3.d** Strengthen the capacity of all countries, particularly developing countries, for early warning, risk reduction and management of national and global health risks |
| 4 | **4.a** Build and upgrade education facilities that are child, disability and gender sensitive and provide safe, non-violent, inclusive and effective learning environments for all |
| 6 | **6.6** By 2020 protect and restore water-related ecosystems, including mountains, forests, rivers and lakes |
| 9 | **9.1** Develop quality, reliable, sustainable and resilient infrastructure, including regional and trans-border infrastructure, to support economic development and human well-being, with a focus on affordable and equitable access for all. **9.a** Facilitate sustainable and resilient infrastructure development in developing countries through enhanced financial, technological and technical support to African countries, LDCs, LLDCs and SIDS |
| 11 | **11.4** Strengthen efforts to protect and safeguard the world's cultural and natural heritage. **11.6** By 2030, reduce the adverse per capita environmental impact of cities, including by paying special attention to air quality, municipal and other waste management. **11.c** Support least-developed countries, including through financial and technical assistance, for sustainable and resilient buildings utilizing local materials |
| 13 | **13.2** Integrate climate change measures into national policies, strategies and planning. **13.3** Improve education, awareness raising and human and institutional capacity on climate change mitigation, adaptation, impact reduction and early warning |
| 14 | **14.2** By 2020, sustainably manage and protect marine and coastal ecosystems to avoid significant adverse impacts, including by strengthening their resilience, and take action for their restoration, to achieve healthy and productive oceans |
| 15 | **15.1** By 2020 ensure conservation, restoration and sustainable use of terrestrial and inland freshwater ecosystems and their services, in particular forests, wetlands, mountains and drylands, in line with obligations under international agreements. **15.3** By 2020, combat desertification, and restore degraded land and soil, including land affected by desertification, drought and floods, and strive to achieve a land-degradation neutral world |

## 18.4 Challenges of Mainstreaming DRR in SDGs

Mainstreaming disaster risk reduction in itself is a challenge for sustainable development goals. There are many hurdles in achieving this mainstreaming. Some of them are discussed here:

Link to the Global Disaster Reduction Framework

The global development agenda and the DRR agenda should be linked and aligned for sustaining the growth. Interactive and cross reference would enable the governments to take notice of DRR framework. Sendai Framework for disaster risk reduction 2015–2030 has many references stressing on the relevance of inclusion of DRR in development plans, policies and programmes. It also identifies the relevance of action-oriented framework for governments and other stakeholders to work with.

The Sendai Framework was accepted on 18 March 2015, and the consultation for revision of SDGs was accomplished at the end of March 2015. Goal 11.5 was revised to 'By 2030, substantially reduce the number of deaths, the number of affected people and the direct economic losses relative to global gross domestic product caused by disasters, including water-related disasters, with a focus on protecting the poor and people in vulnerable situations'. Similarly, 11.b was revised to 'By 2020, substantially increase the number of cities and human settlements adopting and implementing integrated policies and plans towards inclusion, resource efficiency, mitigation and adaptation to climate change, resilience to disasters, develop and implement, in line with the forthcoming Hyogo Framework, holistic disaster risk management at all levels'. The revisions were made 'in line with Sendai Framework, para 18'.

Unifying Indicators for Disaster Risk and Development

The measurement of vulnerability, resilience and risk has always been very difficult. However, these factors can be associated with development measures, for example, vulnerability lines can be established alongside poverty lines. Both can be represented by variables such as household income or community-level capacities (Birkmann 2013). This can be made possible only with the improved data systems at local and national levels which will generate data sets and indicators and would be able to measure long-term changes.

Enhancement of Disaster Knowledge and Access to Such Information

Knowledge of the driving factors in disaster risk is the essential basis for preemptive policy and action to reduce the risks. However, yet again, only few countries are technically equipped to research and utilize such studies for risk reduction. More platforms need to be organized to enable sharing of such information. This information would enable hazard-sensitive investments, vulnerability and exposure mapping and risk reduction actions.

Better International and National Coordination

Countries share natural resources and consequently share vulnerabilities. Effective international coordination is required not only to share these resources but also to work in unison to reduce the effects of the hazards. Similarly, coordination within a country is required for countries with governments at various levels. The world already has a global framework for disaster risk reduction, and SDGs would take the final shape soon. The national and local governments would be required to implement the strategies from both these documents in agreement with each other. This sounds too simplistic, but it would be challenging and would prepare the countries to mainstream DRR into their own development agenda.

Private Sector's Role

Rio+20 outcome document and OWG SDGs report, both, highlight the growth of green economy with the help of corporate sector. However, they are silent on the involvement of the private sector in disaster risk reduction. It is widely acceptable that most of the development would be driven with the corporate support all over the world. Nevertheless, the DRR initiatives need to be assigned, in partial terms, to the private sector to sustain the development attained. It would be a challenge to involve the private sector, especially the big players, in this herculean task. It would also encourage investment for DRR.

Investment for DRR

It is a big challenge as only few countries are capable of investing in DRR while pursuing their developmental goals. For others, it is a choice between the two. Moreover, there is lack of intent and willingness to invest in the wake of limited knowledge about the benefits of investing in DRR.

## 18.5 Conclusion

The impacts of disasters have risen rapidly over recent decades, affecting almost all sectors and developed and developing countries alike. The Sendai Framework for Action 2015–2030, which seeks to build the resilience of nations and communities to disasters, includes the integration of disaster risk considerations into sustainable development processes as a key strategy. Denotive recognition and reference of disaster risk reduction in the sustainable development goals will provide a vital push to the substantial work on underlying disaster risk.

A geophysical hazard event's impacts devolve upon the existing conditions of the people, households and societies, in turn a reflection of overall macro economic, political and social conditions of the country. Both the development state and the development course of the country play a role in dealing with disasters. Moreover, disasters are linked with undermining poverty eradication, unsustainable growth, impacts on cities and globalization and cascading risks. All these denote towards better linking of development agenda with disaster risk reduction. Integrated approaches will improve outcomes and opportunities for both disaster risk reduction and sustainable development.

**Acknowledgements** The first three authors acknowledge the receipt of MEXT Scholarship from the Government of Japan for their studies in Kyoto University.

## References

Birkmann J (2013) Measuring vulnerability to natural hazards – towards disaster resilient societies, 2nd edn. UNU Press, Tokyo

CRED (2009) Annual disaster statistical review 2008: the numbers and trends. CRED, Geneva

Davies M, Oswald K, Mitchell T, Tanner T (2009) Climate change adaptation, disaster risk reduction and social protection. Promote Pro-Poor Growth Soc Prot 3:201–217

DFID (2004) Disaster risk reduction: a development concern. A scoping study on links between disaster risk reduction, poverty and development. DFID, London

Inter- Parliamentary Union (IPU) (2010) Disaster risk reduction: an instrument for achieving the millennium development goals: advocacy kit for parliamentarians. In: Disaster risk reduction: an instrument for achieving the millennium development goals: advocacy kit for parliamentarians. UN International Strategy for Disaster Reduction (UNISDR); IPU Headquarters

Open Working Group (OWG) (UN) (2014) Full report of the Open Working Group of the General Assembly on Sustainable Development Goals, issued as document A/68/970. Retrieved 17 January, 2015 from http://undocs.org/A/68/970UNDESA

UN Millennium Project (2005) Investing in development: a practical plan to achieve the millennium development goals. United Nations, New York

UNDP (United Nations Development Programme) (2004) Reducing disaster risk: a challenge for development. UNDP, New York

UNISDR (UN International Strategy for Disaster Reduction) (2005a) UN Millennium Declaration (A/RES/55/2). Retrieved 29 March, 2015 from http://www.unisdr.org/2005/mdgs-drr/link-mdg-drr.htm

UNISDR (UN International Strategy for Disaster Reduction) (2005b) Road map towards the implementation of the United Nations Millennium Declaration Secretary-General Report to GA (A/56/326). Retrieved 28 March, 2015 from http://www.unisdr.org/2005/mdgs-drr/pdf/a-56-326.pdf

UNISDR (UN International Strategy for Disaster Reduction) (2015) Disaster risk reduction in sustainable development outcome documents. UNISDR Geneva. Retrieved 31 March, 2015 from http://www.preventionweb.net/files/42613_drrinsustainabledevelopmentoutcomed.pdf

United Nations Statistical Commission (UNSC) (2015) Technical report by the Bureau of the United Nations Statistical Commission (UNSC) on the process of the development of an indicator framework for the goals and targets of the post-2015 development agenda (Working draft). Retrieved 28 March, 2015 from http://bit.ly/StatComm_Indicators_Technical_Report_March_2015

FSC
www.fsc.org
MIX
Papier | Fördert
gute Waldnutzung
FSC® C083411

Kolibri 360 GmbH
Ferdinand-Jühlke-Straße 7
99095 Erfurt, Deutschland
produktsicherheit@kolibri360.de